BIOPROSPECTING ETHNOMEDICINAL PLANT RESOURCES

Sustainable Utilization and Restoration

Cover photo credits:

Top left: *Caralluma chung*. Photo by Marie Portas.
https://commons.wikimedia.org/wiki/File:Caralluma_chung_Paris_3.jpg
https://creativecommons.org/licenses/by-sa/4.0/deed.en

Top right: *Vachellia nilotica* (babul, thorn mimosa, or Egyptian acacia). Photo by J.M.Garg.
https://en.wikipedia.org/wiki/Vachellia_nilotica#/media/File:Babool_(Acacia_nilotica)_flowers_at_Hodal_W_IMG_1163.jpg
https://creativecommons.org/licenses/by-sa/3.0/

Bottom left: *Galega officinalis* (goat's rue). Photo by JoJan.
https://en.wikipedia.org/wiki/Bioprospecting#/media/File:Galegaofficinalis03.jpg
https://creativecommons.org/licenses/by-sa/3.0/deed.en

Bottom right: *Taxus brevifolia* (Pacific yew). Photo by Jason Hollinger.
https://en.wikipedia.org/wiki/Taxus_brevifolia#/media/File:Taxus_brevifolia_Blue_Mts_WA.jpg
https://creativecommons.org/licenses/by/2.0/

BIOPROSPECTING ETHNOMEDICINAL PLANT RESOURCES

Sustainable Utilization and Restoration

Edited by

Gopal Shukla, PhD

Jahangeer A. Bhat, PhD

A. P. Das, PhD

Sumit Chakravarty, PhD

APPLE ACADEMIC PRESS

First edition published 2025

Apple Academic Press Inc.
1265 Goldenrod Circle, NE,
Palm Bay, FL 32905 USA

760 Laurentian Drive, Unit 19,
Burlington, ON L7N 0A4, CANADA

CRC Press
2385 NW Executive Center Drive,
Suite 320, Boca Raton FL 33431

4 Park Square, Milton Park,
Abingdon, Oxon, OX14 4RN UK

Library and Archives Canada Cataloguing in Publication

CIP data on file with Canada Library and Archives

Library of Congress Cataloging-in-Publication Data

CIP data on file with US Library of Congress

ISBN: 978-1-77491-618-6 (hbk)
ISBN: 978-1-77491-619-3 (pbk)
ISBN: 978-1-00345-148-8 (ebk)

About the Editors

Gopal Shukla, PhD

Professor and Head, Department of Forestry, North Eastern Hill University, Tura Campus, Tura, Meghalaya, India

Gopal Shukla, PhD, is currently working as Professor in Forestry at North Eastern Hill University, Tura Campus, Tura, Meghalaya, India. Before joining as Professor, he was working as Senior Assistant Professor in Forestry at Uttar Banga Krishi Viswavidyalaya, Pundibari, Cooch Behar, West Bengal, India. Dr. Shukla has teaching and research experience of more than 12 years. He has worked on different projects, including the National Agricultural Innovation Project (ICAR); National Innovations in Climate Resilient Agriculture (ICAR); and also as a young scientist on the project on carbon stock of different land uses (Science and Engineering Research Board, Department of Science and Technology, Government of India). The focus of his research and development work is on agroforestry, medicinal plants, forest ecology, and conservation. He has national and international publications that include 90 research papers, 30 book chapters, and eight books. He has successfully completed several research projects on forest ecology, medicinal plants, and agroforestry. Dr. Shukla also guided four doctoral and six master's students.

Jahangeer A. Bhat, PhD

Piramal Foundation for Education Leadership (State Transformation Program), Jammu and Kashmir, India

Jahangeer A. Bhat, PhD, is currently associated with the Piramal Foundation for Education Leadership at Piramal University, Mumbai, India. Before taking on his current position, he was associated with the College of Horticulture and Forestry at Rani Lakshmi Bai Central Agricultural University, India. Earlier, he served as Head of the Department of Forestry at the College of Agriculture, Fisheries and Forestry at Fiji National University, Republic of Fiji Islands. His research work in India and Fiji Islands spans 12 years and encompasses emerging issues in forestry, including conservation of biodiversity, traditional knowledge of plants, and sustainable management of forest resources, particularly in the light of climate change scenarios. He has worked as counselor, mentor, and coordinator for forestry academic programs and has been instrumental in developing higher education and technical and vocational education and training streams of forestry and allied programs. He worked closely for accreditation with the Fiji Higher Education Commission and forestry stakeholders.

A. P. Das, PhD

A. P. Das, PhD is a retired Professor of Botany, University of North Bengal, Siliguri, West Bengal, India. Dr. Das is now working as an Adjunct Professor in the Department of Botany, Rajiv Gandhi University, Rono Hills, Arunachal Pradesh, India. He is teaching plant taxonomy for over 45 years. Basically, a student of plant taxonomy, later on diversified his fields of work and that includes Biodiversity, Conservation, Ethnobotany, Ecology, Weed Biology, Palynology, Remote Sensing, and Aerobiology. He has published more than 240 research papers, three textbooks, three research based books, two edited books and some popular science books. So far, 27 students have been awarded PhD under his supervision. In different times he acted as Head of the Botany Department, Director of the Center for Life Science, Coordinator for the Department of Biotechnology (DBT), and Head of the Department of Tea Science, NBU. Dr. Das has established the East Himalayan Society for Spermatophyte Taxonomy in the year 2007 and initiated the publication of the journal Pleione. He has also developed the NBU Garden of Medicinal Plants and NBU-Herbarium.

Sumit Chakravarty, PhD
Professor, Department of Forestry, Uttar Banga Krishi Viswavidyalaya (UBKV), Pundibari, Cooch Behar, West Bengal, India

Sumit Chakravarty, PhD has a wide experience of more than 20 years in forestry training, research, and development. He is currently working as a Professor in the Department of Forestry, Uttar Banga Krishi Viswavidyalaya, Pundibari, Cooch Behar, West Bengal, India. He has conducted research on several aspects of forestry, agroforestry, medicinal plants, and climate change with national and international reputed publications. This includes 120 research papers, 48 book chapters and eight books. He has successfully completed and continuing 15 research projects on forest ecology, medicinal plants, and agroforestry. He has trained more than 50 master and doctoral students in the field of forestry. Dr. Chakravarty has attended more than 50 national and international conferences in the capacity of presenter and invited speaker. He holds a master's degree in forestry and doctoral degree in Agronomy from Punjab Agricultural University, Ludhiana.

Contents

Contributors

Md. Shamsher Ahmad
Department of Food Science and Postharvest Technology, BAC, Sabour (Bihar), India

Firoz Ahmed
Vindhyan Ecology and Natural History Foundation, Mirzapur, Uttar Pradesh, India

Rofik Ahmed Barbhuiya
Wildlife Conservation Research Center, Department of Ecology and Environmental Sciences, Assam University, Silchar, Assam, India; Center for Biodiversity and Climate Change Research, Udayan, Hailakandi, Assam, India

Jahangeer A. Bhat
Piramal Foundation for Education Leadership (State Transformation Program), Jammu and Kashmir, India

B. P. Bhatt
ICAR Research Complex for Eastern Region, Patna, Bihar, India

Saroj Biswakarma
Uttar Banga Krishi Viswavidyalaya, Pundibari, West Bengal, India

Rainer W. Bussmann
Department of Ethnobotany, Institute of Botany, Ilia State University, Tbilisi, Georgia

Moumita Chakraborty
Department of Genetics and Plant Breeding, Uttar Banga Krishi Viswavidyalaya, Pundibari, West Bengal, India

Sumit Chakravarty
Department of Forestry, Uttar Banga Krishi Viswavidyalaya, Pundibari, Cooch Behar, West Bengal, India

Kuntal Narayan Chaudhuri
Department of Botany, Vivekananda College, Thakurpukur, Kolkata, West Bengal, India

Sangram Chavan
ICAR–Central Agroforestry Research Institute, Jhansi, Uttar Pradesh, India

Parthankar Choudhury
Wildlife Conservation Research Center, Department of Ecology and Environmental Sciences, Assam University, Silchar, Assam, India; Center for Biodiversity and Climate Change Research, Udayan, Hailakandi, Assam, India

Sayan Chowdhury
Uttar Banga Krishi Viswavidyalaya, Pundibari, Cooch Behar, West Bengal, India

S. S. Dahiya
Division of Animal Nutrition and Feed Technology, ICAR–Central Institute for Research on Buffaloes, Hisar, Haryana, India

Bapi Das
MARKFED, RGM Hospital, Kailashahar, Unakoti, Tripura, India

Bikash Das
ICAR Research Complex for Eastern Region, Farming System Research Center for Hill and Plateau Region, Plandu, Ranchi, Jharkhand, India

Litan Das
Uttar Banga Krishi Viswavidyalaya, Pundibari, Cooch Behar, West Bengal, India

Subrata Das
Department of Forestry, Uttar Banga Krishi Viswavidyalaya, Pundibari, Cooch Behar, West Bengal, India

Inder Dev
ICAR–Central Agroforestry Research Institute, Jhansi, Uttar Pradesh, India

M. P. Devi
Department of Pomology and Postharvest Technology, Faculty of Horticulture, UBKV, Pundibari, Cooch Behar (WB), India

Yanglem Lakshimai Devi
Department of Agricultural Extension, Uttar Banga Krishi Viswavidyalaya, Pundibari, Cooch Behar, West Bengal, India

Avijit Dey
Division of Animal Nutrition and Feed Technology, ICAR–Central Institute for Research on Buffaloes, Hisar, Haryana, India

Mahesh Kumar Dhakar
ICAR Research Complex for Eastern Region, Farming System Research Center for Hill and Plateau Region, Plandu, Ranchi, Jharkhand, India

S. Eda
Department of Pomology and Postharvest Technology, Faculty of Horticulture, UBKV, Pundibari, Cooch Behar (WB), India

K. A. Geetha
ICAR–Directorate of Medicinal and Aromatic Plants Research, Boriavi, Anand, Gujarat, India

Srinjoy Ghosh
Uttar Banga Krishi Viswavidyalaya, Pundibari, Cooch Behar, West Bengal, India

L. A. Gonzalez
School of Life and Environmental Sciences, The University of Sydney, NSW, Australia

Satyajit Hembram
Regional Research Station (Terai Zone), Uttar Banga Krishi Viswavidyalaya, Pundibari, Cooch Behar, West Bengal, India

Lakshmi Hijam
Department of Genetics and Plant Breeding, Uttar Banga Krishi Viswavidyalaya, Pundibari, West Bengal, India

Md. Azizul Hoque
Department of Horticulture, Bangabandhu Sheikh Mujibur Rahman Agricultural University, Salna, Gazipur, Bangladesh

Thadiyan Parambil Ijinu
Naturæ Scientific, Kerala University-Business Innovation and Incubation Centre, Kariavattom Campus, Thiruvananthapuram 695581, Kerala, India; The National Society of Ethnopharmacology, VRA 179, Mannamoola, Thiruvananthapuram 695005, Kerala, India

T. Kalaivani
Department of Biotechnology, School of Bio-Sciences and Technology, Vellore Institute of Technology, Vellore, Tamil Nadu, India

N. Kanagam
Department of Biotechnology, School of Bio-Sciences and Technology, Vellore Institute of Technology, Vellore, Tamil Nadu, India

Dhiraj Kumar
ICAR–Central Agroforestry Research Institute, Jhansi, Uttar Pradesh, India

Sutapa Kumar (Rai)
Department of Botany, Vivekananda College, Thakurpukur, Kolkata, West Bengal, India

Dipa Kundu
Uttar Banga Krishi Viswavidyalaya, Pundibari, Cooch Behar, West Bengal, India

P. C. Lailer
Division of Animal Nutrition and Feed Technology, ICAR–Central Institute for Research on Buffaloes, Hisar, Haryana, India

Rosalin Laishram
Uttar Banga Krishi Viswavidyalaya, Pundibari, Cooch Behar, West Bengal, India

Rupsanatan Mandal
Department of Genetics and Plant Breeding, Uttar Banga Krishi Viswavidyalaya, Pundibari, West Bengal, India

Somnath Mandal
Uttar Banga Krishi Viswavidyalaya, Pundibari, Cooch Behar, West Bengal, India

K. Abha Manohar
Department of Agronomy, MS Swaminathan School of Agriculture, Centurion University of Technology and Management, Odisha, India

Bilin Maying
Department of Genetics and Plant Breeding, Uttar Banga Krishi Viswavidyalaya, Pundibari, West Bengal, India

Hossain Ali Mondal
School of Crop Improvement, College of Postgraduate Studies in Agricultural Sciences, Umiam, Meghalaya, India

Prithusayak Mondal
Uttar Banga Krishi Viswavidyalaya, Pundibari, Cooch Behar, West Bengal, India

Kumaresh Pal
Uttar Banga Krishi Viswavidyalaya, Pundibari, Cooch Behar, West Bengal, India

Prabhat Kumar Pal
Department of Agricultural Extension, Uttar Banga Krishi Viswavidyalaya, Pundibari, Cooch Behar, West Bengal, India

Subhrajyoti Panda
Department of Agricultural Extension, Uttar Banga Krishi Viswavidyalaya, Pundibari, Cooch Behar, West Bengal, India

Rakesh Pandey
Microbial Technology and Nematology Department, CSIR–Central Institute of Medicinal and Aromatic Plants, P.O. CIMAP, Lucknow, India

Vineeta Pandey
National Medicinal Plants Board, Ministry of Ayush, New Delhi, India

Goutam Kumar Pandit
Uttar Banga Krishi Viswavidyalaya, Pundibari, Cooch Behar, West Bengal, India

Bablu Paul
Department of Genetics and Plant Breeding, Uttar Banga Krishi Viswavidyalaya, Pundibari, West Bengal, India

P. K. Paul
Department of Pomology and Postharvest Technology, Faculty of Horticulture, UBKV, Pundibari, Cooch Behar (WB), India

Shiva Pokhrel
International Center for Integrated Mountain Development (ICIMOD), Khumaltar, Lalitpur, Kathmandu, Nepal

Sumita Pradhan
Regional Research Station, Hill Zone, Uttar Banga Krishi Viswavidyalaya, Kalimpong, West Bengal, India

Rajendra Prasad
ICAR–Central Agroforestry Research Institute, Jhansi, Uttar Pradesh, India

Md. Masud Karim Prodhan
Department of Horticulture, Bangabandhu Sheikh Mujibur Rahman Agricultural University, Salna, Gazipur, Bangladesh

Palpu Pushpangadan
The National Society of Ethnopharmacology, VRA 179, Mannamoola, Thiruvananthapuram 695005, Kerala, India; Amity Institute for Herbal and Biotech Products Development, Thiruvananthapuram 695005, Kerala, India

Tata Raghu
Uttar Banga Krishi Viswavidyalaya, Pundibari, Cooch Behar, West Bengal, India

C. Rajasekaran
Department of Biotechnology, School of Bio-Sciences and Technology, Vellore Institute of Technology, Vellore, Tamil Nadu, India

Asha Ram
ICAR–Central Agroforestry Research Institute, Jhansi, Uttar Pradesh, India

R. Nagaraja Reddy
ICAR–Directorate of Medicinal and Aromatic Plants Research, Boriavi, Anand, Gujarat, India

Satyajit Roy
ICAR–Directorate of Medicinal and Aromatic Plants Research, Boriavi, Anand, Gujarat, India

Nandita Sahana
Uttar Banga Krishi Viswavidyalaya, Pundibari, Cooch Behar, West Bengal, India

Biplov Chandra Sarkar
Department of Forestry, Uttar Banga Krishi Viswavidyalaya, Pundibari, Cooch Behar, West Bengal, India

Pradip Kumar Sarkar
ICAR Research Complex for Eastern Region, Farming System Research Center for Hill and Plateau Region, Plandu, Ranchi, Jharkhand, India

Pranati Sarkar
ICAR Central Potato Research Station, Shillong, Meghalaya, India

N. N. Shahina
Department of Forestry, Uttar Banga Krishi Viswavidyalaya, Pundibari, Cooch Behar, West Bengal, India

Anjali Sharma
Uttar Dinajpur Krishi Vigyan Kendra, Uttar Banga Krishi Viswavidyalaya, Chopra, West Bengal, India

Pushkar Sharma
Regional Research Station, Hill Zone, Uttar Banga Krishi Viswavidyalaya, Kalimpong, West Bengal, India

Ashok Shukla
ICAR–Central Agroforestry Research Institute, Jhansi, Uttar Pradesh, India

Gopal Shukla
Department of Forestry, Uttar Banga Krishi Viswavidyalaya, Pundibari, Cooch Behar, West Bengal, India; Department of Forestry, North Eastern Hill University, Tura Campus, Tura, Meghalaya, India

Md. Wasim Siddiqui
Department of Food Science and Postharvest Technology, BAC, Sabour (Bihar), India

Manendra Singh
Department of Forestry, Uttar Banga Krishi Viswavidyalaya, Pundibari, Cooch Behar, West Bengal, India

Manendra Singh
Uttar Banga Krishi Viswavidyalaya, Pundibari, West Bengal, India

Prashant Singh
ICAR–Central Agroforestry Research Institute, Jhansi, Uttar Pradesh, India

Sajitha Siril
Uttar Banga Krishi Viswavidyalaya, Pundibari, West Bengal, India

Kasireddy Sivasankarreddy
Department of Genetics and Plant Breeding, Uttar Banga Krishi Viswavidyalaya, Pundibari, West Bengal, India

Nazimur Rahman Talukdar
Wildlife Conservation Research Center, Department of Ecology and Environmental Sciences, Assam University, Silchar, Assam, India; Center for Biodiversity and Climate Change Research, Udayan, Hailakandi, Assam, India

Bandan Thapa
Regional Research Station, Hill Zone, Uttar Banga Krishi Viswavidyalaya, Kalimpong, West Bengal, India

Vineeta
Department of Forestry, Uttar Banga Krishi Viswavidyalaya, Pundibari, Cooch Behar, West Bengal, India

Abbreviations

1,3,5-triCQA	1,3,5-tricaffeoylquinic acid
3,4-diCQA	3,4-dicaffeoylquinic acid
3,5-diCQA	3,5-dicaffeoylquinic acid
3-CQA	3-caffeoylquinic acid
3-CSA	3-caffeoylshikimic acid
4,5-diCQA	4,5-dicaffeoylquinic acid
4-C,5FQA	4-caffeoyl-5-feruloylquinic acid
4-CQA	4-caffeoylquinic acid
4-CSA	4-caffeoylshikimic acid
5-CQA	5-caffeoylquinic acid
5-CSA	5-caffeoylshikimic acid
5-FQA	5-feruloylquinic acid
5-p-CoQA	5-p-coumaroylquinic acid
a cis-3,5-diCQA	A cis-3,5-dicaffeoylquinic acid
a cis-4,5-diCQA	A cis-4,5-dicaffeoylquinic acid
AFS	agroforestry system
AICRP	All-India Coordinated Research Project
AICRP-MAP	All-India Co-ordinated Research Project on Medicinal and Aromatic Plants
AICRP-MAP&B	All-India Co-ordinated Research Project on Medicinal and Aromatic Plants and Betelvine
AICTE	All-India Council for Technical Education
ANOVA	analysis of variance
AS	average score
AU	alternative utilization
BARC	Bangladesh Agricultural Research Council
BARI	Bangladesh Agricultural Research Institute
BAS	Bangladesh Academy of Sciences
BBS	Bangladesh Bureau of Statistics
BFRI	Bangladesh Forest Research Institute
BIS	Bureau of Indian Standards
BSMRAU	Bangabandhu Sheikh Mujibur Rahman Agricultural

	University
CARE	Cooperative for Assistance and Relief Everywhere
CBD	convention on biological diversity
CFQA	caffeoyl-feruloylquinic acid
CIMAP	Central Institute of Medicinal and Aromatic Plants
cis-4,5-diCQA	cis-4,5-dicaffeoylquinic acid
CMI	cell mediated immune
CSIR	Council of Scientific and Industrial Research
CT	condensed tannins
CT	conventional treatment
DBT	department of biotechnology
DMAPR	Directorate of Medicinal and Aromatic Plants Research
DNA	deoxyribonucleic acid
DRDO	Defense Research and Development Organization
DST	Department of Science and Technology
ECO	eucalyptus oils
Ef	effectiveness
EN	endangered
e-NAM	electronic national agriculture marketing
e-NWRs	electronic negotiable warehouse receipts
EO	essential oil
ET	ethnomedicinal treatment
EVM	ethnoveterinary medicine
F&F	flavor and fragrance
FAO	Food and Agricultural Organization
FDA	Food and Drug Administration
FDC	Forest Development Corporation
FMCG	fast moving consumer goods
FPO	Farmer Producer Company
FQA	feruloylquinic acid
FRLHT	Foundation for Revitalization of Local Health Tradition
FYM	farmyard manure
GACP	good agriculture collection practices
GAPs	good agricultural practices
GATT	general agreement on tariffs and trade
GBPIHED	Gobind Ballav Pant Institute of Himalayan Environment and Development

GHG	greenhouse gas
GIP	gastrointestinal parasites
GSCS	Girijan Service Cooperative Society
GSH	glutathione
HAP	hyper-ammonia producing bacteria
HDL	high density lipoprotein
HK	having knowledge
HKI	Helen Killer International
HRDI	Herbal Research and Development Institute
HT	hydrolysable tannin
IARI	Indian Agricultural Research Institute
ICAR	Indian Council of Agricultural Research
ICAR-DMAPR	ICAR–Directorate of Medicinal and Aromatic Plants Research
ICMR	Indian Council of Medical Research
IFS	integrated farming system
IIHR	Indian Institute of Horticultural Research
IIIM	Indian Institute of Integrative Medicine
INR	Indian National Rupee
IPCC	intergovernmental panel on climate change
IPM	integrated pest management
IPR	intellectual property right
IRRI	International Rice Research Institute
IUCN	International Union for Conservation of Nature and Natural Resources
JNTBGRI	Jawaharlal Nehru Tropical Botanic Garden and Research Institute
KGF	Krishi Gobeshona Foundation
LA	locally available
LAE	less adverse effects
LC	low cost
LDL	low density lipoprotein
LGED	local government and engineering department
LPO	lipid peroxidation
MAP	medicinal and aromatic plant
MAS	marker-assisted selection
MD	market demand

MDA	malondialdehyde
MFP	minor forest product
MFPD	minor forest produce division
MIS	marketing information system
MoEF	Ministry of Environment and Forest
MoU	memorandum of understanding
NABARD	National Bank for Agriculture and Rural Development
NATP	National Agricultural Technology Project
NBPGR	National Bureau of Plant Genetic Resources
NBRI	National Botanical Research Institute
NCR	National Capital Region
NGO	non-government organization
NMITLI	New Millennium India Technology Leadership Initiative
NMPB	National Medicinal Plant Board
NRC-MAP	National Research Center for Medicinal and Aromatic Plants
NTFP	non-timber forest product
ORS	oral rehydration solution
PA	proanthocyanidins
PCT	patent cooperation treaty
PHA-P	phytohaemagglutinin-P
Pr	preference
QPM	quality planting material
RG	ranking groups
RRS (HZ)	regional research station (hill zone)
SBT	single bud technology
SDG	sustainable development goals
SHG	self-help group
SOD	superoxide dismutase
SOPs	standard operating protocols
TDCC	Tribal Development Cooperative Corporation
TEK	traditional ecological knowledge
TFC	total flavonoids content
TK	traditional knowledge
TKDL	traditional knowledge digital library
TKRC	traditional knowledge resource classification
TM	treatment methods
TNAU	Tamil Nadu Agricultural University

TPC	total phenol content
TRI	turmeric redness index
triCQA	tricaffeoylquinic acid
TRIFED	tribal federation
TRIPS	trade-related aspects of intellectual property rights
TS	total score
UBKV	Uttar Banga Krishi Viswavidyalaya
UGC	University Grant Commission
UNEP	United Nations Environment Program
UNESCO	United Nation Education, Scientific, and Cultural Organization
USAID	United States Agency for International Development
VDVK	Van Dhan Vikash Kendra
VU	vulnerable
WCED	World Commission on Environment and Development
WHO	World Health Organization
WIPO	World Intellectual Property Organization
WTO	World Trade Organization
WWF	World Wildlife Fund

Preface

Medicinal plants are gaining popularity as herbal raw medications, and the majority of modern drugs are based on phytochemicals produced from the ethnomedicinal plants. The natural products generated from medicinal plants are gaining popularity, as can be seen by the increased acceptance of plant-derived healthcare products. The importance of traditional knowledge (TK) and biodiversity resources in healthcare management, revived interest in global bioprospecting, and the use of indigenous information about medicinal plants to help in the search for novel chemical compounds and new drugs. Bio-prospection is a pressing problem for biodiversity-rich countries in order to catalog their bio-resources and identify valuable plants. The production of medicinal plants should be prioritized in order to support rural populations and produce healthy products. Due to the global usage and demand for medicinal herbs, cultivation and bioprospecting of these plants has enormous potential for the future.

This book titled *Bioprospecting of Ethnomedicinal Plants: Sustainable Utilization and Restoration* is focused on sustainable utilization, management of ethnomedicinal plants and their bioprospecting. The understanding of this aspect is of utmost importance to formulate modern medicine for raw materials from ethnomedicinal plants cultivated and collected from the wild for healthcare practices. The depletion of ethnomedicinal plant resources has forced scientists and policy developers to strategize the future possibilities for sustainable utilization and restoration of these resources. Therefore, proper and systematic identification, collection, and conservation of these resources is very important. If serious and concrete steps are not considered for their sustainable utilization and conservation, then this wealth of plant resources with medicinal properties could vanish forever.

This edited book contains extensive information on medicinal plant diversification, cultivation techniques, biotechnological approaches, modern, and traditional usage of different resources to contribute to current knowledge and improve management of plant resources sustainably. It also provides information for the development of further strategies and the formulation of policies for the systematic management and conservation of medicinal resources for future use.

The book has addressed all of the important features of the subject for a large audience of academics, researchers, policy planners, and other stake-holders. The book's compilation of data and contents will provide in-depth information to users/readers from the relevant subject and other linked disciplines worldwide.

We would like to express our heartfelt gratitude and acknowledgment to all contributors. The AAP team deserves credit for their ideas and for releasing this edited book.

—*Editors*

Introduction

Traditional systems of medicine have wide acceptance with in rural areas, particularly in developing nations. In addition to serving the healthcare needs of human and livestock ailments, ethnic plants are the exclusive source of raw drugs even in modern medical treatment. Plant products are used as daily food and nutrition supplements by the forest adjacent communities, and is also marketed as industrial raw material in pharmaceutical, cosmetic, food flavoring, medical industries.

Increasing popularity of plant-based drugs has put a lot of pressure on the existing plant resources; therefore, conservation strategies and systematic research are needed for better management of their natural habitat and is a great concern in the present scenario. This book is focused on ethnomedicinal plants, traditional system of medicine, cultivation, scope of research and development. The brief highlights of chapters are described below.

Chapter 1 discusses the development of traditional system of medicine in the present scenario through different collaborative programs initiated for new discoveries in bioprospecting of ethnomedicinal plants.

Chapter 2 emphasizes on the nutritionally and medicinally rich ethnic plants which are less known to modern science, impacts of ethic plants on global nutraceutical market and economic development, modulation of metabolism and lifestyle of people, misconceptions and duplication of products, and scope of this newly emerging science.

Chapter 3 discusses the integration of medicinal plants under multitier based agroforestry system (AFS) and could be a viable option for farmers to improve their livelihood by adopting Horti-medicinal and Silvi-medicinal systems. The insights of such agroforestry systems are recommended for wasteland management.

Chapter 4 discusses different bioactive compounds produced under agroforestry based medicinal plants and their significant implications to improve livestock health. Therefore, feeding of bio-active compounds containing a blend of different plants is beneficial to control inhabitation of methanogenic archaea in rumen, gastro-intestinal parasites, increase milk production, feed efficiency and utilization with reduction in methane emission growth performance and health status of animals.

Chapter 5 discusses the ethnobotanical uses by forest dependent people and how they transmit their traditional knowledge and culture to future generations, and the need of conservation with systematic research and development through modern technology.

Chapter 6 highlights the importance of traditional healthcare systems after the COVID-19 pandemic, thereby increasing the consumption of herbal medicine and focussing on the opportunity to generate income, employment, and foreign exchange across the world. It also addresses the prospects and problems, emerging research needs and collaborative efforts of different organizations related to MAPs to convert the research achievements into deliverables for the benefit of all the stakeholders.

Chapter 7 discusses the traditional system of medicine, medicinal uses, marketing, habitat destruction, and conservation of medicinal plants in Bangladesh.

Chapter 8 highlights the geographic distribution, ecological niche, and medicinal values of parasitic and carnivores' plants of the Indian Himalayas, as these unique groups of plants have antioxidant, anti-fertility, and anthelmintic properties, and are efficiently used as a traditional medicine by tribal communities.

Chapter 9 discusses a case study of medicinal plants and NTFPs of a forest fringe area in North Bengal. The strategies to enhance cultivation and entrepreneurial scope through ethnic plants nearby forest is also explored.

Chapter 10 emphasizes on ethnomedicinal plant species used by the Reang tribe of Northeast India. It also lists plants' family, genera, and species along with their medicinal properties.

Chapter 11 discusses the role of sacred groves of India in *in situ* conservation of ethnomedicinal plants and also checks the degradation of natural habitats and their protection by different autochthonous communities in the Eastern Himalayas and Western Ghats.

Chapter 12 highlights the scope of medicinal plants and their proliferation in a very well-mannered agroforestry system as demand of raw drugs is increasing in pharmaceutical and medical industries, etc. It also discusses, the needs of conservation, maintenance, and up-scaling of threatened plants species in homestead-based agroforestry.

Chapter 13 discusses the diversity of plant genetic resources as an integral part of treatment from commencement of human memory. Diversification in locally adapted niche and human diets has increased utilization and substantial consumption of indigenous plants as vegetables; many underutilized plants are vital in alleviating hunger, malnutrition, and improving health, thereby these are commonly marketed by rural people as a part of their socio-economy.

Chapter 14 discusses nursery management, productive agro-technology through quality planting material, quality and quantity of chemical constituents of *Stevia rebaudiana* as a sugar substitute.

Chapter 15 emphasizes on *Oroxylum indicum* as a folklore medicinal plant in India and Southeast Asia. Its traditional medicinal formulations, therapeutic properties, phytochemical compounds, pharmacological properties, and commercial exploitation are discussed.

Chapter 16 highlights the medicinal importance of turmeric as its applications in food and condiment industries. The multi-benefit crop is suitable for intercropping and this rhizome extract has antifungal and antibacterial properties, thereby, making this crop of diverse potential a low cost to high impact industrial crop for socio-economically poor farming communities of West Bengal region.

Chapter 17 discusses a medicinal orchid *Malaxis acuminata* that plays an important role in Ayurvedic formulations; the species has a wide range of healing properties used in treatment of male sterility, rheumatism dysentery, fever, emaciation, and general debility through antioxidant, anti-aging, UV-A blocking, anti-inflammatory, antiproliferative, and antimicrobial activities. *In situ* conservation and *in vitro* propagation technique envisaged for potential conservation and management is also detailed.

Chapter 18 focusses on the genetic diversity in *Oryza* species and its role in the world food security and socio-economic development. Among the cereals, the rice has unique quality in terms of mineral content, starch quality, glycemic index, and antioxidant activity. The rice-based ORS (oral rehydration solution) have been proved effective in decreasing stool output and improving intestinal absorption in acute diarrhea. The promotion and conservation of this national heritage as a health food is critical in order to stem the onslaught of lifestyle-related diseases.

Global ethnic population, mainly in the developing nations, still depends upon medicinal and aromatic plants for treating various diseases. COVID-19 outbreak was an opportunity to again look into many forgotten medicinal and aromatic plants which have reflections over the global market as well. To ace the upcoming demand of disease-free medicinal and aromatic plants, disease management of these plants is inevitable to improve yield and quality of plant derived products. The focus of chapter 19 is largely on disease management practices of medicinal plants, its importance, and commercial value.

Medicinal and aromatic plants are one of the important segments of herbal based pharmaceuticals industries. There is an increase in the demand of herbal products due to consciousness among the masses that herbal treatment are known to be healthy. While the demand is increasing the middleman in the marketing chain benefit more than the growers. Slowly, the situation is changing as of now since the growers are trying to directly link their produce with consumers and retail chains, etc., and this shift is what we can say as the marketing revolution. This important aspect is detailed in chapter 20.

CHAPTER 1

Bioprospecting of Ethnomedicinal Plants: Basic Issues

THADIYAN PARAMBIL IJINU[1,2] and PALPU PUSHPANGADAN[2,3]

[1]*Naturæ Scientific, Kerala University-Business Innovation and Incubation Centre, Kariavattom Campus, Thiruvananthapuram, Kerala, India*

[2]*The National Society of Ethnopharmacology, Mannamoola, Thiruvananthapuram, Kerala, India*

[3]*Amity Institute for Herbal and Biotech Products Development, Thiruvananthapuram, Kerala, India*

ABSTRACT

The use of plants for medicine can be traced all the way back to the history and development of the human race. The desire to alleviate pain and discomfort and to have better health prompted humans to search for therapeutic or health-promoting agents from nature. Medicinal plants serve as the primary resource base of most traditional healthcare systems. During the 19th century, plant-based medicines faced a declining trend and almost disappeared in most developed countries. New discoveries of sulfa drugs, synthetics, antibiotics, cortisones, and other therapeutic agents emerged in quick succession and swept all other systems of medicine off their feet. Almost 35–40% of the medicinal plants found in tropical regions are cross-pollinated species, and there exists extensive genetic variability particularly in the secondary metabolites in these species. CSIR has initiated a coordinated program on drug discovery in the area of traditional systems of medicines. CSIR has commenced a number of bioprospecting programs with specified targets/

Bioprospecting Ethnomedicinal Plant Resources: Sustainable Utilization and Restoration.
Gopal Shukla, Jahangeer A. Bhat, A. P. Das, and Sumit Chakravarty (Eds.)

goals – such as the inter-laboratory collaborative programs on biomolecules/drugs and drug prospecting.

1.1 INTRODUCTION

The 21st century is known as the 'century of Biology' which is powered and instigated by technological expertise and scientific knowledge. Herbal technology, biotechnology, nanotechnology, and information technology (bioinformatics) will be the four major powerful technologies in the world influencing the prosperity and well-being of the populations across the globe (Pushpangadan & George, 2010). It was a logical outcome of the intellectual triumphs of the 20th century – now called as the 'Century of Physics' which was triggered by the outstanding contribution of a few scientists like Roentgen, Mendeleev, Rutherford, Albert Einstein, and Max Planck. 'Century of Physics' stemmed from the quantum physics of Max Planck in 1900 who described that energy is dispensed in discrete packets called 'quanta' and not as a continuous stream. This led to the understanding of the relationship between matter and energy. Einstein took Max Planck's theory further in his Special Theory of Relativity in 1905 and General Theory of Relativity in 1916 (Pushpangadan et al., 2014). Einstein combined space and time as a continuum and explained mass and energy are inter-convertible ($E = mc^2$). High temperature can create matter and temperature is inseparable from radiation. Electromagnetic radiation turns into masses. The inverse of this process, turning mass into energy, makes a nuclear bomb. The advances in physics led to the development of sophisticated instruments for experimentation that gave a better understanding of the structure and functional dynamics of natural objects. The previous century's physics thus split atoms, made atom bombs, airplanes, rockets, television, telecommunications, X-rays, electron microscopes, analytical tools, computers, and satellites. These advances in material sciences of physics have immensely contributed to the understanding of the structure and functional dynamics of life at the cellular level, mechanized diagnosis, spliced genes, and entering into genomic or proteomic therapies.

1.2 TRADITIONAL KNOWLEDGE (TK)

The Convention on Biological Diversity (CBD) defined biodiversity as the variability among living organisms from different sources such as terrestrial, marine, and other aquatic ecosystems. It also refers to the diversity existing within species, between species, as well as ecosystems. The continual interactions, experimentation, and observations of indigenous and local communities with the surrounding environment developed the community-based functional knowledge which has been preserved and maintained over many generations. The traditional system of knowledge developed as a result of the co-evolution and co-existence of traditional cultures and resource utilization patterns. In this context, 'access' denotes the capacity of individual(s) or organizations to receive, exchange, or utilize biological resources in nature for multiple purposes including commercial applications.

1.3 GENESIS OF THE GLOBAL CONCERN ON BIODIVERSITY/ BIOTECHNOLOGY

The United Nations Environment Programme (UNEP) identified nature, wildlife, and genetic resources as important areas of conservation. In 1987, the World Commission on Environment and Development (WCED) came up with a report called 'Our common future' which highlighted the conservation of biological diversity for sustainable development. Later in 1992, the CBD was created with 186 member countries as a result of the constitution of an ad-hoc working group of technological and legal experts by the UNEP. CBD operates as an international law enforcement agency for the conservation and sustainable use of the environment and biodiversity (Rao et al., 2004).

1.4 PATENTS AND INTELLECTUAL PROPERTY RIGHTS (IPRS)

A patent is a collection of exclusive rights given by the government to an inventor or applicant for a period of 20 years normally from the filing date. It is basically a monopolistic right that excludes third parties from activities such as making, selling, using, and importing patented inventions. The chronology of patent or monopolistic rights is given in Table 1.1. The Statute of Monopolies Act, 1623 was initially proposed by Great Britain through a parliamentary act, which granted exclusive rights over novel inventions for a term of 14 years. Later at the end of the 18th century and during the 19th century, various countries including USA and France started formalizing comprehensive patent statutes. The chronology of property rights conventions/agreements/treaties formation of the TRIPS (trade-related aspects of intellectual property rights) regime is given in Table 1.2.

TABLE 1.1 Chronologies of Patent or Monopolistic Rights

Year	Country	Patenting Event
700 BC	Greece	Idea of monopolistic rights/privilege.
15th century	Italy	Patent introduced.
1421	Republic of Florence	First known patent grant by a "state" to an inventor.
1474	Venice	Patent ordinance.
1623	Great Britain	Statute of monopolies proposing grant of exclusive rights to inventors for new inventions for a period of 14 years.

TABLE 1.2 Chronology of Property Right Conventions/Agreements/Treaties

Year	Conventions/Agreements/Treaties
1883	Paris convention for the protection of industrial property.
1886	Berne convention for the protection of literary and artistic works.
1891	Madrid agreement concerning the international registration of trademarks.
1947	Formation of GATT (general agreement on tariffs and trade).
1958	Lisbon agreement for the protection of appellations of origin and their international registration.
1961	Rome convention for the protection of performers, producers of phonograms and broadcasting organizations.
1967	Decision to form WIPO (world intellectual property organization) by UN.
1970	PCT (patent cooperation treaty).
1994	WIPO established WTO (World Trade Organization) which administered the TRIPS.
2005	TRIPs came into full force.

During the 19th and 20th centuries, IPR remains the primary steering force for industrial growth and development. However, this didn't recognize the informal system of innovations by the indigenous communities. The provisions of mechanisms for the compensation and sharing of benefits with indigenous communities are not laid out. IPR laws in general ignored the interest of traditional/local communities. This is due to the fact that the principles followed by the societies of the developed and developing countries are quite different from the concept of intellectual property and other resources followed by the indigenous communities. But there is great scope to protect the interests of such communities through various laws. However, this requires access to information, better legal advice, and monetary resources, which most of the indigenous people couldn't afford.

1.5 IPR ISSUES AND BENEFIT SHARING STRATEGIES

1.5.1 CONTRIBUTION TO TKDL (TRADITIONAL KNOWLEDGE DIGITAL LIBRARY) AND TKRC (TRADITIONAL KNOWLEDGE RESOURCE CLASSIFICATION)

Value addition to TK and indigenous medicinal plants is required through scaling up of IPRs, pharmaceuticals, herbal drugs, natural products and byproducts, nutraceuticals, and other functional foods. India should develop capabilities in the sectors of biotechnology, bioinformatics, and bioprospecting via national, regional, and global 'biopartnership' programs while addressing and resolving the issue of access to and transfer of genetic resources as well as technologies between North-South countries. Steps to check bio-piracy and misappropriation of genetic resources and TK are required. The development of an international system for the protection of TK is urgently needed (Pushpangadan & Pradeep, 2008).

1.5.2 BIOTECHNOLOGY

Manipulation of biological organisms, systems, or biological materials with the objective of value addition for human use is called biotechnology. Modern biotechnology is a collection of powerful tools employed to understand the structure and functional dynamics of living organisms at the cellular or even up to the molecular level so as to manipulate them to the best advantage of humankind. Biotechnology is becoming increasingly trans-disciplinary and one of the most powerful technologies of the 21st century which has revolutionized the life sciences and has grown as a separate discipline encompassing a number of other disciplines like cell biology, immunology, molecular biology, recombinant DNA (deoxyribonucleic acid) technology and bioinformatics. The development of biotechnology (human genome projects, animal and plant genome projects, genomics, proteomics, and metabolomics) is making visible impacts in varied areas such as biology, medicine, agriculture, and the environment (Pushpangadan et al., 2013).

1.5.3 BIOPROSPECTING

A systematic exploration of genes, non-synthetic compounds, designs, and whole organisms, which possess the potential for the development of value-added products through biological observations and biochemical, biophysical, and genetic methods without causing disturbance to the environment.

1.5.4 SYSTEM BIOLOGY APPROACH

A biological system not only refers to the mere assembly of cells, tissues, genes, and proteins, rather it is significant in terms of the various interactions between these units-systems biology. Understanding the structural and functional dynamics of the cell system and its fine-tuned regulation is a necessity. The challenge is to design systems to deal with the extraction, separation, and refinement of chemistries and elements of interest. The strength of the Indian biological landscape is enlisted below:

- Rich in biodiversity at all levels, i.e., species, genes, and habitat.
- Rich cultural diversity that resulted in diverse indigenous systems of knowledge.
- Humanity has accessed only a small portion of nature's enormous genetic library. About 70-80% of the genetic potential of India and other South Asian nations is yet to be tapped.
- Excellent opportunities to discover new genes, medicines, pharmaceuticals, and chemicals/raw materials for emerging industries.

1.6 ETHNOPHARMACOLOGY

Modern healthcare systems are largely market-driven by the pharmaceutical industries, which mainly focus on curing illness rather than the maintenance of health. The economic foundation of this sector is undermined by disease prevention and eradication. There are no effective treatments for the majority of degenerative disorders common in the growing population and for resurgent resistant infections. The clinical efficacy of several commonly used contemporary drugs has also not been adequately validated. The USA has the highest per capita yearly expenditure on healthcare ($3,600), but still, 15% of the population lacks access to even the most basic medical facilities (the best Indian state like Kerala has an annual per capita expenditure of ~ $15). The majority of potent phytomedicines on the market are available as whole plant extracts. Practitioners believe that the synergistic interactions between the components of an individual or herbal mixtures are a key component of therapeutic efficiency. The mechanism of action of numerous phyto-medicines still lacks a clear understanding of how they work, and in certain cases, a total extract performs better than an equivalent dose of an isolated ingredient. In order to demonstrate conclusively the concept of synergism unequivocally, chemical and pharmacological evidence is required.

The global health market is currently experiencing the repolarization of products. There is a shift in people's attitude toward buying more healthy foods, preventive vitamins, and herbs and less dependent on pharmaceutical drugs. Products using agrotechnology are greatly preferred by the people and they believe that good herbal remedies will reduce the frequency of illnesses (Rao et al., 2004). Reverse pharmacology refers to the science of pursuing clinical or experiential leads through transdisciplinary exploratory and experimental studies, at different levels of biological organization; this is for understanding the mechanisms of action or optimizing efficacy, safety, and acceptability of the lead-modality. What is required is 'Scientific evidence.' The answer lies in providing scientific validation for efficacy and safety. The solution is to offer scientific proof of efficacy and safety. The objective is more likely to be hit based on Ethnomedical and Ethnobotanical screening. According to Hansen et al. (1995), applied research aims to transform traditional medicine into contemporary medicine (pharmacotherapy) or to expand its original application using modern technologies. Basic research aims to provide a rational explanation for how a traditional medicine works (phytotherapy). Small-scale *in vitro* biological test systems, which can test multiple samples for activity in a short amount of time, provide data from replicate tests to enable sound statistical analyses, and are based on cultured cells, isolated enzymes, or cloned receptors, have become a major feature of the drug discovery process over the past 30 years. These methods are now often used by the pharmaceutical industry for a diverse array of fractions for leads in the process of drug development. Ethnopharmacology in India was started by Dr. C. K. Atal and Dr. P. Pushpangadan in 1985. Dr. P. Pushpangadan was the Chief Coordinator of All-India Coordinated Research Project (AICRP) on Ethnobiology. Dr. Pushpangadan along with Prof. M. G. K. Menon, Dr. S. Rajasekharan, Dr. V. George, Dr. B. G. Nagavi and others registered a society in 1986. The first ethnopharmacology laboratory started at IIIM (Indian Institute of Integrative Medicine), Jammu, the second one at JNTBGRI, Trivandrum, and the third one at CSIR-NBRI, Lucknow. Later many ethnopharmacology laboratories were started in different parts of the country. For the discovery of bioactive molecules. The NMITLI (New Millennium India Technology Leadership Initiative) program and 20 other laboratory networked programs on the discovery of bioactive compounds were started by the CSIR. These programs mainly derived their cues from traditional remedies. Many laboratories outside of the CSIR and the pharmaceutical industry are working together on this endeavor. The initiative aims to generate standardized herbal formulations, practice Ayurvedic medicine, and establish a pharmacoepidemiologic data basis. Traditional and integrative health sciences are increasingly being used in research and clinical practice. Common approaches to drug discovery are chemical biology, chemical synthesis, serendipity, combinatorial chemistry, genomics, and proteomics. It also involves ethnopharmacology, reverse pharmacology, holistic approach, systems biology, and personalized medicine. Research and development work on medicinal plants in India is now focused on the following aspects:

- Expertise in resource survey on availability and distribution of medicinal plant species in natural habitat.
- Genetic enhancement of high-yielding and high-quality medicinal plants using conventional and modern techniques including biotechnological interventions.
- Packages of practice for agro-techniques such as organic cultivation and use of marginal/wasteland, as well as pre- and post-harvest cultivation.
- Standardization and authentication of compound herbal formulations and crude/unprocessed drugs.
- Developing standardized chemical and biological profiles for herbal formulations so that the pharmaceutical industry may easily refer to them and receive a comprehensive package in accordance with the guidelines of WHO.
- Evaluation of pesticide residue, heavy metals, and microbial contamination in crude samples or finished products.

1.7 CONCLUSION

The driving force of the 21st century is knowledge and a nation's future is determined based on its ability to convert knowledge into wealth and social welfare. A novel concept of 'knowledge engineering' for building up future 'knowledge societies' and 'knowledge industries' is now gaining popularity and acceptance both nationally and internationally. Modern biotechnology has an important role to play in nation-building through bioprospecting. Healthcare systems are greatly influenced by phytonutrients and phytochemicals.

They have tremendous potential in providing a multitude of health benefits such as the prevention and treatment of diseases and physiological disorders. New technologies are constantly being developed to isolate and identify the constituents responsible for the various plant activities.

ACKNOWLEDGMENT

The author expresses his sincere thanks to Dr. Ashok K. Chauhan, Founder President of RBEF and Amity Group of Institutions and Shri. Atul Chauhan, President, Ritnand Balved Education Foundation for facilities and encouragement. The author also expresses his gratitude to Smt. Sunija for helping in preparing the manuscript.

KEYWORDS

- **bioprospecting**
- **ethnomedicine**
- **knowledge societies**
- **pesticide residue**
- **phytochemicals**
- **phytonutrients**
- **traditional knowledge**
- **traditional rights**

REFERENCES

Hansen, K., Nyman, U., Smitt, U. W., Adsersen, A., Gudiksen, L., Rajasekharan, S., &Pushpangadan, P., (1995). *In-vitro* screening of traditional medicines for antihypertensive effect based on inhibition of the angiotensin converting enzyme (ACE). *J. Ethnopharmacology*, *48*, 43–51.

Pushpangadan, P., & George, V., (2010). Ethnomedical practices of rural and tribal populations of India with special reference to the mother and childcare. *Indian Journal of Traditional Knowledge*, 9–17.

Pushpangadan, P., & Pradeep, P. R. J., (2008). *A Glimpse at Tribal India-An Ethnobiological Enquiry*. Amity Institute for Herbal and Biotech products Development, Thiruvananthapuram.

Pushpangadan, P., Ijinu, T. P., & George, V., (2014). Origin, growth and development of ethnobiology/ethnobotany, human culture, and civilization. In: Kumar, S., (ed.), *Ethnobotanical Studies in India* (pp. 1–11). Deep Publications, New Delhi.

Pushpangadan, P., Vipin, M. D., Ijinu, T. P., & George, V., (2013). Phytochemical characterization of medicinal plant in herbal drugs. In: Tripathi, I. P., (ed.), *Chemistry, Biochemistry and Ayurveda of Indian Medicinal Plants* (pp. 1–12). Mahaveer Publishers.

Rao, C. V., Ojha, S. K., Radhakrishnan, K., Govindarajan, R., Rastogi, S., Mehrotra, S., &Pushpangadan, P., (2004). Antiulcer activity of *Ulteriasalicifolia* rhizome extract. *Journal of Ethnopharmacology*, *91*, 243–249.

CHAPTER 2

Ethnomedicinal Plants: Medicine and Nutrition for Future Generations

K. ABHA MANOHAR,[1] N. N. SHAHINA,[2] BIPLOV CHANDRA SARKAR,[2] GOPAL SHUKLA,[2,3] VINEETA,[2] SUBRATA DAS,[2] MANENDRA SINGH,[2] SHIVA POKHREL,[4] and SUMIT CHAKRAVARTY[2]

[1]*Department of Agronomy, MS Swaminathan School of Agriculture, Centurion University of Technology and Management, Odisha, India*

[2]*Department of Forestry, Uttar Banga Krishi Viswavidyalaya, Pundibari, Cooch Behar, West Bengal, India*

[3]*Department of Forestry, North Eastern Hill University, Tura Campus, Tura, Meghalaya, India*

[4]*International Center for Integrated Mountain Development (ICIMOD), Khumaltar, Lalitpur, Kathmandu, Nepal*

ABSTRACT

The concept of nutritionally rich and medicinally important ethnic plants is valued like never before in this era. Ethnic knowledge harbors many different plants which are less known to modern science. Food can be beneficially used as medicine if it's been taken wisely. The global nutraceutical market is expanding at a high pace while in India it is evolving and gaining momentum. Nutraceutical helps in reducing the risk of various chronic lifestyle-related diseases as it modulates metabolism. It is very relevant in present times as globalization and economic development have damaged our eating habits thus questioning healthy living. A systematic key approach is needed as many misconceptions and duplicate products surround this industry. For the well-being of humankind and to regulate and minimize

Bioprospecting Ethnomedicinal Plant Resources: Sustainable Utilization and Restoration.
Gopal Shukla, Jahangeer A. Bhat, A. P. Das, and Sumit Chakravarty (Eds.)

those misconceptions it is essential to raise scientific backup and popularization of this idea. People nowadays embrace living naturally and healthily. Owing to the high acceptability of this term, it is widely used in marketing varied products. Thus, there is greater scope for this newly emerging science along with incorporating traditional knowledge (TK) in research and quality improvement.

2.1 INTRODUCTION

Recent years have witnessed sudden progress in the use of plant resources in healthcare. The age-old relationship between human beings and plants is irreplaceable. Plant diversity is vital for India's ecological, environmental, and healthcare sectors. The ancient relationship between man and plants is ethnobotany. It is also known as aboriginal botany. Indigenous people play a pertinent role in the usage of traditional ways of using plants as a healing remedy for various diseases, culinary purposes, and other construction and craft works too. Plants possess an array of secondary metabolites in them which can be carefully utilized by humans for different applications like antibiotics, antimycotics, animal nutrition, cosmetics, food additives, and bio-repellents. It is very relevant in present times as globalization and economic development have damaged our eating habits thus questioning healthy living. Foods with high calories, carbohydrates, and unsaturated fats have made human health terrible in the present times.

Back to nature is the only way to combat most of these lifestyle-related diseases/disorders. In ancient times, people were using only natural products for nutrition and medical purposes as well. But modernization has totally destroyed most of the plant resources and thus such traditional practices. Still, in some places of the world, people are relying on ethnic or folk medicine mostly because of their poor economic status or remoteness. The sudden surge in medicinal and aromatic plants (MAPs) in healthcare is mainly due to the various side effects created by unsafe synthetic pharmacological agents. People are more concerned about maintaining their health with innocuous and effective natural products. Going back to nature is the only way to achieve this.

2.2 ETHNOMEDICINAL PLANTS

Recently, ethnobotany and ethnomedicinal plants have gained wide acceptance by the public. According to the International Union for Conservation of Nature (IUCN) and the World Wildlife Fund (WWF), about 50,000 to 80,000 flowering plant species are used for medicinal purposes. Approximately 20,000 or possibly more of the 60,000 plant species classified as threatened are from India alone (Kala & Sajwan, 2007). Traditional medicine has been used by approximately 60% of the world's population and 80% of the people residing in developing nations. According to Bhat et al. (2013), medicinal plants represent an important part of healthcare for more than 4.5 billion people in the developing world. Due to the high cost of allopathic medications and their negative side effects, medicinal plants have gained more popularity in rural areas (Marwat et al., 2008). Ethnic people still use medicinal plants for their medical purposes, primarily because they are easily available and secondly because of the poor economic conditions prevailing in such communities. There are many unknown sources and uses of plant species among tribal people. The knowledge source is depleting as the younger generation doesn't care about such things and they are moving towards cities. Ethnobotany can be defined as the entire natural and traditional relationship, as well as interactions, between man and his surrounding plant wealth (Das & Rahman, 2014). It has been estimated that 5-10% of the plant species in India have been identified for the presence of biologically active compounds (Singh et al., 2007). However, only 25% of medical drugs are based on plant sources in developed countries (Cragg et al., 1997).

The daily rigorous life has made people focus on ready-to-eat or ready-made things. Now people have realized the usefulness of dietary supplements, nutritional therapy, phytotherapy, and nutraceuticals. Nutraceuticals (nutrients + pharmaceuticals) are the nonspecific biological therapies used to enhance wellness, prevent malignancy, and manage symptoms. Based on their chemical components, they are divided into the following categories: (i) nutrients such as vitamins, minerals, amino acids, etc.; (ii) herbs or botanicals used in Ayurveda; and (iii) dietary supplements like probiotics, prebiotics, antioxidants, and enzymes (Hathcock, 2001).

2.3 NUTRACEUTICALS

In 1989, Dr. Stefan DeFelise coined the word "Nutraceutical," which is a combination of "Nutrition and Pharmaceutical" (Table 2.1). A food or component of food that offers medical or health benefits, such as the prevention and treatment of disease can be referred to as a nutraceutical. These products may range from simply isolated minerals or nutritional supplements to genetically engineered foods and herbal products. Natural functional/medical foods or bioactive phytochemicals that are curative, disease-preventive, and health-promoting are also referred to as nutraceuticals. They may cover functional foods and beverages that contain particular ingredients like vitamins, lipids, proteins, carbohydrates, minerals, or other essential nutrients that have health benefits. Dietary supplements deliver nutraceuticals in a nonfood matrix form like tablets or capsules with higher dosages than what is found in normal food (Brower, 1998; Zeisel, 1999). Such nutraceuticals may be referred to as drugs when they are employed in the treatment or prevention of diseases. The plethora of phenolic chemicals, terpenoids, sulfur compounds, pigments, and other naturally occurring antioxidants found in vegetables, fruits, whole grains, herbs, nuts, and seeds serve as therapeutic agents for a variety of medical conditions (Prabu et al., 2012). Thus, the nutraceutical is a very broad term that includes a variety of substances/compounds ranging from essential metals, large polymers and even bacteria.

TABLE 2.1 Nutraceuticals versus Pharmaceuticals

Nutraceuticals	Pharmaceuticals
Nutraceuticals are highly concerned with the prevention and ensuring wellness.	Pharmaceuticals are majorly focused on illness and treatment.
Nutraceuticals neither require medical supervision nor the approval of the FDA (Food and Drug Administration).	Pharmaceuticals are designed for medical use under a physician's supervision and are subjected to the approval of the FDA.
Nutraceuticals are sold without any prescription at health food stores, drug stores, grocery stores, the internet, etc.	Pharmaceuticals require a prescription from a certified doctor for purchase.
Nutraceuticals function on the root cause of the problem and take a long time in showing results. The products do not have any side effects on the human body.	Pharmaceuticals work on symptoms in a particular part or whole body, providing instant results along with or without side effects.

2.3.1 CATEGORIES OF NUTRACEUTICAL

Several studies have highlighted the importance of nutraceuticals in numerous ailments including diabetes. The majority of nutraceutical products, such as cardiovascular agents, anti-obesity agents, anti-diabetic agents, anti-cancer agents, immune boosters, substances that manage chronic inflammatory disorders, and formulations to treat degenerative diseases, have been shown to have multidimensional therapeutic benefits as well as physiological benefits (Rajasekaran et al., 2008). They can be categorized into three broad classes (Craig & Beck, 1999; Dureja et al., 2003) as nutrients (substances with recognized nutritional functions, like vitamins, minerals, amino acids, and fatty acids), herbals (herbs or botanical products as concentrates and extracts), and dietary supplements (reagents derived from other sources like pyruvate, chondroitin sulfate, steroid hormone precursors serving defined functions such as sports nutrition, weight-loss supplements, and meal replacements).

2.3.2 POTENTIAL OF NUTRACEUTICALS

Nutraceuticals are found in a mosaic of products ranging from - (i) the food industry; (ii) the herbal and dietary supplement market; (iii) the pharmaceutical industry; and (iv) the newly merged pharmaceutical/agribusiness/nutrition conglomerates (Dureja et al., 2003). The most well-known nutrients are antioxidants, water-soluble, and fat-soluble vitamins, and they are now frequently used to treat many serious illnesses (Table 2.2).

TABLE 2.2 Nutraceuticals in Major Health Issues/Diseases

Disease	Nutraceuticals Used
Cardiovascular	• Dietary fibers, poly-unsaturated fatty acids, polyphenols, flavonoids, vitamins, omega-3, and minerals.
Diabetes	• Lipoic acid and dietary fibers.
Obesity	• Buckwheat seed proteins, antioxidants in caffeine and green tea, Vitamin C, chitosan, and glucomannan.
Cancer	• Flavonoids, saponins, and lycopene. • Spinach, potatoes, clover, alfalfa, cucumber, and turmeric contain anti-tumor and anti-mutagenic substances.
Allergy	• Quercetin onions, green tea and red wine reduces the inflammation resulting from hay fever, gout, asthma, and arthritis.

Source: Baldi et al. (2013).

Antioxidant intake, be it through dietary intake or supplementation, has been associated with numerous potential advantages. In general, antioxidants may be helpful in the prevention of cerebrovascular disease and cancer (Keservani et al., 2010). A high vitamin E diet may fend against Parkinson's disease. Dehydro-ascorbic acid, the oxidized form of vitamin C, is shown to easily cross the blood-brain barrier, according to the findings by Agus et al. (2000). These findings have significance in enhancing the intake of antioxidants in the central nervous system, which may improve Alzheimer's disease treatment. The reduction of low-density lipoprotein oxidation and ensuing atherosclerosis has been made possible by the combination of vitamins E, C, and beta carotene (Jialal & Fuller, 2010). As vitamin supplements stimulate macrophage and T cell activity, they are related to the improved antibody response to the hepatitis B and tetanus vaccines. Selenium serum levels are low in those who are genetically prone to pancreatic cancer; it is therefore considered that selenium supplementation may help to avoid this disease. Selenium has also been tested for potential benefits in the treatment of skin cancer and asthma, but the results are still speculative (Iyen et al., 2019). More than 100 enzymes that are involved in digestion, metabolism, and wound healing have zinc as an essential constituent. L-arginine is a semi-essential amino acid and a substrate for the synthesis of nitric oxide (Keservani et al., 2010).

Flavonoids are the primary active nutraceutical components of Ayurvedic medicines. In addition to their considerable effects on the central nervous system, phenolic compounds typically have antioxidant, antimicrobial, anti-ulcer, hepatoprotective, anti-inflammatory, antidiabetic, vasorelaxant, anti-atherosclerotic, anti-thrombogenic, cardioprotective, and anti-neoplastic activity. It implies that they could play a crucial role in managing the complex pathophysiology of diabetes and its associated consequences (Tapas et al., 2008). A phytochemical is a naturally occurring bioactive substance that can be found in plants in the form of vegetables, fruits, medicinal plants, flowers, leaves, and roots. It functions with nutrients and fibers to protect against disease by acting as a defensive system (Sudhanshu et al., 2012). Various plant species are frequently combined to create herbal medicines. The effects of such mixtures may result from the synergistic impact of several phytochemicals (Williamson, 2001). Commercial antimicrobial chemotherapy frequently uses these synergistic combinations to treat a number of infections.

2.3.3 PLANT-DERIVED NUTRACEUTICALS

Nutraceuticals derived from plants are very essential in the contemporary medical and healthcare systems. Plant foods are edible plant parts (either as-is or after cooking, processing, etc.) that have been recognized as appropriate, desirable foods or diet ingredients by any community by custom, habit, and tradition. Food products typically give the body nourishment without having any negative short- or long-term effects on health or wellbeing. Nutraceuticals made from plants have both therapeutic and nutritional benefits (Table 2.3) due to the active ingredients or chemical compounds present in them (Table 2.4). Pharmaceutical and biotech industries are increasingly emphasizing the development and manufacture of nutraceuticals over pharmaceuticals.

TABLE 2.3 Functional and Nutraceutical Components in Common Herbals

Common Name	Scientific Name	Constituent	Health Benefits
Garlic	*Allium sativum*	Alliin, alicin	Anti-inflammatory, antibacterial, anti-gout, nervine tonic.
Maidenhair tree	*Ginkgo biloba*	Ginkolide, bilobalide	PAF antagonist, memory booster, antioxidant.
Ginger	*Zingiber officinale*	Zingiberene, gingerols	Stimulant, chronic bronchitis, hyperglycemia, throat pain.
Echinacea	*Echinacea purpurea*	Alkyl amide, echinacoside	Anti-inflammatory, anti-viral, immuno-modulatory.
Ginseng	*Panax ginseng*	Ginsenosides, panaxinosides	Immune and nervous system stimulator.
Liquourice	*Glycirrhyza glabra*	Glycirrhizine, liquirtin	Anti-inflammatory, anti-allergic, expectorant.
St. John's wort	*Hypericum perforatum*	Hypeicin, hyperforin	Anti-depressant, anti-HIV, anti-Hepatitis C.
Turmeric	*Curcuma longa*	Curcumin	Anti-inflammatory, anti-arthritic, anticarcinogenic, antiseptic.
Onion	*Allium cepa*	Allicin, allin	Hypoglycemic promoter, antibiotic, anti-atherosclerosis.
Valeriana	*Valeriana officinalis*	Valerenic acid, valerate	Tranquilizer, relieve migraine, menstrual pain, intestinal cramps, and bronchial spam.
Aloes	*Aloe barbadensis*	Aloins, alosin	Dilates capillaries, anti-inflammatory emollient, wound healing properties.
Goldenseal	*Hydrastis canadensis*	Hydrastine, berberine	Antimicrobial, astringent, anti-hemorrhagic, treatment of mucosal inflammation.
Senna	*Cassia angustifolia*	Sennosides	Purgative
Bael	*Aegle marmelos*	Marmelosin	Digestive, appetizer, treatment of diarrhea and dysentery.
Brahmy	*Centella asiatica*	Asiaticoside, madecosside	Nervine tonic, spasmolytic, anti-anxiety.

TABLE 2.4 Some Common Chemical Compounds Isolated from Plants Used as Nutraceuticals

Compound/Source	Properties	References
Allicin *Allium sativum*	Powerful antibiotic and antioxidant treat arteriosclerosis and serum cholesterol.	Block (1985)
Betaine (trimethyl glycine) Green leafy vegetables and germinated grains.	Reduce toxicity of homocysteine.	Pulliainen et al. (2010)
Bromelain (protease enzyme) *Ananas* sp.	Prevent heart disease, anti-aging, immune booster, anti-arthritis, anti-inflammatory, used as an inhalant to treat cold and flu.	Chen et al. (2002); Hakim et al. (2012)
Capsaicin or trans-8-methyl N-vanillyl-5 nonenamide-*Capsicum annum.*	Topical pain reliever, digestive, and antioxidant but in higher doses allergic and damage eyes and skin.	Dray (1992)
Carnitine or L-Carnitine asparagus.	Long-chain fatty acids transport to mitochondria.	Steiber et al. (2004)
Ellagic acid Strawberries, raspberries	Anticarcinogenic	Vattem& Shetty (2005)
Ricinoleic acid *Ricinus communis*	Dermal problems, constipation, lungs, liver, and kidney problems.	Okui et al. (1963)
Flavonoids, serotonin, theo-bromine, phenylethylamine. Chocolate (*Theobroma cacao*, vanilla orchid).	Cardiac and blood pressure regulator, neurotransmitter, and stimulator, antidepressant.	Taubert et al. (2007)
Curcumin *Curcuma longa*	Might clear brain plaque caused by Alzheimer's disease.	Mohan et al. (2000)
Glutathione (GSH)	Antioxidant	Mullineaux & Creissen (1996)
Hesperitin	Anti-inflammatory	Chauhan et al. (1978)
Hydroxy citric acid *Garcinia cambogia*	–	Lee & Lee (2007)
Isoquercitin *Mangifera indica* *Rheum nobile*	Increases blood flow for varicose veins, arterial flow, increase brain function and might treat progressive Alzheimer's disease.	Hiroyuki et al. (2003)
Glycirrhizic acid Liquorice	Decongestant, anti-inflammatory, stomach ulcers.	Stormer et al. (1993)
Lignan (phytoestrogens) Rye, soybean, broccoli	Antioxidants, anti-aging	Prasad (2005)

TABLE 2.4 *(Continued)*

Compound/Source	Properties	References
Lutein, lutein esters (carotenoid) – marigold, kale, spinach, rosemary.	Ophthalmic health	Berendschot et al. (2000)
Nattokinase (enzyme) Soybeans	Fermentation (*Bacillusnatto*)	Chang et al. (2008)
Olive oil (monounsaturated fat) *Olea europaea*	Regulate good cholesterol level.	Machowetz et al. (2007)
Omega 3 fatty acids *Linum usitatissimum*	Ophthalmic health.	Holman (1998)
Phloretin Apple	Malaria as a quinine replacer, inhibits protein kinase C and regulates cellular translocation of sodium/potassium.	MacDonald & Bishop (1952)
Phytosterol Corn (germinated)	Regulate cholesterol absorption.	St-Onge et al. (2003)
Proanthocyanins Grapes	Prevent urinary tract infections.	Valverde et al. (2011)
Resveratrol Grapes	Anti-inflammatory, anticarcinogenic (breast).	Gehm et al. (1997)
Tall oil (derived phytosterols)	Reduce arteriosclerosis.	De Graaf et al. (2002)
Zeaxanthin (carotenoid) Marigold	Antioxidant, ophthalmic health, retards effect of aging eyesight or age-related macular degeneration (AMD).	Toyoda & Thomson (2002)

Non-edible herbal medicines are not considered nutraceuticals. Pharmaceuticals are substances produced from modified plants or animals that have medicinal use (usually produced through biotechnological intervention). Ayurvedic and other traditional medicines from plants are not regarded as a nutraceutical if they are derived from non-edible parts of the plant. Ayurveda and traditional medicinal plants should be supported by cutting-edge scientific research and technological advancements. For instance, it has been documented that several medicinal plants utilized in Ayurveda and indigenous healing customs contain hepatotoxic pyrrolizidine alkaloids. Some pharmacies and biotech companies incorrectly widened the definition of "nutraceutical" to include isolated components from non-edible wild plants (Kamboj, 2000). This is a dangerous trend considering human health. The term "nutraceutical" is also applied to a bioactive substance that is present in food but is isolated from a non-edible (but non-toxic) plant. Nutraceuticals, along with natural supplements and herbal products, are natural health products that enhance health (Shahidi, 2012). Botanicals are frequently used to refer to herbal products. A legal distinction between plant nutraceuticals and botanicals or herbs with pharmacological activity is required (Koch et al., 2014).

2.3.3.1 ETHNOMEDICINAL PLANTS-DERIVED NUTRACEUTICALS

Plants have been used to treat both acute and chronic diseases for as long as human civilization has existed. The primary ingredients of many of these medicinal herbs are nutraceuticals (Prabu et al., 2012). People who seek conventional healthcare frequently employ a significant category of nutraceuticals called ethnomedicinal plant bioactive components. These contain a wide range of active phytochemicals, such as phthalates, lignin, sulfides, polyphenols, carotenoids, coumarins, saponins, plant sterols, and terpenoids (Craig, 1999). For the treatment of benign prostatic hyperplasia, herbal extracts such as beta-sitosterols (found in Saw Palmetto berry), Cernilton (pollen extract), and *Pygeum africum* (African plum) have undergone clinical evaluation (Whitman, 2001; Villasenor et al., 2002).

Ethnic plants have been utilized for medicinal uses throughout the entire human history. Numerous chemical components, such as vitamins, flavonoids, terpenoids, carotenoids, phytoestrogens, and minerals are associated with their beneficial effects on health (Calucci et al., 2003; Suhaj, 2006). Plants including basil, lemon balm, marjoram, mint, oregano, rosemary, sage, savory, and thyme are abundant in bioactive terpenoids, phenolic acids, and flavonoids. Other herbs with high flavonoid content include parsley, chervil, coriander, dill, fennel, and lovage. Drumsticks, also known as *Moringa oleifera*, have been shown to contain 25 times more iron than spinach, 17 times more calcium than milk, 15 times more potassium than bananas, 9 times more protein than yogurt, and 10 times more vitamins than carrots. In addition to its abundance in these micronutrients, various experimental and clinical research has indicated that drumstick has the ability to treat hyperglycemia and dyslipidemia (William et al., 1993; Ndong et al., 2007; Jaiswal et al., 2009; Kumari, 2010; Giridhari et al., 2011; Mbikay, 2012). Researchers discovered that *Aloe vera* extract contains lectins, polysaccharides, and anthraquinones that have antidiabetic action (Eshun & He, 2004). *Acacia arabica* seeds and *Carallumaedulis* roots have also been shown to cause hypoglycemia by triggering the release of insulin from pancreatic cells (Wadood et al., 1989; Vats et al., 2002). A plant-based diet contains a variety of key non-nutritive components that play important roles in enhancing health in addition to the macro- and micronutrients like proteins, lipids, carbs, vitamins, or minerals required for regular metabolism (Halliwell, 1996; Dulloo et al., 1999; Bell & Goodrick, 2002).

2.3.3.2 SPICES AS A NUTRACEUTICAL

The majority of spices contain various amounts of anti-inflammatory, hypolipidemic, and antioxidant qualities. These qualities enhance cardiovascular health and aid in preventing and/or treating diabetes and cancer. Most of these spices are only used in trace amounts when preparing food. Some spices may also be hazardous when consumed in large quantities. The acceptable dose should be evaluated based on clinical trials when purified nutraceutical substances from food, particularly spices, are utilized; too high doses may not deliver intended health advantages and in rare situations, toxicity may also result. Scientific validation of the function of various spice-based nutraceuticals in different diseases requires long-term clinical trials. The adverse effects of some climatic circumstances can be managed with the use of certain spice-based nutraceuticals, such as thermogenic spices. For instance, black pepper is found more favorable in cold climatic conditions (Majeed & Subramoniam, 2017).

2.4 NUTRACEUTICAL: HEALTH BENEFITS

According to a recent study by Mapping Global Malnutrition Trends, the obesity rate in India has increased by around 20%, and a significant amount of health concerns associated with cardiovascular disease have come to light. One of the most serious growing health concerns in India is diabetes. The urban lifestyle and increasing influence of Western culture have been the major reasons for inclining cases of obesity and diabetes. Humans regularly consume a variety of foods that are high in phenolics. There are only a few negative side effects, but their half-life is comparatively long. Following consumption, they can be quickly absorbed in the intestine. They are freely accessible without a prescription and there is no need for an appointment with a doctor. Many people think that this strategy is more natural than utilizing prescription medication. They believe that taking nutritional supplements will make them feel better physically and mentally, give them more energy, and stave off disease. When they assume that conventional therapies have failed to cure their particular conditions, some people resort to these goods (Prabu et al., 2012).

2.4.1 RETARDS THE PROGRESSION OF DISEASE

Nutraceuticals may be utilized to enhance health, prevent chronic diseases, and delay the aging process (increasing life expectancy). They work on the principle of retarding the progression of diseases by acting upon the root cause of the diseases.

2.4.2 PREVENTION OF DISEASE

Nutraceuticals are good sources for preventing diseases including diabetes, kidney, gastrointestinal problems, as well as various other infections.

2.4.3 COMBATING LIFESTYLE-RELATED HEALTH ISSUES

Nutraceuticals are receiving appreciation as being advantageous in treating obesity, coronary heart diseases, cancer, diabetes, and other chronic and degenerative diseases.

2.4.4 REDUCES TREATMENT COST

Nutraceuticals offer low-cost and early control of risk factors for a number of diseases, e.g., heart disease. Nutraceuticals improve the immune system of a diseased person thus reducing treatment costs.

2.4.5 DELAYS AGING AND INCREASES LIFE EXPECTANCY

Better control of infectious diseases and access to a more nutritious diet has increased life expectancy in the 20th century. According to many reports, oxidative stress and certain nutritional deficiencies may contribute to the aging process, so intake of nutraceuticals in the form of antioxidants can delay the aging process.

2.4.6 SAFETY

As these are obtained from natural sources, they have little or no negative side effects and are safe to use. Due to their alleged safety and potential medicinal and nutritional advantages, nutraceuticals have recently attracted a lot of attention. The notion of nutraceuticals stemmed from a survey conducted in the UK, Germany, and France that suggested consumers place a higher value on nutrition than on exercise or inherited factors for maintaining good health. Nutraceuticals, which offer health advantages and serve as an alternative to contemporary medicine, have attracted increasing attention in recent years. It may be possible to reduce the consumption of conventional medications through nutraceuticals and thereby minimizing the chances of any adverse impacts. Nutraceuticals often exhibit distinctive chemical actions that are lacking in pharmaceuticals. Obesity, osteoporosis, cancer, diabetes, allergies, and dental issues are a few of the modern-day ailments that are expanding widely. Exercise and nutrition both play important roles in the prevention and treatment of obesity, which is becoming more prevalent throughout the world. Major components of nutraceuticals include nutrients, herbs, and dietary supplements, making them essential for maintaining health, fighting off various disease problems, and enhancing the quality of life. It is well known that food products can be used to treat diseases and promote health. The majority of drug compounds utilized today in formulations were once used in their basic form (Gupta et al., 2010). The term "nutraceutical" is used to describe a variety of items including medicines, functional foods, medicinal plants, and dietary supplements. The key components of nutraceuticals were found to have a wide range of positive health and medical advantages (Table 2.5). These compounds increase the health value of our diet and help us to live a long and healthy life by avoiding particular medical conditions (Elhassan et al., 2016). They provide psychological advantages from doing something for oneself as they were perceived as more natural than traditional medications and are less likely to produce adverse side effects. Moreover, they supplement specific food for the population with special requirements (e.g., nutrient-dense foods for the elderly).

TABLE 2.5 Nutraceuticals, Their Major Components and Their Benefit

Class/Components	Sources	Benefits
Beta-carotene	Carrots, tomato, sweet potato, spinach, and broccoli.	Neutralizes free radicals, strengthen antioxidant defense mechanism of cells, and maintains prostate health.
Lycopene	Tomatoes, apricot, grapes, and guava.	Enhances immunity.
Dietary fiber	Wheat bran, peas, beans, and lentils.	Good for digestive tract health.
Monosaturated fatty acids.	Tree nuts	Reduce the risk of coronary heart disease.
Flavonoids and flavonols	Onion, berry, cherry, soybean, pear, apple, tea, and broccoli.	Neutralize free radicals, strengthening of cellular antioxidant defenses.
Isothiocyanates. Sulforaphane.	Cauliflower, cabbage, broccoli, kale, horse radish.	Detoxification of undesirable compounds and antioxidant properties.
Phenols – caffeic and ferulic acid.	Apple, potato, yam, citrus fruits, and vegetables	Cardiac health and vision maintenance, antioxidant.
Stanol/sterol	Peanut, butter, soybean, corn oil.	Coronary heart disease risk reduction.
Prebiotics and probiotics.	Yoghurt, milk, and other dairy products.	Gastrointestinal health and systematic immunity.
Phytoestrogens, isoflavones	Soy-based food products.	Bone and brain health, immune system, and menstrual health.
Sulfides	Cruciferous vegetables.	Immunoregulatory properties.
Thiols and dithiolthiones		

Source: Modified from Dev et al. (2014).

2.5 ADVANTAGES OF NUTRACEUTICALS

- Nutraceuticals have the potential to promote healthy eating and aid in disease prevention as well as treatment.
- There is undoubtedly a very large and expanding market for nutraceuticals because it is more expensive, safe, and difficult to discover new molecules.
- Consumers believe that these "food-like compounds" are less hazardous or innocuous than conventional medicines.

- Nutraceuticals play an important role in treating improper dietary practices, which are thought to be a major cause of fatalities from coronary heart disease, certain types of malignancies, etc. (Golani et al., 2012).

2.6 DISADVANTAGES OF NUTRACEUTICALS

- They are not subjected to the same testing and regulations as pharmaceuticals.
- Nutraceuticals do not have any regulatory definition and the majority of them are not regulated by the regulatory bodies (Keservani et al., 2010).
- Manufacturers create unregulated products to create a high-profit margin.
- Bioavailability of nutrients is lower.

2.7 NUTRACEUTICAL MARKET

The market today consists of both traditional foods and non-traditional foods. Traditional nutraceuticals are only natural, whole foods that have been given fresh insight into their potential health benefits. Most fruits, grains, vegetables, fish, meat, and dairy consist of numerous natural components that provide several advantages beyond basic nutrition. For instance, tomatoes contain lycopene, salmon has omega-3 fatty acids, and soy contains saponins. According to certain studies, even tea and chocolate have health-promoting properties. Contrarily, non-traditional nutraceuticals are foods produced through agricultural breeding or the addition of nutrients and/or additives. The advent of pharmaceutical and large food companies into the nutraceutical market has increased competition in the sector (Table 2.6). Nutraceutical divisions have been developed by many food firms with the goal of offering a diverse product line. Pharmaceutical firms have also entered the arena by purchasing producers of dietary supplements. Major food and pharmaceutical firms, including Kellogg, Heinz, M& M, Quaker Oats, Unilever, Cargill, Hormel, GlaxoSmithKline, Warner-Lambert, Johnson & Johnson, and Wyeth, have recently entered the nutraceutical sector (Faisal & Varma, 2009).

TABLE 2.6 List of Marketed Nutraceutical Products and Their Manufacturer

Product	Category	Manufacturers
Garcinia cambogia capsules	Appetite supplement.	Bionova Lifesciences, India
Soya protein hydrolysate	Protein supplement.	Chaitanya Agrobiotech Group, India
Banana leaf extract	Nutritional supplement.	LactonovaNutripharm, India
Jetex F capsules	Nutritional supplement.	Sydler, India
Multivitamins capsules with essential trace elements	Vitamin supplements.	Zeon life sciences, India Ltd.
Niswell women	Health supplement.	Neiss wellness (India) Ltd.
Coral calcium	Calcium supplement.	Nature's Answer Hauppauge, USA
Rox	Energy drink.	Rox America, USA
Mushroom optimize	Immune supplement.	Jarrow formulas, USA
Calcirol	Calcium supplement.	Cadilla healthcare LTD, India
Proteinex	Protein supplement.	Pfizer Ltd., India
Threptin diskettes	Protein supplement.	Raptakos, Brett & Co. Ltd., Mumbai

Source: Adapted from Aditi (2019); Patil (2011).

2.7.1 GLOBAL NUTRACEUTICAL MARKET

In 2010, the global market for nutraceutical components increased by 5.8% annually to $15.5 billion, supporting a $197 billion global nutritional product industry. Due to their ability to improve and diversify their food, beverage, and pharmaceutical manufacturing capacities due to robust economic growth, China and India will emerge as the nutraceutical markets with the fastest rate of growth. The demand for herbal and non-herbal extracts increased by 6.5% annually to $1.85 billion in 2010 due to consumers' and doctors' growing acceptance of these products. Demand for nutrients, minerals, and vitamins reached $9.5 billion in 2010, an increase of 6.3% from 2005. Nutraceutical vitamin components had an increase in demand of up to 4.6% yearly, reaching $4.2 billion in 2010. Based on the effectiveness (Ef) benefits of nutraceuticals over synthetic compounds for adult and pediatric nutrition, natural vitamin E formulations and beta carotene (vitamin A) also do well in the global market (Prabu et al., 2012).

2.7.2 *INDIAN NUTRACEUTICAL MARKET*

In the past few years, nutraceuticals have attracted substantial interest among the Indian population due to their potential nutritional, safety, and therapeutic effects, and hence Indian companies are also entering the nutraceutical market (Table 2.7). The Indian nutraceutical market was estimated to be US $1 billion. The international market is expanding at a CAGR of 7%, and the Indian market is growing at a CAGR of 18% for the last three years, mainly in the functional food and beverages categories. However, the latent market in India is 2–4 times the current market size (Radhika et al., 2011). The Indian market scenario is mixed with both big and regional players, with exports as well. Both the Indian and global markets are attracting interest from the population. There are numerous issues that exist despite the ongoing growth. Cultivating medicinal plants is crucial for the nutraceutical industry. Over time, Americans and Europeans have begun to recognize that Ayurveda can be a source of materials of extraordinary quality. Increasing demand has caused many medicinal plants to be threatened with extinction in India. As a result, maintaining the cultivation of medicinal plants is crucial to the long-term viability of the nutraceutical industry.

TABLE 2.7 Leading Indian Companies in Nutraceuticals

Companies	Product Category	Products
Himalaya	Protein	HiownaKidz
Dabur	Herbal	Chyawanprash
	Functional beverage	Nutritionally improved fruit juices (real fruit power).
Parry nutraceutical	Vitamins and minerals.	Multigrain products.
Britannia	Functional food.	Multigrain bread.
Lifegenix	Antidepressants.	Medicinal drugs.
ITC	Functional food.	Multigrain flour.
NutraCulture	Health supplement.	Spirulina.
Xena Bioherbals Private Ltd.	Herbal extracts.	Soft gel capsules.
Amul	Functional food.	Probiotic Dhahi.

Source: Company Websites.

2.8 REGULATORY ASPECTS OF NUTRACEUTICALS

Some ingredients in nutraceutical products may have direct toxic effects and result in health issues. It is often easier to prove a nutraceutical product's safety than its effectiveness. Numerous nutraceutical items have been employed as nutritional and medicinal substitutes. Without adequate data to substantiate the safety and effectiveness of their products, many manufacturers make illegal claims. Customers, therefore, require confirmation that a product is secure and ideally, capable of performing as promised. Above everything, nutraceuticals should be safe (Patil, 2011). The appropriate authorities must pay attention to the nutraceutical's regulatory framework. In other countries, regulatory authorities are proactive in protecting consumers by amending existing laws to reflect changes, but in India, laws like the Prevention of Food Adulteration Act (1954), the Bureau of Indian Standards (BIS) Act (1998), and many others need to be reviewed in light of the production and marketing of nutraceuticals (Gupta et al., 2010; Palthur et al., 2010).

Numerous studies have evaluated the efficiency and safety of herbal treatments or dietary supplements, and their adverse effects also have been documented. To help patients and doctors in choosing the optimal risk-benefit ratio before using any herbal or conventional treatment (CT), this data must be combined with information gathered from adverse reaction monitoring systems. Through appropriate manufacturing practices, regulatory oversight, research initiatives, and the reporting of adverse events, definitive and greater control over the safety and quality of nutraceutical products might be attained (Koh & Woo, 2000). Regulators need to take measures to reduce these dangers. In order to increase the safety and effectiveness of dietary supplements, including medicinal herbs, more studies are required to characterize the pharmacology, stability, and bioavailability of these products (Bent & Ko, 2004).

2.9 CURRENT AND FUTURE DEVELOPMENTS IN NUTRACEUTICALS

Although nutraceuticals hold great promise for improving human health and preventing disease, regulatory toxicologists and health professionals could strategically collaborate to establish the right policies to deliver the ultimate therapeutic advantages to humanity. Quality controls, such as composition and concentrations of active ingredients in natural plants, and maintenance, are crucial for nutraceutical production processes. Extensive safety studies, such as acute, subacute, subchronic, chronic, and long-term toxicity studies, as well as supplementation studies in animals and clinical trials in people, are required to establish a product's safety and efficacy (Schilter et al., 2003; Kroes et al., 2004). Utilizing DNA microarray technology, researchers may also assess the effectiveness and safety of pharmaceuticals, chemicals, food supplements, and nutraceuticals (Roy et al., 2007).

2.10 CONCLUSION

It has been noted that very few nations have created regulatory criteria and adopted the word "nutraceutical." The lack of regulatory standards for nutraceuticals may cause a rush of fake and subpar goods onto the healthcare industry. It is past due for the Indian government to recognize and classify nutraceuticals, functional foods, and nutritional supplements. Additionally, the government must provide appropriate regulatory rules to oversee the healthcare industry and prevent the entry of counterfeit and subpar nutraceutical products. Nutraceuticals may be advantageous to our health, but we are still learning about their advantages and possible adverse effects (Chauhan et al., 2013). Government regulatory authorities also face several difficulties in this new category of health products, which exists between food and drugs. Health professionals, nutritionists, and regulatory toxicologists should strategically work together to plan appropriate regulations to provide the utmost health and therapeutic benefits to humankind.

KEYWORDS

- **conventional treatment**
- **diseases**
- **ethnomedicine**
- **healthy living**
- **herbal treatments**
- **medicines**
- **nutraceuticals**
- **nutrition**
- **pharmaceuticals**

REFERENCES

Aditi. *List of Nutraceutical Manufacturers in India*. Pharma Adda. Available at: https://www.pharmaadda.in/list-of-nutraceutical-manufacturers-in-india (accessed on 18 December 2023).

Agus, D. B., Gambhir, S. S., Pardridge, W. M., Spielholz, C., Baselga, J., Vera, J. C., & Golde, D. W., (2000). Vitamin C crosses the blood-brain barrier in the oxidized form through the glucose transporters. *J. Clin. Invest.*, *100*, 2842–2848.

Baldi, A., Choudhary, N., & Kumar, S., (2013). Nutraceuticals as therapeutic agents for holistic treatment of diabetes. *Int. J. Green Pharm.*, *7*, 278–287.

Bell, S. J., & Goodrick, G. K., (2002). A functional food product for the management of weight. *Critical Reviews in Food Science and Nutrition*, *42*(2), 163–178.

Bent, S., & Ko, R., (2004). Commonly used herbal medicines in the United States: A review. *Am. J. Med.*, *116*(7), 478–485.

Berendschot, T. T., Goldbohm, R. A., Klöpping, W. A., Van De, K. J., Van, N. J., & Van, N. D., (2000). Influence of lutein supplementation on macular pigment, assessed with two objective techniques. *Invest. Ophthalmol. Vis. Sci.*, *41*(11), 3322–3326.

Bhat, J. A., Kumar, M., Negi, A. K., & Todaria, N. P., (2013). Informants consensus on ethnomedicinal plants in Kedarnath Wildlife Sanctuary of Indian Himalayas. *J. Med. Plant Res.*, *7*(4), 148–154.

Block, E., (1985). The chemistry of garlic and onions. *Scientific American*, *252*, 114–119.

Brower, V., (1998). Nutraceuticals: Poised for a healthy slice of the healthcare market? *Nat. Biotechnol.*, *16*, 728–731.

Calucci, L., et al., (2003). Effects of gamma-irradiation on the free radical and antioxidant contents in nine aromatic herbs and spices. *J. Agric. Food Chem.*, *51*(4), 927–934.

Chang, Y. Y., Liu, J. S., Lai, S. L., Wu, H. S., & Lan, M. Y., (2008). Cerebellar hemorrhage provoked by combined use of nattokinase and aspirin in a patient with cerebral micro-bleeds. *Intern. Med.*, *47*(5), 467–469.

Chauhan, B., Kumar, G., Kalam, N., & Ansari, S. H., (2013). Current concepts and prospects of herbal nutraceutical: A review. *J. Adv. Pharm. Technol. Res.*, *4*(1), 4–8.

Chauhan, J. S., Srivastava, S. K., & Sultan, M., (1978). Hesperetin 7-rhamnoside from *Cordia obliqua*. *Phytochemistry*, *17*(2), 334–335.

Chen, P. F., Chang, S. T., & Wu, H. H., (2002). Antimite activity of essential oils and their components from *Cinnamomum osmophloeum*. *Quart. J. Chin. For.*, *35*, 397–404.

Cragg, G. M., Newman, D. J., & Snader, K. M., (1997). Natural products in drug discovery and development. *J. Nat. Prod.*, *60*, 52–60.

Craig, W. J., (1999). Health-promoting properties of common herbs. *Am. J. Nutr.*, *70*(suppl.), 491S–9S.

Craig, W., & Beck, L., (1999). *Can. J. Diet. Pract. Res.*, *60*, 78–84.

Das, U., & Rahaman, C. H., (2014). An observation on the phyto-therapeutic uses of plants by the tribal people of some forest areas in Birbhum and Burdwan districts, West Bengal. *Indian Journal of Applied Pharmaceutical Science*, *4*(4), 72–78.

De Graaf, J., De Sauvage, N. P. R., Van, D. M., Belsey, E. M., Kastelein, J. J., Haydn, P. P., & Stalenhoef, A. F., (2002). Consumption of tall oil-derived phytosterols in a chocolate matrix significantly decreases plasma total and low-density lipoprotein cholesterol levels. *Br. J. Nutr.*, *5*, 479–488.

Dev, R., Kumar, S., Singh, J., & Chauhan, B., (2014). Potential role of nutraceuticals in present scenario: A review. *Journal of Applied Pharmaceutical Science, 1*(4), 26–28.

Dray, A., (1992). Mechanism of action of capsaicin like molecules on sensory neurons. *Life Sci., 51*(23), 1759–1765.

Dulloo, A. G., Duret, C., Rohrer, D., Girardier, L., Mensi, N., Fathi, M., Chantre, P., & Vandermander, J., (1999). Efficacy of a green tea extract rich in catechin polyphenols and caffeine in increasing 24-h energy expenditure and fat oxidation in humans. *The American Journal of Clinical Nutrition, 70*(6), 1040–1045.

Dureja, H., Kaushik, D., & Kumar, V., (2003). Developments in nutraceuticals. *Ind. J. Pharmacol., 35*, 363–372.

Elhassan, G. O., Khan, M. M. U., Khalilullah, H., & Eid, E. E. M., (2016). Nutraceuticals- an opportunity of healthcare: A review. *Tropical Journal of Pharmaceutical Research., 5*(11), 229–239.

Eshun, K., & He, Q., (2004). *Aloe vera*: A valuable ingredient for the food, pharmaceutical and cosmetic industries: A review. *Crit. Rev. Food Sci. Nutr., 44*, 91-96.

Faisal, N., & Varma, K. S., (2009). *Nutraceuticals and its Impact on Health Care.* https://www.scribd.com/document/102377477/Nutraceuticals-and-Its-Impact-on-Health-Care (accessed on 18 December 2023).

Gehm, B. D., McAndrews, J. M., Chien, P. Y., & Jameson, J. L., (1997). Resveratrol, a polyphenolic compound found in grapes and wine, is an agonist for the estrogen receptor. *Proceedings of the National Academy of Sciences of the United States of America, 94*(25), 14138–14143.

Giridhari, V. V., Malhati, D., & Geetha, K., (2011). Anti-diabetic properties of drumstick (*Moringa oleifera*) leaf tablets. *Int. J. Health Nutr., 2*, 1-5.

Golani, S., Jain, G., Jadon, A. S., & Kaushik, A., (2012). Nutraceuticals: Modern era's nutritional pharmaceuticals. *Current Research in Pharmaceutical Sciences, 2*, 69–72.

Gupta, S., Chauhan, D., Mehla, K., Sood, P., & Nair, A., (2010). An overview of nutraceuticals: Current scenario. *J. Basic & Clin. Pharm., 2*, 55–62.

Hakim, M. A., et al., (2012). Role of health hazardous ethephon in nutritive values of selected pineapple, banana, and tomato. *J. Food Agric. Environ., 10*(2), 247–251.

Halliwell, B., (1996). Antioxidants in human health and disease. *Annual Review of Nutrition, 16*(1), 33–50.

Hathcock, J., (2001). Dietary supplements: How they are used and regulated. *J. Nutr., 131*, 1114–1117.

Hiroyuki, S., Yoshinori, H., Satoshi, N., Hitoshi, A., & Kazuki, K., (2003). Simultaneous determination of all polyphenols in vegetables, fruits, and teas. *Journal of Agricultural and Food Chemistry, 51*(3), 571–581.

Holman, R. T., (1998). The slow discovery of the importance of omega 3 essential fatty acids in human health. *J. Nutr., 128*(2), 427–433.

Iyen, Sandra, I., Ogofotha, Oghenekeno, G., Oko, & Odiba, J., (2019). Advances in nutraceuticals. *International Journal of Medicinal Plants and Natural Products, 5*(2), 8–22.

Jaiswal, D., Kumar, R. P., Kumar, A., Mehta, S., & Watal, G., (2009). Effect of *Moringa oleifera* Lam. leaves aqueous extract therapy on hyperglycemic rats. *J. Ethnopharmacol., 123*, 392–396.

Jialal, I., & Fuller, C. J., (2010). Effect of vitamin E, vitamin C and beta-carotene on LDL oxidation and atherosclerosis. *Can. J. Card., 1*(11), 97–103.

Kala, C. P., & Sajwan, B. S., (2007). Revitalizing Indian systems of herbal medicine by the National Medicinal Plants Board through institutional networking and capacity building. *Curr. Sci.*, *93*(6), 797–806.

Kamboj, V. P., (2000). Herbal medicine. *Curr. Sci.*, *78*(1), 35–39.

Keservani, R. K., Kesharwani, R. K., Sharma, A. K., Vyas, N., & Chadokar, A., (2010). Nutritional supplements: An overview. *International Journal of Current Pharmaceutical Review and Research*, *1*(1), 59–97.

Koch, A., Brandenburger, S., Türpe, S., & Birringer, M., (2014). The need for a legal distinction of nutraceuticals. *Food and Nutrition Sciences*, *5*, 905–913.

Koh, H. L., & Woo, S. O., (2000). Chinese proprietary medicine in Singapore: Regulatory control of toxic heavy metals and undeclared drugs. *Drug Safety*, *23*(5), 351–361.

Kroes, R., Renwick, A. G., Cheeseman, M., Kleiner, J., Mangelsdorf, I., Piersma, A., Schilter, B., et al., (2004). Structure-based thresholds of toxicological concern (TTC): Guidance for application to substances present at low levels in the diet. *Food and Chemical Toxicology*, *42*(1), 65–83.

Kumari, D. J., (2010). Hypoglycemic effect of *Moringa oleifera* and *Azadirachta indica* in type-2 diabetes. *Bioscan*, *5*, 211–214.

Lee, K. H., & Lee, B. M., (2007). Evaluation of the genotoxicity of (–)-hydroxy citric acid (HCASX) isolated from *Garcinia cambogia*. *J. Toxicol. Environ. Health*, *70*(5), 388–392.

MacDonald, R. E., & Bishop, C. J., (1952). Phloretin: An antibacterial substance obtained from apple leaves. *Canadian Journal of Botany*, *30*, 486–489.

Machowetz, A., Poulsen, H. E., Gruendel, S., Weimann, A., Fitz, M., & Marrugat, J., (2007). Effect of olive oils on biomarkers of oxidative DNA stress in Northern and Southern Europeans. *FASEB J.*, *21*, 45–52.

Majeed, M., & Subramoniam, A., (2017). Nutraceuticals in health care. In: *Insights on Global Challenges and Opportunities for the Century Ahead* (p. 279).

Marwat, S. K. M. A., Khan, M., Ahmad, M., Zafar, & Rehman, F. U., (2008). Ethnophytomedicines for treatment of various diseases in D. I. Khan District. *Sarhad J. Agric.*, *24*(2), 305–315.

Mbikay, M., (2012). Therapeutic potential of *Moringa oleifera* leaves in chronic hyperglycemia and dyslipidemia: A review. *Front. Pharmacol.*, *3*, 1–12.

Mohan, R. S., Anderson, A. M., & Mitchell, M. S., (2000). Isolation of curcumin from turmeric. *Journal of Chemical Education*, *77*(3), 359–360.

Mullineaux, P., & Creissen, G., (1996). Opportunities for the genetic manipulation of antioxidants in plant foods. *Biochem. Soc. Trans.*,*24*, 829–835.

Ndong, M., Uehara, M., Katsumata, S., & Suzuki, K., (2007). Effects of oral administration of *Moringa oleifera* Lam on glucose tolerance in goto-kakizaki and Wistar rats. *J. Clin. Biochem. Nutr.*, *40*, 229–233.

Okui, S., Uchiyama, M., & Mizugaki, M., (1963). Metabolism of hydroxy fatty acids. II. Intermediates of the oxidative breakdown of ricinoleic acid by genus Candida. *J. Biochem.*, *54*, 536–540.

Palthur, M. P., Palthur, S. S. S., & Chitta, S. K., (2010). Nutraceuticals: Concept and regulatory scenario. *Int. J. Pharm. Sci.*, *2*(2), 14–20.

Patil, C. S., (2011). Current trends and future prospective of nutraceuticals in health promotion. *Bioinfo. Pharmac. Biotech.*, *1*(1), 1–7.

Prabu, S. L., Suriyaprakash, T. N., Kumar, C. D., & Kumar, S. S., (2012). Nutraceuticals and their medicinal importance. *Int. J. Health Allied Sci.*, *1*, 47–53.

Prasad, K., (2005). Hypocholesterolemic and anti-atherosclerotic effect of flax lignan complex isolated from flaxseed. *Atherosclerosis, 179*, 269–275.

Pulliainen, K., Nevalainen, H., Vakevainen, H., Jutila, K., & Gummer, C. L., (2010). An analytical method for the determination of betaine (trimethylglycine) from hair. *International Journal of Cosmetic Science, 32*, 135–138.

Radhika, R. P., Singh, R. B. M., & Sivakumar, T., (2011). Nutraceuticals: An area of tremendous scope. *IJRAP, 2*(2), 410–415.

Rajasekaran, A., Sivagnanam, G., & Xavier, R., (2008). Nutraceuticals as therapeutic agents: A review. *Res. J. Pharm. Tech., 1*, 328–340.

Roy, S., Shah, H., Rink, C., Khanna, S., Bagchi, D., Bagchi, M., & Sen, C. K., (2007). Transcriptome of primary adipocytes from obese women in response to a novel hydroxycitric acid–based dietary supplement. *DNA and Cell Biology, 26*(9), 627–639.

Schilter, B., Andersson, C., Anton, R., Constable, A., Kleiner, J., O'Brien, J., Renwick, A. G., et al., (2003). Guidance for the safety assessment of botanicals and botanical preparations for use in food and food supplements. *Food and Chemical Toxicology, 41*(12), 1625–1649.

Shahidi, F., (2012). Nutraceuticals, functional foods and dietary supplements in health and disease. *Journal of Food and Drug Analysis, 20*(1), 226–230.

Singh, S., Taneja, M., & Majumdar, D. K., (2007). Biological activities of *Ocimum sanctum* L. fixed oil- an overview. *Ind. J. Exp. Biol., 45*, 403–412.

Steiber, A., Kerner, J., & Hoppel, C., (2004). Carnitine: A nutritional, biosynthetic, and functional perspective. *Mol. Aspects Med., 25*(5, 6), 455–473.

St-Onge, M. P., Lamarche, B., Mauger, J. F., & Jones, P. J. H., (2003). Consumption of a functional oil rich in phytosterols and medium-chain triglyceride oil improves plasma lipid profiles in men. *Journal of Nutrition, 133*(6), 1815–1820.

Stormer, F. C., Reistad, R., & Alexander, J., (1993). Glycyrrhizic acid in liquorice – evaluation of health hazard. *Food Chem. Toxicol., 31*(4), 303–312.

Sudhanshu, M. S., Rao, N. I., & Menghani, E. K., (2012). Phytochemical and antimicrobial activity of *Withania coagulans* (Stocks) Dunal (Fruit). *International Journal of Pharmacy and Pharmaceutical Sciences, 4*, 387–389.

Suhaj, M., (2006). Spice antioxidants isolation and their antiradical activity: A review. *Journal of Food Composition and Analysis, 19*, 531–537.

Tapas, A. R., Sakarkar, D. M., & Kakde, R. B., (2008). Flavonoids as nutraceuticals: A review. *Trop. J. Pharm. Res., 7*, 1089–1099.

Taubert, D., Roesen, R., Lehmann, C., Jung, N., & Schömig, E., (2007). Effects of low habitual cocoa intake on blood pressure and bioactive nitric oxide. *The Journal of the American Medical Association, 298*(1), 49–60.

Toyoda, Y., & Thomson, L. R., (2002). Effect of dietary zeaxanthin on tissue distribution of zeaxanthin and lutein in quail. *Invest. Ophthalmol. Vis. Sci., 43*(4), 1210–1221.

Valverde, N. B., Plaza, E. G., Roca, J. M. L., Muoz, R. G., & Bautista-Ortn, A. B., (2011). The extraction of anthocyanins and proanthocyanidins from grapes to wine during fermentative maceration is affected by the enological technique. *J. Agric. Food Chem., 59*(10), 5450–5455.

Vats, V., Grover, G. K., & Rathi, S. S., (2002). Evaluation of anti-hyperglycemic and hypoglycaemic effect of *Trigonella-foenumgraecum* Linn, *Ocimum sanctum* Linn and *Acacia arabica* in normal and alloxanized diabetic rats. *J. Ethnopharmacol., 79*, 95–100.

Vattem, D. A., & Shetty, K., (2005). Biological function of ellagic acid: A review. *Journal of Food Biochemistry, 29*(3), 234–266.

Villasenor, I. M., Simon, M. K., & Villanueva, A. M., (2002). *Nutr. Cancer*, *44*, 66–70.

Wadood, A., Wadood, N., & Shah, S. A., (1989). Effects of *Acacia arabica* and *Caralluma edulis* on blood glucose levels of normal and alloxan diabetic rabbits. *J. Pak. Med. Assoc.*, *39*, 208–212.

Whitman, M., (2001). Understanding the perceived need for complementary and alternative nutraceuticals: Lifestyle issues. *Clin. J. Oncol.Nurs.*, *5*, 190–194.

William, F., Lakshminarayanan, S., & Chegu, H., (1993). Effect of some Indian vegetables on the glucose and insulin response in diabetic subjects. *Int. J. Food Sci. Nutr.*, *44*, 191–196.

Williamson, E. M., (2001). Synergy and other interactions in phytomedicines. *Phytomedicine, 8*(5), 401–409.

Zeisel, S. H., (1999). Regulation of "nutraceuticals."*Science*, *285*, 185–186.

CHAPTER 3

Medicinal Plant-Based Agroforestry Systems

PRADIP KUMAR SARKAR,[1] MAHESH KUMAR DHAKAR,[1]
BIKASH DAS,[1] BAPI DAS,[2] PRANATI SARKAR,[3] and B. P. BHATT[4]

[1]*ICAR Research Complex for Eastern Region, Farming System Research Center for Hill and Plateau Region, Plandu, Ranchi, Jharkhand, India*

[2]*MARKFED, RGM Hospital, Kailashahar, Unakoti, Tripura, India*

[3]*ICAR Central Potato Research Station, Shillong, Meghalaya, India*

[4]*ICAR Research Complex for Eastern Region, Patna, Bihar, India*

ABSTRACT

The forests are rich in medicinal plant diversity in the world. The plants like Tulsi, Arjuna, Drumstick, Amla, and Bael grown under traditional agroforestry systems (AFS) are having both medicinal as well as religious value. Commercial cultivation of medicinal plants under multitier based AFS can be a good option for an entrepreneur or a farmer to get economic benefit. Some of the AFS like Horti-medicinal and Silvi-medicinal AFS are the best options for the farmers to improve their livelihood by increasing the level of income. Moreover, those AFS can play a key role in preserving biodiversity as well as reclamation of wastelands and problem soils.

3.1 INTRODUCTION

Medicinal plants are one of the major NTFPs (non-timber forest products) or MFPs (minor forest products) which primarily supply 80% of the total net forest export earnings in India annually (Jain, 1996). The Indian system

Bioprospecting Ethnomedicinal Plant Resources: Sustainable Utilization and Restoration.
Gopal Shukla, Jahangeer A. Bhat, A. P. Das, and Sumit Chakravarty (Eds.)

of medicines derives many of its curative products from medicinal plants (Kumar et al., 2005) and the information about the plants used as drugs is also available in old Indian literature such as Charaka Samhita, Atharvaveda, and Sushruta Samhita (Tomar, 2009). Around 45,000 plant species are being reported from the Indian subcontinent, out of which approximately 6,500 species of higher plants in India have diverse medicinal uses (Kala et al., 2006). Many tree species are being used by many industries to formulate various herbal drugs in this country, which constitutes 11% of the total known world flora having medicinal properties (Sati et al., 2010; Balasundaram et al., 2011). Additionally, the contemporary Ayurvedic industries are using more than 500 medicinal plant species, of which 80% are obtained extremely from the forest (Thakur et al., 2017).

Species like *Origanum vulgare, Saussurea obvallata, Ocimum sanctum, Cedrus deodara, Cynodon dactylon, Aegle marmelos, Juniperus communis, Nardostachys grandiflora, Zanthoxylum armatum, Ficus benghalensis*, and *Ficus religiosa* are few examples of medicinal plants having both medicinal as well as religious values. In India, the fruits of *Terminalia chebula* are well considered as medicinal as well as sacred fruits by both Hindu and Buddhist (Kala, 2005, 2006). Likewise, there are many such examples of medicinal plant species having area or region-specific diversified values. Hence, medicinal plants are attracting attention from all parts of the world due to their multiple uses as well as untapped economic potential for use by many herbal industries. Most of the people in Asian countries depend on local traditional medicines, particularly plant-originated raw medicine, for the betterment of their health (Kumar & Chandrashekar, 2011). Nowadays, due to the enduring pandemic situation of COVID-19, the medicinal plant growing sectors have drawn the attention of most people towards them and a few have been inclined towards developing their own medicinal plant-based enterprises. Recognizing the importance of medicinal plants, an entrepreneur can choose a sustainable integrated farming system (IFS) where either the medicinal plant component can be integrated in the alleys of trees or all components (i.e., trees, shrubs, herbs, and grasses) can be grown together, as they are medicinally very important. Water scarcity during the summer season has a negative impact on the livelihood of many people in India. Hence, farmers have to choose rainfed farming as an option for their livelihood. For rainfed farming, suitable species or location-specific varieties have to be chosen to maximize land productivity and net return (Kalaichelvi & Swaminathan, 2009). Some of the identified suitable medicinal plants for rainfed farming systems are mentioned in Table 3.1. Growing medicinal plants as intercrops under IFS or AFS systems would be a better option for an entrepreneur to develop an enterprise rather than growing them as sole crops.

TABLE 3.1 Suitable Medicinal Plants for Rainfed Farming

Scientific Name	Common Name	Plant Type
Acacia rugata	*Shikai*	Shrub
Aegle marmelos	*Bael/Bel*	Tree
Aloevera	Aloe vera/*Khorpad*	Succulent
Argemone mexicana	Mexican poppy	Herb
Asparagus racemosus	*Shatavari*	Climber/shrub
Camphora officinarum	*Camphor*	Tree
Chlorophytum borivilianum	*Safed musli*	Herb
Datura stramonium	*Datura*	Shrub
Eclipta prostrate	*Bhringraj*	Creeping herb
Jatropha curcas	Jatropha	Shrub
Lawsonia inermis	*Henna*	Tree
Ocimumba silicum	*Tulsi*	Herb erect plant
Phylanthus emblica (syn. *Emblica officinalis*)	*Aonla*	Tree
Plantago ovate	*Isabgol*	Herb
Rauwolfia serpentina	*Sarpagandha*	Shrub
Tridax procumbens	Jhakhamjudi/coat buttons	Herb erect plant
Vinca rosea	*Sadaphuli/Sadabahar*	Herb
Withania somnifera	*Ashwagandha*	Herb

3.2 MARKETING SCOPE

In the world market of herbal medicine, Europe plays a significant role (Figure 3.1) in global market share with respect to importing and exporting of raw or processed products of medicinally important plants followed by Asia, America, and Japan (BFTI, 2014).

Around 35,000 plant species are being used by almost all parts of the world for medicinal purposes (Sukumaran & Raj, 2010). Among these, the species like *Aconitum heterophyllum* (Aconites, Indian Atees), *Adhatoda vasica* (Vasaka), *Andrographis paniculata* (Kalmegh), *Azadirachta indica* (Neem), *Moringa oleifera* (Drumstick), *Ocimum sanctum* (Tulsi), *Rauwolfia serpentina* (Sarpagandha), *Stevia rebaudiana* (Stevia), *Terminalia arjuna* (Arjuna), *Terminalia chebula* (Hara, Haritaki), *Withania somnifera* (Ashwa-gandha) and some climber type medicinal plants like *Tinospora cordifolia*

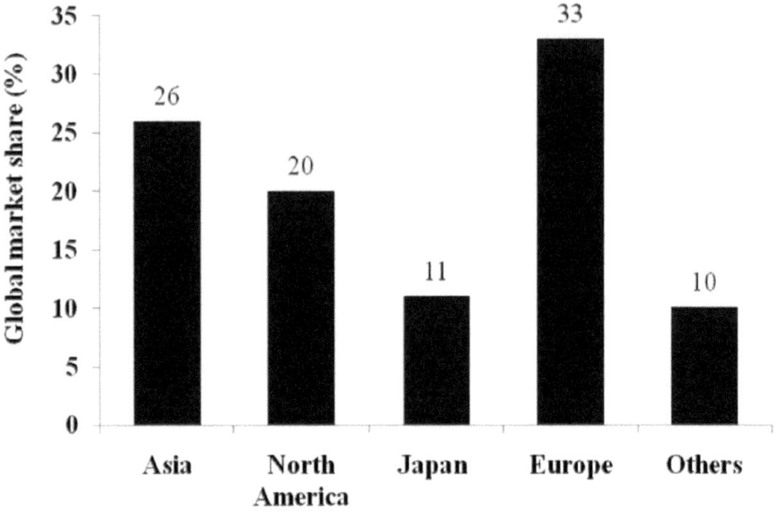

FIGURE 3.1 Global market share of herbal medicine.
Source: BFTI (2014).

(Giloe), *Mucuna pruriens* (Kwanch), *Asparagus racemosus* (Shatavari), *Piper cubeba* (Tailed pepper) and *Piper nigrum* (Black pepper) are in high market demand (MD) (Table 3.2 and Plate 3.1).

TABLE 3.2 Average Market Price of Important Medicinal Plants During 2020–2021

Species	Plant Parts	Average Price (Rs./kg)
Aconitum heterophyllum	Root	7143.94
Adhatoda vasica	Leaf	40.33
Andrographis paniculata	Upper part	47.12
Asparagus racemosus	Root	311.16
Azadirachta indica	Leaf	43.81
Moringa oleifera	Leaf	249.28
Ocimum sanctum	Whole plant	119.60
Piper cubeba	Fruit	1818.42
Rauwolfia serpentina	Root	699.38
Terminalia arjuna	Bark	40.03
Terminalia chebula	Fruit	201.90
Tinospora cordifolia	Stem	34.48
Withania somnifera	Roots	273.77

Source: www.e-charak.in (accessed on 22 December 2023).

PLATE 3.1 Some important medicinal plants as intercrops.

India is the greatest exporter of crude drugs and exports around 75–80% of the total crude products to the six most developed countries like USA, Germany, France, Switzerland, UK, and Japan (Quraishi et al., 2011). As per the International market, the World Health Organization (WHO) had estimated the projected demand of INR (Indian National Rupee) Rs. 25,000,000 crores by the year 2050 for medicinal plants and its products (Purohit & Vyas, 2005). More than Rs. 800 crores of herbal extracts, crude drugs or finished products are being exported from India, with a share in the global market (export) of about 2.5% only (Mishra et al., 2013). Many drug industries of India are importing raw materials like artemisinin (Cyranoski, 2004; Enserink, 2005) which demands sustainable cultivation of these crops for which WHO and many other organizations are highly concerned. Hence, these gaps between demand and supply bring about the opportunities for the growers or entrepreneurs to think about the selection and growing of demand-specific plants to get additional income. Moreover, species selection should also be based on the prevailing market prices (Table 3.2) and the microclimatic condition of that particular growing area. Nowadays, trading of the economic produces is not so difficult. The National Medicinal Plant Board (NMPB) had launched an online virtual platform (www.e-charak.in) and a mobile application e-Charak, for easy trading of medicinal plants, where farmers and collectors can display their produces and the buyers like traders, manufacturers, and exporters can be able to look into their requirements and trade. Though this sector is more remunerative to the farmers in terms of their livelihood, nutritional, and health security, but then there is ample scope in promoting the exports of medicinal plants and its products in an organized manner right from the stage of cultivation, collection, storage, processing, value addition, packaging, and distribution following an integrated approach (Mishra et al., 2013).

3.3 STATE-WISE PRODUCTION OF MEDICINAL PLANTS

The major medicinal plants growing states of India are Madhya Pradesh, followed by Rajasthan and Tamil Nadu (Table 3.3) with an average production of 359,870 tons, 156,500 tons and 132,940 tons, respectively during 2011–2017 (Chowti et al., 2018).

TABLE 3.3 State-Wise Average Area and Production of Aromatic and Medicinal Plants in India (2011–2017)

States/UTs	Area (000' ha)	Production (000' tons)
Andhra Pradesh	3.83	13.54
Arunachal Pradesh	3.07	62.67
Assam	2.60	0.09
Bihar	3.55	0.49
Chhattisgarh	9.58	63.14
Karnataka	3.22	20.36
Madhya Pradesh	59.25	359.87
Mizoram	1.12	2.59
Odisha	1.91	0.62
Punjab	11.04	2.03
Rajasthan	303.63	156.50
Tamil Nadu	13.45	132.94
Uttar Pradesh	133.89	13.42
Others	552.45	829.67

The species like Psyllium (*Plantago ovata*) is cultivated in more than one lakh hectares of areas covering the states of Rajasthan and Gujarat. Similarly, the species like Opium poppy (*Papaver somniferum*) and Senna (*Cassia angustifolia*) are commercially grown in 20,000 hectares areas in the states of Madhya Pradesh, Uttar Pradesh, and Rajasthan. A very common species, i.e., *Rauwolfia serpentina* (Indian snakeroot) is majorly grown over 2,500 hectares areas in the state of Madhya Pradesh (Ved & Goraya, 2007).

3.4 INDUSTRIES DEALING WITH MEDICINAL PLANTS

The major problem in the medicinal plant sector is the marketing. The farmers or the entrepreneurs are mostly unaware of the channel through which they can access the market for business. Hence, there is a need to provide them awareness through training by various institutions or organizations. Moreover, the farmers or the entrepreneurs can approach directly the industries that deal with herbal medicine formulation. Some of the important industries dealing with medicinal plants are listed in Table 3.4.

TABLE 3.4 Industries Dealing with Medicinal Plants in India

SL. No.	Name of Industries	Location
1.	Ansar Drug Laboratories	Surat
2.	Acis Laboratories	Kanpur
3.	Amil Pharmaceutical	New Delhi
4.	Allen Laboratories	Kolkata
5.	Bharti Rasanagar	Kolkata
6.	Dabur India Ltd.	Ghaziabad
7.	Dattatraya Krishan Sandu Bros.	Mumbai
8.	Herbals Pvt. Ltd.	Patna
9.	Herbo-med (P) Ltd.	Kolkata
10.	The Himalaya Drug Co.	Bangaluru
11.	Indian Herb and Research Supply Co.	Saharanpur
12.	J&J Dechane Laboratories Pvt. Ltd.	Hyderabad
13.	Madona Pharmaceutical Research Pvt. Ltd.	Kolkata
14.	Kruzer Herbals	New Delhi
15.	Shilpachem	Indore
16.	Hamdard (Wakf) Laboratories	Delhi
17.	Zandu Pharmaceutical Works Ltd.	Mumbai
18.	Baidyanath Ayurveda Bhavan	Jhansi
19.	Charak Pharmaceuticals	Mumbai

Source: http://www.pharmacy180.com/article/medicinal-plant-based-industries-in-indigenous-system-of-medicine-52 (accessed 22 December 2023).

3.5 PLANTS OF MEDICINAL AND RELIGIOUS VALUES

In India, there are many plants which are being conserved by the rural community in their tradition, either in the form of sacred groves or in the form of scattered patches, considering them as religious values. Such plants are also being exploited for ethno-medicinal uses. Some of the important plants of medicinal and religious values have been enlisted in Table 3.5.

3.6 MEDICINAL PLANT-BASED AGROFORESTRY SYSTEMS (AFS)

India, with more than 1.30 billion people, requires food grains, pulses, fruits, and other agricultural commodities for sustaining their life and for nutritional security. Hence, to secure the health of the people, medicinal plants can be grown integrated with other essential commodities in the same

TABLE 3.5 Plant Species of India Used as Medicine as well as Religious Purposes

Species	Religious Uses	Medicinal Uses
Aegle marmelos (L.) Corr.	To worship Lord Shiva and Goddess Lakshmi.	Digestive disorder, fever.
Artemisia nilagirica (Cl.) Pamp	To worship local deities.	Asthma, skin ailment, headache, antiseptic.
Betula utilis D. Don	Sacred plant	Cut, wounds, hysteria.
Cannabis sativa L.	Associated with Lord Shiva.	Delirium, eye disorder.
Cedrus deodara (Roxb.) Loud.	Associated with Gods and Goddesses.	Ulcer, rheumatic pain.
Datura stramonium L.	Associated with Lord Shiva.	Asthma, joint pain.
Dendrocala musstrictus Nees	Used at various religious ceremonies performed right from the birth to death.	Tonic, fever.
*Ficus benghalensis*L.	Worship as symbol of Lord Shiva.	Dysentery, scabies, tonic.
Ficus religiosa L.	Worship as Lord Vishnu.	Dysentery, cholera, scabies.
Hedychium spicatum Ham. ex Smith	Used on the marriage ceremony.	Stomachache, liver disorder.
Juniperus communis L.	Its incense used for worshipping Gods and Goddesses.	Kidney trouble.
Musa balbisiana Cola	Used on the marriage ceremony.	Indigestion.
Nardostachys jatamansi (Don) DC.	Its incense used for worshipping Gods and Goddesses.	Blood purification, cough.
Ocimum sanctum L.	Worship as a local deity.	Bronchitis, constipation.
Origanum vulgare L.	Used in worshipping the Lord Vishnu.	Diarrhea, bronchitis.
Pinus roxburghii Sarg.	Used on the marriage ceremony.	Cut, wounds, bone fracture.
Prunus cerasoides D. Don	Used in some special religious occasion.	Body swelling.
Saussureaobvallata (DC.) Edgew	Worship as Lord Brahma.	Cuts, aphrodisiac.
Sesamum orientale L.	Worship as Lord Saturn.	Aphrodisiac, body ache.
Zanthoxylum armatum DC.	Used in worshipping as Lord Shiva, Narsingh, Vairoun.	Toothache

Source: Kala (2009).

piece of land without neglecting either of these. Moreover, the fallow land or waste land can be brought under plantation of trees, fruit orchard considering medicinal plants as intercrops. Hence, commercial cultivation of medicinal plants under agroforestry system (AFS) can be a good alternate option for the entrepreneur and for the farmers. Many studies explored the possibilities for growing medicinal plants under fruit trees (like mango, litchi, Bael, Jack-fruit, etc.), timber trees (Teak, Gamhar, Poplar, etc.) and fodder tree based AFS, considering it as economically and ecologically sound and sustainable (Chauhan, 2000; Verma & Thakur, 2011; Thakur et al., 2014). Homegarden is a typical example for growing different plants under different strata, more particularly the medicinal plants in the lower strata (Rao et al., 2004). There are many such Horti-medicinal plant-based systems existing in India, which in other Asian countries like China are known by the name of Silvi-medicinal system (Zou & Sanford, 1990). Medicinal plants like *Decalepishamiltonii*, *Gloriosa superba* and *Hemidesmus indicus* were reported to grow well under Casuarinas based AFS (Selvam et al., 2014). The economic analysis of medicinal plants integrated under fruit based AFS revealed that many of the medicinal plants can be grown at different ages of orchards and the economic benefits are invariably higher than the sole orchard practice. Fruit orchards were reported to provide simulated conditions for medicinal plants as exists in the forest (Vyas, 2001). Maheswari et al. (1985, 1995) reported that growing of Rauwolfia and Patchouli as intercrops can provide extra income for coconut growers. The *Ocimum* species can be intercropped with commercially economic important tree species (Singh et al., 2002; Suvera et al., 2015). Some of the profitable AFS of India are mentioned in Table 3.6. Some profitable AFS for livelihood, nutritional, and health security have been mentioned hereunder.

3.6.1 TRADITIONAL AGROFORESTRY SYSTEMS (AFS)

Generally, many of the plants grown under traditional AFS possess ethno-medicinal and religious values. Plants are either grown in home gardens or in the form of scattered trees planted in agricultural fields or grazing lands or on field bunds or planted as live fences. The species like *Terminalia arjuna* (Arjuna), *Ocimum sanctum* (Tulsi), *Moringa oleifera* (Drumstick), and *Murrayakoenigii* (curry leaf) are some common backyard plants which are generally used for household purposes for routine healthcare, religious needs while some are used in food preparation (Kikon & Angami, 2018). In traditional farming systems, the trees are primarily grown for the purpose of

TABLE 3.6 Some Profitable Agroforestry Systems in India

Medicinal Agroforestry System or Sole Crop	B:C Ratio	References
Horti-Medicinal Plant-based System		
Emblica officinalis + *Andrographis paniculata*	2.47	Madhavi et al. (2012)
Coconut + Guava + *Aloe vera*	3.54	Bari &Rahim (2012a)
Coconut + Lemon + *Aloe vera*	2.80	
Coconut + Guava + *Asparagus racemosus*	3.21	Bari &Rahim (2012b)
Coconut + Lemon + *Asparagus racemosus*	3.08	
Coconut + Guava + *Kaemferia angustifolia*	3.83–5.06	
Tamarind + *Safed musli*	1.94	Kumar et al. (2010)
Coconut + *Aloe vera*	3.80	Mohandas (2011)
Coconut + *Alpina galangal*	3.20	
Coconut + *Ocimum sanctum*	3.00	
Coconut + *Pogestemon patchouli*	2.10	
Mango + Guava + Aloe vera	7.21	Bari &Rahim (2012c)
Mango + Guava + Asparagus	6.65	
Silvi-Medicinal Plant-based System		
Terminalia chebula + *Ashwagandha*	8.41	Madhavi et al. (2012)
Terminalia chebula + *Andrographis paniculata*	11.11	

shade, fodder, fuelwood, timber, and fruits and at the same time meeting the secondary needs as medicine for their health security.

3.6.2 HORTI-MEDICINAL AGROFORESTRY SYSTEMS (AFS)

Fruit plants are commonly grown in India, covering a vast area, and hence there is scope for integration of medicinal plants under fruit-based AFS, also known as Horti-medicinal AFS. Patil et al. (2005) stated that medicinal plants can be grown under young fruit orchards based on the size and intensity of the canopy and the plant spacing. Even the shaded conditions under Horti-medicinal systems help in the process of formation of secondary products and alkaloids in medicinal plants (Ram et al., 1999). Thus, there is ample scope for farmers or entrepreneurs to select medicinal plants as intercrops under fruit tree-based systems without deteriorating the active constituents in the intercrops (Basavaraju & Nanjappa, 2011a). Hence, medicinal plants can be grown successfully as intercrops under fruit trees (Table 3.7) without much effect on their growth, yield, and quality (Korwar et al., 2006; Thakur & Verma, 2012; Thakur et al., 2014).

TABLE 3.7 Tree-Crop Interactions Under Horti-Medicinal Agroforestry Systems

Agroforestry Systems	Performances	References
Peach + Grewia + Setaria + *Mucuna*/Holy basil/*Withania somnifera* Peach + Morus + Setaria + *Mucuna*/Holy basil/*Withania somnifera*	• No effect on growth and yield of both the components. • No effect on active constituents of medicinal plants.	Thakur et al. (2009, 2011, 2014); Verma et al. (2010)
Aonla + Turmeric (var. Rajendra Sonia) *Aonla* + Ginger (var. Nadia)	• Significant enhancement in height and collar girth of tree over sole tree. • Fruit weight, fruit size and pulp weight improved significantly due to growing of turmeric as intercrop under *Aonla* based system.	Das et al. (2011)
Aonla/Tamarind + Castor/ Green gram	No antagonistic effect on growth of fruit trees.	Korwar et al. (2006); Awasthi et al. (2009)
Papaya + *Pogostemoncablin*	Patchouli alcohol increased under papaya.	Ram et al. (1999)
Prosopis cineraria + Senna	Increased growth and yield of senna under *Prosopis cineraria*.	Vyas (2001)
Tamarind + *Safed musli*	Better plant growth and yield in *Safed musli* under tamarind.	Kumar et al. (2010)
Aonla/*Terminalia chebula*+ *Ashwagandha*/*Andrographis paniculate*	Less with anolide content in *Ashwagandha* under *Aonla.*	Madhavi et al. (2012)
Coconut + Guava + *Aloe vera* Coconut + Lemon + *Aloe vera*	Plant height increased.	Bari & Rahim (2012a)
Coconut (30 years old) + *Adhatodabeddomei*/ *Nilgirianthus ciliates*/ *Baliospermum montanum*/ *Vetiveriazi zanoides*/*Piper longum*	Coconut equivalent of intercrops yield was maximum in case of *Nilgirianthus ciliates.*	Maheswarappa et al. (2008)
Coconut + Long pepper/ *Vetiveriazizanoides*	No adverse impact of intercrops on yield of coconut.	George et al. (2010)
Coconut + *Alpina galangal*/ *Aloe vera*/*Ocimum sanctum*/ *Cymbopogan flexuosus*/ *Pogestemon patchouli*	Mean annual nut yield of coconut increased to the tune of 18% due to intercropping of medicinal plants.	Mohandas (2011)

TABLE 3.7 *(Continued)*

Agroforestry Systems	Performances	References
Coconut + *Andrographis paniculata/Solanum nigrum/ Coleus forskolii/Ruta graveolens/Lepidium sativum/ Ocimum sanctum/Curcuma Amanda/Maranta arundinacea/ Kaemferia galanga/Mucuna pruriens/Hibiscus sabdariffa/ Abelmoschus moschatus*	• Growth of medicinal plants was affected when intercropped except *Mucuna pruriens* and *Kaemferia galanga.* • Plant height of *Solanum nigrum, Ocimum sanctum* and *Maranta arundinacea*increased under coconut. • Fresh rhizome yield of *Kaemferia galangal* was increased. • Forskolin in *Coleus forskohlii*and essential oil content in *Abelmoschus moschatus*increased significantly.	Basavaraju & Nanjappa (2011a, b)
Coconut + *Kaemferia galangal/ Maranta arundinacea*	Increased in yield of coconut.	Ghosh et al. (2007)
Mangifera indica + Aloe vera/ Catharanthus roseus	• Yield of both the intercrops was not affected. • Yield of mango was increased due to addition of fertilizers/organic manures applied to intercrops.	Ravitchandirane & Haripriya (2011)

Growing medicinal plants under fruit orchards provides beneficial advantages to fruit plants (Das et al., 2011; Basavaraju & Nanjappa, 2011a). Though there may be instances of lower yield and active constituents of medicinal plants, the overall returns from the systems are greater compared to sole crops or orchards (Madhavi et al., 2012; Bari & Rahim, 2012a). In states like Kerala and Karnataka, medicinal plants like Ginger, Turmeric, Black pepper, and Cardamom are successfully grown as intercrops under Arecanut palm-based systems (Korikanthimath & Hegde, 1994). Moreover, medicinal plants like black pepper can also be successfully intercropped with mango-based AFS (Plate 3.2).

3.6.3 SILVI-MEDICINAL AGROFORESTRY SYSTEM (AFS)

The Medicinal plants like Coleus grown under Teak based AFS had induced an increase in herbage yield by 12.93% over sole crop (Nagarajaiah et al., 2012). The tree species like *Aegle marmelos* (Bael), *Anona squamosa* (Custard Apple), *Emblica officinalis* (Amla), *Prunus africana* (African cherry), *Santalum album* (Sandalwood), *Saraca indica* (Asoka tree), Moringa (Drumstick), *Sapindus mukorossi* (Soapnut), *Azadirachta indica*

PLATE 3.2 Mango + black pepper agroforestry system.

(Neem), *Terminalia chebula* (Haritaki/Chebuli), *Terminalia arjuna* (Arjun tree), and *Jatropha curcas* (Physic nut) can be intercropped with annual medicinal plants in early years until the tree canopy covers the ground. A good amount of yield from *Rauwolfia serpentina*, *Curcuma aromatica*, *Chlorophytum arundinaceum* and *Curculigo orchioides* was obtained when it was intercropped with *Eucalyptus tereticornis*, *Acacia auriculiformis,* and *Leucaena leucocephala* (Chadhar & Sharma, 1996). Medicinal plants like Safed Musli (*Chlorophytum borivilianum*), Sarpagandha (*Rauwolfia serpentina*), Turmeric (*Curcuma longa*), wild Turmeric (*C. aromatica*) and Ginger (*Zingiber officinale*) have been successfully intercropped with fuel wood trees like *Albizia lebbeck*, *Eucalyptus tereticornis*, *Gmelina arborea*, and *Leucaena leucocephala* (Mishra & Pandey, 1998; Prajapati et al., 2003). In Assam, medicinal plants like *Dioscorea floribunda* have been successfully intercropped under Rubber tree (*Hevea brasiliensis*) based system (Kikon & Angami, 2018). Climber type medicinal plants like *Asparagus racemosus* (Shatavari) and *Tinospora cordifolia* (Giloe) can be grown better under poplar based AFS, particularly under close spacing of poplar (i.e., 5 m × 4 m and 10 m × 2 m) as compared to the wider spacing and sole crop without compromising their yield (Bimlendra & Madan, 2017).

3.6.4 HOMEGARDENS

Homegarden is a mixed cropping system where fruits, vegetables, spices, timber trees, ornamental, medicinal plants, other plants as well as livestock, poultry, and fishery are grown and managed within the home premises for supplementary source of food, nutrition, and income. The plants like *Adhatoda vasica, Aloe vera, Asparagus racemosus, Chlorophytum tuberosum, Curcuma angustifolia, Dioscorea bulbifera, Dioscorea hispida, Gymnema sylvestre, Rauwolfia serpentina,* and *Tinospora cordifolia* are grown in harmony with several fruit and timber trees (Bajpai et al., 2013). Medicinal plants in 80 different homegardens of upper Assam of northeastern region of India was found to comprise with 15–17% of the total economically important plants grown, having diverse curative properties (Saikia et al., 2012). Likewise, Mohan et al. (2006) also reported the availability of diverse medicinal plants in homegardens of Thrissur district in Kerala. In traditional homegarden of northeast India, medicinally important plants like *Curcuma longa, Clerodendron viscosum* and *Spilanthes* sp. were successfully cultivated with *Areca catechu, Musa* sp., and *Artocarpus heterophyllus* (Tangjang & Arunachalam, 2009).

3.6.5 MEDICINAL PLANT-BASED AGROFORESTRY SYSTEM (AFS) FOR RECLAMATION OF PROBLEM SOILS AND WASTELANDS

Medicinal plants have been proven to be successfully grown on degraded or wastelands. Cultivation of medicinal plants on degraded or wastelands will not only help in reducing the pressure on croplands but also will help to reclaim such lands and thereby will help in conserving the diversity of medicinal plants therein. The plantations of Aonla (*Phyllanthus emblica*) and Neem (*Azadirachta indica*) were found to improve the soil fertility as well as biological diversity of such degraded lands (Verma, 2004). Under sodic wasteland management, growing aromatic plants *viz.,* Palmorosa (*Cymbopogon martini*), Lemon grass (*Cymbopogon flexuosus*), Vetiver (*Vetiveria zizanioides*) and German chamomile (*Matricaria chamomilla*) (Singh, 1997) and among fruits plants like Aonla, Bael, Ber, Guava, and Karonda on sodic soils (Pandey & Singh, 1997) were reported as most suitable. Moreover, the medicinal plants like Ashwagandha (*Withania somnifera*), Kalmegh (*Andrographis paniculata*), Senna (*Cassia angustifolia*), Makoi (*Solanum nigrum*), Tulsi (*Ocimum sanctum*), Sarpagandha (*Rauwolfia serpentina*),

Giloe (*Tinospora cordifolia*), Gudmar (*Gymnema sylvestre*), Shatavari (*Asparagus racemosus*) and Langali (*Gloriosa superba*) can suitably be cultivated under dry land conditions (Narashima, 2006; Kikon & Angami, 2018). Under waterlogged conditions, the medicinal plants like Bacha (*Acorus calamus*), Brahmi/Indian pennywort (*Bacopa monnieri*), Umbrella sedge (*Cyprus scariosus*) and Vetiver (*Vetiver zizanoides*) were found most suitable (Singh et al., 2003). Cultivation of medicinal and aromatic plants (MAPs) in degraded land will not only help minimize the pressure on croplands and reclaim the degraded lands but will also help in saving the diversity of these herbal plants at the same space and time.

3.7 CONCLUSION

In traditional AFS, the woody perennials are mainly exploited by the farmers for fodder, fuelwood, timber, and fruits, whereas, exploiting the plants for medicinal uses are of secondary requirements. But growing medicinal plants to obtain quality and quantity productions are getting more attention in the world market. Horti-medicinal and Silvi-medicinal AFS are the economically viable options for entrepreneurs or farmers to fulfill the demands of industries dealing with medicinal plants. The Horti-medicinal AFS like Aonla + Turmeric and Aonla + Ginger can improve the livelihood of farmers. With the help of medicinal plants based multitier AFS, entrepreneurs or the farmers can convert their wastelands or fallow lands into an economically viable, ecologically sound, sustainable, and productive land use system.

KEYWORDS

- **Agroforestry**
- **Horti-medicinal agroforestry system**
- **Livelihood**
- **Medicinal plants**
- **Reclamation**
- **Silvi-medicinal agroforestry system**
- **Tree crop interaction**

REFERENCES

Awasthi, O. P., Singh, I. S., & More, T. A., (2009). Performance of intercrops during establishment phase of aonla (*Emblica officinalis*) orchard. *Indian J. Agric. Sci., 79*(8), 587–591.

Bajpai, S., Sharma, A. K., & Kanungo, V. K., (2013). Traditional homegardens: A preserve of medicinal plants. *International Journal of Herbal Medicine, 1*(2), 152–161.

Balasundaram, A., Kumari, P., John, G., & Selvakumar, B. N., (2011). Antimicrobial activity of the leaf extracts of two medicinal plants against MRSA (Methicilin resistant *Staphylococcus aureus*) from human urinary tract pathogens. *Research Journal of Microbiology, 6*, 625–631.

Bari, M. S., & Rahim, M. A., (2012a). Bio-economic evaluation of *Aloe vera* in coconut based multistoried agroforestry systems. *J. Agrofor. Environ.,3*(2), 61–63.

Bari, M. S., & Rahim, M. A., (2012b). Economic evaluation and yield performance of some medicinal plants in coconut based multistoried agroforestry systems. *The Agriculturists,10*(1), 71–80.

Bari, M. S., & Rahim, M. A., (2012c). Performance of medicinal plants under mango based multistoried agroforestry system. *J. Agrofor. Environ., 3*(2), 99–102.

Basavaraju, T. B., & Nanjappa, H. V., (2011a). Growth, yield, and attributing characters of medicinal and aromatic crops as intercrops in coconut garden. *Mysore J. Agric. Sci.,45*(2), 332–341.

Basavaraju, T. B., & Nanjappa, H. V., (2011b). Yield and economics of medicinal and aromatic crops as intercrops in coconut garden. *Mysore J. Agric. Sci., 45*(1), 74–82.

BFTI, (2014). *A Study on Sector-based Need Assessment of Business Promotion Council-Herbal Products and Medicinal Plants*. Bangladesh Foreign Trade Institute, Kawran Bazar, Dhaka –1215.

Bimlendra, K., & Madan, V. K., (2017). Performance of climber type of medicinal plants in poplar-based agroforestry system. *Indian Journal of Agroforestry, 18*(2), 47–52.

Chadhar, S. K., & Sharma, M. C., (1996). Survival and yield of four medicinal plant species grown under tree plantations of bhataland. *Vaniki Sandesh, 20*(4), 3–5.

Chauhan, H. S., (2000). Performance of poplar (*Populus deltoides*) based agroforestry system using aromatic crops. *Indian Journal of Agroforestry, 2*, 17–21.

Chowti, S. P., Rudrapur, S., & Naik, B. K., (2018). Production scenario of medicinal and aromatic crops in India. *Journal of Pharmacognosy and Phytochemistry, SP3*, 274–277.

Cyranoski, D., (2004). Campaign to fight malaria hit by surge in demand for medicine. *Nature, 18*(7015), 259.

Das, D. K., Chaturvedi, O. P., Jha, R. K., & Kumar, R., (2011). Yield, soil health and economics of aonla (*Emblica officinalis* Gaertn.)-based agrihorticultural systems in eastern India. *Current Science, 101*(6), 786–790.

Enserink, M., (2005). Source of new hope against malaria is in short supply. *Science, 307*, 33.

George, V., Thomas, V., Krishanakumar, H. P., & Maheswarappa, P. C., (2010). *Coconut Based Cropping/Farming Systems* (p. 231). Central Plantation Crops Research Institute, Kasaragod.

Ghosh, D. K., Bandhopadhyay, A., Mahapatra, S., & Maji, M. K., (2007). Studies on the performance of medicinal plants under coconut plantation in West Bengal. *Indian Coconut Journal, 38*, 15–18.

Jain, A., (1996). *Current Trends: Marketing of Minor Forest Produce* (Vol. 6, No. 1). MFP News.

Kala, C. P., (2005). Health traditions of Buddhist community and role of *Amchis* in trans-Himalayan region of India. *Current Science, 89*, 1331–1338.

Kala, C. P., (2006). Problems and prospects in the conservation and development of the Himalayan medicinal plants sector. *International Journal of Sustainable Development, 9*(4), 370–389.

Kala, C. P., (2009). Medicinal plants conservation and enterprise development. *Medicinal Plants, 1*(2), 79–95.

Kala, C. P., Dhyani, P. P., & Sajwan, B. S., (2006). Developing the medicinal plants sector in Northern India: Challenges and opportunities. *Journal of Ethnobiology and Ethnomedicine, 2*, 1–15.

Kalaichelvi, K., & Swaminathan, A. A., (2009). Alternate land use through cultivation of medicinal and aromatic plants: A review. *Agric. Rev., 30*(3), 176–183.

Kikon, N., & Angami, T., (2018). Medicinal and aromatic plants for climate change adaptation: Scope and prospects with special reference to North Eastern Himalayas. *Journal of Medicinal Plants Studies, 6*(1), 150–155.

Korikanthimath, V. S., & Hegde, R., (1994). Cardamom and arecanut mixed-cropping systems. *Indian Cocoa Arecanut and Spices Journal, 18*(4), 109–112.

Korwar, G. R., Pratibha, G., Ravi, V., & Kumar, D. P., (2006). Performance of castor (*Ricinus communis*) and green gram (*Vigna radiata*) in agroforestry systems in semi-arid tropics. *Indian Journal of Agronomy, 51*(2), 112–115.

Kumar, R. D., Sreenivasulu, G. B., Prashanth, S. J., Jayaprakashnarayan, R. P., Nataraj, S. K., & Hegde, N. K., (2010). Performance of Safed musli (*Chlorophytum borivilianum* L.) in tamarind plantation as intercrop and as sole crop in open area. *International Journal of Agricultural Science, 6*(1), 359, 360.

Kumar, S., Parveen, F., Goyal, S., & Chouhan, A., (2005). Trading of ethnomedicinal plants in the Indian arid zone. *Indian Forester, 131*, 371–378.

Kumar, T., & Chandrashekar, K. S., (2011). *Bauhinia purpurea* Linn: A review of its ethnobotany, phytochemical and pharmacological profile. *Research Journal of Medicinal Plant, 5*, 420–431.

Madhavi, L. A., Srinivasa, R. M., Joseph, B., Siva, S. A., & Rao, C. P., (2012). Performance of medicinal herbs in tree-based cropping system under dry land conditions. *Journal of Pharmacognosy, 3*(2), 92.

Maheswarappa, H. P., Reddy, S., Krishnakumar, D. V., Subramanian, V. P., & Dhanapal, R., (2008). Organic cultivation of medicinal plants as inter/mixed crops in coconut gardens. In: *Proc. National Conference on Organic Farming in Horticultural Crops with Special Reference to Plantation Crops, Program and Abstracts*. CPCRI, Kasaragod.

Maheswari, S. K., Dhantonde, B. N., Yadav, S., & Gangrade, S. K., (1985). Intercropping of *Rauwolfia serpentina* for higher monetary return. *Indian J. Agric. Sci., 58*, 487, 488.

Maheswari, S. K., Sharma, R. K., & Gangarade, S. K., (1995). Studies on spatial arrangements in palmarosa-pigeon pea intercropping in black cotton soil. *Agron. J., 92*, 812–818.

Mishra, D., Singh, R. K., & Srivastava, R. K., (2013). Agribusiness and entrepreneurship development through medicinal and aromatic plants: An Indian state of affairs. *International Journal of Food, Agriculture and Veterinary Sciences, 3*(1), 238–246.

Mishra, R. K., & Pandey, V. K., (1998). Intercropping of turmeric under different tree species and their planting pattern in agroforestry systems. *Range Management and Agroforestry, 19*(2), 199–202.

Mohan, S., Alavalapati, J. R. R., & Nair, P. K. R., (2006). Financial analysis of homegardens: A case study from Kerala state, India. In: Kumar, B. M., & Nair, P. K. R., (eds.), *Tropical Homegardens: A Time-Tested Example of Sustainable Agroforestry* (pp. 283–296). Springer Science, Dordrecht, The Netherlands.

Mohandas, S., (2011). Prospects of intercropping medicinal and aromatic plants in coconut garden. *Madras Agric. J., 98*(1–3), 82, 83.

Nagarajaiah, C., Kittur, B. H., Mukthamath, U., & Venkatesh, L., (2012). Evaluation of medicinal and aromatic crops under teak based agroforestry system. *Environment and Ecology, 30*(1), 221–225.

Narashima, R. K., (2006). *Agri and Herbal Vision*, 13–17.

Pandey, G., & Singh, R. P., (1997). Appropriate technology for sodic-wasteland management. *Advances in Wastelands Development*, 105–111.

Patil, D. R., Patil, H. B., Patil, S. N., & Prashanth, J. M., (2005). *Studies on Evaluation of Intercrops in Mango Orchards.* In:National seminar on commercialization of Horticulture in non-traditional areas held at CIAH, Bikaner from 5–6 February, Abstract 2005, 81.

Prajapati, N. D., Purohit, S. S., Sharma, A. K., & Kumar, T., (2003). *A Handbook of Medicinal Plants* (p. 553). Agrobios, India.

Purohit, S. S., & Vyas, S. P., (2005). *National Consultative Workshop on Medicinal and Aromatic Plants*. Held at G B Pant University of Agriculture and Technology, Pantnagar, Uttarakhand.

Quraishi, A., Jadhav, S. K., & Gupta, S., (2011). *In vitro* clonal propagation of *Cassia tora* L. (Coffee Pod): A medicinal plant. *Biotechnology, 10*, 546–550.

Ram, M., Ram, D., Singh, S., Naqvi, A. A., & Kumar, S., (1999). Studies on intercropping of patchouli (*Pogostemon patchouli*) with papaya (*Carica papaya*). *Indian Journal of Medicinal and Aromatic Plants, 21*, 358–360.

Rao, M. R., Palada, M. C., & Becker, B. N., (2004). Medicinal and medicinal plants in agroforestry systems. *Agroforestry Systems, 61*(1), 107–122.

Ravitchandirane, V., & Haripriya, K., (2011). Intercropping with medicinal plants in mango cv. Alphonso. *Plant Archives, 11*(1), 413–416.

Saikia, P., Choudhury, B. I., & Khan, M. L., (2012). Floristic composition and plant utilization pattern in homegardens of upper Assam, India. *Tropical Ecology, 53*(1), 105–118.

Sati, S. C., Sati, N., Rawat, U., & Sati, O. P., (2010). Medicinal plants as a source of antioxidants. *Research Journal Phytochemistry, 4*, 213–224.

Selvam, K. P., Sittu, R. T., Saravanan, S., Vijayaraghavan, A., & Aravindhan, V., (2014). Performance of economically important medicinal plants in *Casuarina* based agroforestry systems in Tamil Nadu, India. In: *Proc. Fifth International Casuarina Workshop*. Organized at Mamallapuram, Chennai, India.

Singh, D. V., (1997). Prospects of cultivation of aromatic crops on salt affected wastelands. *Advances in Wastelands Development*, 123–126.

Singh, K., Rajeswara, R. B. R., Rajput, D. K., Bhattacharya, A. K., Chauhan, H. S., Mallavarapu, B. R., & Ramesh, S., (2002). Composition of essential oils of menthol mint (*Mentha arvensis* var. *piperascens*), citronella (*Cymbopogon winterianus*) and palmarosa (*Cymbopogon martinii*var. *motia*) grown in the open and partial shade of poplar (*Populus deltoides*) trees. *J. Med. Aromatic Plant Sci., 24*, 710–712.

Singh, S., Singh, A., & Khanuja, S. P. S., (2003). Medicinal and aromatic plants for crop diversification and economic utilization of waterlogged soils. In: Singh, V. P., & Yadav, R. N., (eds.), *Wastewater Treatment and Waste Management* (pp. 38–44). Allied Publishers, New Delhi.

Sukumaran, S., & Raj, A. D. S., (2010). Medicinal plants sacred groves in Kanyakumari District, Southern Western Ghats. *Indian Journal of Traditional Knowledge, 9*(2), 294–299.

Suvera, A. H., Thakur, N. S., & Jha, S. K., (2015). Herbage and essential oil yield of *Ocimum* spp. Intercropped under *Pongamia pinnata* based Silvi-medicinal systems in Gujarat, India. *The Bioscan, 10*(1), 81–85.

Tangjang, S., & Arunachalam, A., (2009). Role of traditional home garden systems in northeast India. *Indian Journal of Traditional Knowledge, 8*(1), 47–50.

Thakur, N. S., & Verma, K. S., (2012). Financial flows from sacred basil (*Ocimum sanctum*) based agroforestry land use systems in mid hills of Western Himalayas. *Indian Forester, 138*(7), 638–645.

Thakur, N. S., Attar, S. K., & Chauhan, R. S., (2017). Horti-medicinal agroforestry systems: A potential land use for commercial cultivation of medicinal and aromatic plants. In: Gupta, S. K., Panwar, P., & Kaushal, R., (eds.), *Agroforestry for Increased Production and Livelihood Security* (pp. 163–180). New India Publishing Agency, New Delhi.

Thakur, N. S., Verma, K. S., & Rana, R. C., (2009). Effect of tree-crop combinations and nitrogen levels on fresh herbage and oil yield of sacred basil (*Ocimum sanctum*) grown in agrihorti-silvi-pasture system in mid hill Himalayas. *Indian Perfumer, 53*, 39–44.

Thakur, N. S., Verma, K. S., & Rana, R. C., (2011). Effect of tree crop combinations and nitrogen levels on velvet bean (*Mucuna pruriens* L) grown in agri-Horti-silvi-pasture system of agroforestry in mid hill Himalayan region. *Journal of Non-Timber Forest Products, 18*(4), 303–312.

Thakur, N. S., Verma, K. S., & Rana, R. C., (2014). Growth and yield performance of Ashwagandha (*Withaniasomnifera*) under agroforestry. *Indian Journal of Agricultural Sciences, 84*(8), 937–941.

Tomar, A., (2009). Folk medicinal uses of plant roots from Meerut district, Uttar Pradesh. *Indian Journal of Traditional Knowledge, 8*, 298–301.

Ved, D. K., & Goraya, G. S., (2007). *Demand and Supply of Medicinal Plants in India*. NMPB, New Delhi and FRLHT, Bangalore, India.

Verma, K. R., (2004). Evaluation of biological diversity and soil quality under plantations on degraded land. *Indian Forester, 130*(7), 791–799.

Verma, K. S., & Thakur, N. S., (2011). Economic appraisal of Kewach (*Mucuna pruriens*) based agroforestry land-use systems in mid hills of Western Himalayas. *Journal of Non-Timber Forest Products, 18*(1), 47–54.

Verma, K. S., Thakur, N. S., & Rana, R. C., (2010). Effect of tree-crop combinations and nitrogen levels on herbage yield of sacred basil (*Ocimum sanctum* L.) grown in agrihortisilvi-pasture system in mid hill Himalayas. *Indian Journal of Agroforestry, 12*(1), 71–76.

Vyas, S. P., (2001). Effect of *Prosopsis cineraria* (L. MacBride) on growth and productivity of *Cassia angustifolia* (vahl) under medico- forestry system in arid region of Thar Desert. *Journal of Tropical Forestry, 17*(3), 71–73.

Zou, X., & Sanford, R. L., (1990). Agroforestry systems in China: A survey and classification. *Agroforestry Systems, 11*, 85–94.

CHAPTER 4

Plant Bio-Actives from Agroforestry Systems in Improving Livestock Health, Production, and Methane Mitigation

AVIJIT DEY,[1] P. C. LAILER,[1] S. S. DAHIYA,[1] and L. A. GONZALEZ[2]

[1]*Division of Animal Nutrition and Feed Technology, ICAR–Central Institute for Research on Buffaloes, Hisar, Haryana, India*

[2]*School of Life and Environmental Sciences, The University of Sydney, NSW, Australia*

ABSTRACT

Plant bioactive compounds are the secondary metabolites, synthesized as products of biochemical sidetracks in the plant cells, not generally needed for daily performance. Several bioactive compounds are present in agroforestry-based medicinal plants, however, essential oils (EOs), tannins, phenolic acids, saponins, and flavonoids are the most commonly occurring bioactive compounds and have significant implications on animal agriculture not only by improving livestock health and production but also in mitigating environmental pollution through reduction of enteric methane and ammonia emissions. Polyphenols have potent antioxidant properties and prevent cardiovascular, neurodegenerative diseases, and cancer. Condensed tannins (CTs), flavonoids, saponins, and EOs can modify the rumen fermentation towards reduced methanogenesis by altering rumen microbial community and their supplementation reduces nitrogen excretion in ruminants by improving its utilization efficiency. Inhibition of methanogenic archaea in rumen by plant bioactive compounds has attracted their application in enteric methane mitigation from ruminants. Controlling gastrointestinal parasites

Bioprospecting Ethnomedicinal Plant Resources: Sustainable Utilization and Restoration.
Gopal Shukla, Jahangeer A. Bhat, A. P. Das, and Sumit Chakravarty (Eds.)

by feeding of leaves (Ficus infectoria, Ficus glomerata, Psidium guajava, and Ficus bengalensis) containing CT has been advocated to ameliorate drug resistance. Investigation with inclusion of CT in the diet of sheep and crossbred cows through leaves of F. infectoria or F. bengalensis was demonstrated to increase milk production, feed efficiency, growth performance, and health status of animals. A recent study indicated that feeding of plant bioactive compounds containing a blend of F. bengalensis leaves, soapnut fruits (Sapindus mukorossi), and eucalyptus EOs improved growth rate and feed utilization with a reduction in methane emission from buffalo calves. Supplementation of EOs rich mixed feed additives containing poplar and eucalyptus leaves improved antioxidant status and immunity of buffalo calves besides reduction in enteric methane emission and improvement in growth rate. Therefore, plant bioactive compounds from agroforestry-based medicinal plants could be used as feed additives for improving health status and production performances of livestock and mitigation of enteric methane and ammonia emissions for clean animal production.

4.1 INTRODUCTION

Livestock farming in recent years is experiencing dual challenges of reducing its impact on the environment with financial feasibility owing to enhanced cost and scarcity of feedstuffs. Therefore, improvement in production performances with a reduction in environmental pollutants emission from animal production systems is the key requirement for sustainable animal production. The feedstuffs consumed by ruminants are fermented in the rumen by ruminal microbiota and then pass into the intestine for the usual digestion process. The rumen fermentation process enables the animals for suitable digestion of feedstuffs. However, the ruminal fermentation also produces some gaseous products viz. ammonia and methane, which are energy-consuming and harmful to the environment. Most of the feed protein that enters the rumen (60–90%) is deaminated, and rates of ammonia production often exceed the needs of ammonia-utilizing microorganisms. The rumen epithelium absorbs the excess ammonia and subsequently transformed to urinary urea and excreted through urine. Therefore, ruminal degradation of dietary protein beyond the capacity of rumen microbes for microbial protein synthesis results in the wastage of feed protein for productive purposes because it is a nutritionally wasteful process (Firkins et al., 2007) and contributes to environmental pollution (Food and Agricultural Organization

(FAO), 2010). Thus, recently ruminant nutritionists have focused on reducing protein degradation and ammonia genesis in the rumen by inhibiting either the number or action of hyper-ammonia-producing bacteria (HAP) because ammonia is produced at a rapid rate (> 300 nmol NH3/mg cell protein/min) by the HAP (Gano, 2013). HAP has minimal ability to utilize sugars and the majority of them are considered to form an amino acid-utilizing niche. Although the numbers of the HAP inside the rumen are very low (0.01%), the concentration of ammonia they can produce makes them quantitatively important.

Methane is produced mainly in the rumen, the first compartment of the stomach of livestock species, during fermentation of feeds by resident anaerobic microbes and emitted to the environment through burping mostly, contributing to environmental pollution. The IPCC (Intergovernmental Panel on Climate Change, 2019) estimated that 23% of global anthropogenic greenhouse gas (GHG) emissions come from agriculture, forestry, and other land use. These sources contribute 44% of all methane emissions, with enteric methane from ruminants being the highest at 46%, followed by rice cultivation. The livestock production system contributes about 18% to global anthropogenic GHG emissions, mostly due to anaerobic enteric fermentation of feeds (FAO, 2010). Not only does this contribute to environmental pollution, but about 10% of feed energy is wasted due to enteric methane production, which could otherwise be utilized for animal production. Therefore, ruminant nutritionists are searching for suitable mechanisms to modulate rumen functions to increase feed utilization efficiency, not only for enhanced production but also to reduce environmental pollution through livestock farming.

Aromatic plants, their extracts containing secondary metabolites mainly essential oils (EOs), saponins, and tannins are residue-free and generally recognized as safe. They could be a possible way out in improving animal performance and reducing environmental pollution. There are various classes of plant bioactive compounds, but only some have the potential of improving animal health and production performance by increasing antioxidant status and immunity of animals, improving protein utilization, feed efficiency, controlling gastrointestinal parasitism, ruminal bloat, etc. These plant secondary compounds have the capacity to decrease nitrogen emission and enteric methane production from ruminants by modifying rumen microbiome, thereby reducing environmental pollution. The major bioactive compounds, namely phenolic compounds, saponins, and EOs, could be included in the livestock ration as phytogenic feed additives/phytobiotics for the improvement of animal health and production, and environmental remediation through the abatement of enteric methane production and nitrogen excretion from ruminants.

4.2 PHENOLIC COMPOUNDS

Phenolics are the most common secondary metabolites and extensively available in vegetables and plants. They generally protect the plants from enemies and play a significant role in the organoleptic attributes of plants. They abundantly exist in all plant tissues and contribute essentially to the foods of livestock and humans (D'Archivio et al., 2007). Plant phenolics comprise the most widespread phenolic acids, flavonoids, and tannins. However, stilbenes and lignans are also present in some plants, coming under this category. Fruits and vegetables are a rich source of polyphenolics for humans. Whereas the key bioactive compounds present in the tropical tree leaves are phenolic compounds, which have been demonstrated to improve the health status, production, and reproduction performances of animals (Dey et al., 2006). Potent antioxidant features and credible effects in the prevention of various stress-associated diseases have placed polyphenols in the interest of scientists and industry. Modulation of key biological actions by regulating several enzyme activities and cell receptors, polyphenols have been reported to have significant effects in preventing cardiovascular, neurodegenerative diseases, and cancer. Proanthocyanidins (PAs) can modify the rumen functions and reduce enteric methane production by altering rumen microbial community (Carulla et al., 2005). CTs also reduce nitrogen excretion in ruminants by improving its utilization efficiency. Research conducted by several workers (Min et al., 2001; Ramírez-Restrepo et al., 2005; Dey et al., 2008a, b, 2009a, b) revealed improvement in feed intake, growth rate, wool production, reproduction, and milk production in ruminants fed CTs containing diets. However, delivery of plant phenolics to animal systems and their doses are important considerations to achieve maximum beneficial effects. Phenolic acids, flavonoids, and tannins are the most common phenolic compounds present in human and animal diets. Two classes of phenolic acids, namely benzoic acid derivatives (e.g., gallic acid) and cinnamic acid (e.g., coumaric, caffeic, and ferulic acid), are present in foods. Whereas flavonoids are classified into six sub-groups, namely flavonols, flavanols, flavones, isoflavones, flavanones, and anthocyanins, based on the state of oxidation. The degree and form of methoxylation, hydroxylation, or glycosylation mostly determine the structural distinction of compounds in each subgroup. Quercetin is a flavonol abundant in onion, broccoli, and apple, whereas catechin is a common flavanol found in tea. Mustard and soybean contain isoflavones, namely daidzein, genistein, and glycitein (D'Archivio et al., 2007). Tannins are most abundant in tropical tree leaves (Table 4.1) and common feeds for animals (Dey et al., 2006).

TABLE 4.1 Total Phenolics and Tannin Fractions of Common Tropical Tree Leaves

Leaves	Hindi Name	TPh	NTPh	TTPh	CT
Acacia nilotica	*Desi babool*	4.9	1.1	3.8	1.7
Artocarpus heterophyllus	*Kathal*	24.2	1.3	23.0	19.1
Azadirachta indica	*Neem*	2.9	1.3	1.7	0.6
Eugenia jambolana	*Jamun*	8.5	1.1	7.4	4.2
Ficus benghalensis	*Bargad*	19.6	3.5	16.1	12.6
Ficus glomerata	*Gular*	17.5	1.4	16.2	12.1
Ficus infectoria	*Pakar*	19.4	5.5	13.9	9.4
Grewia oppositifolia	*Bhemal*	2.7	1.4	1.3	0.2
Mangifera indica	*Aam*	5.8	1.4	4.4	0.9
Morus alba	*Tut*	2.1	1.4	0.7	0.2
Musa paradisiacal	*Kela*	1.7	0.8	0.9	0.6
Populus fastigiata	*Poplar*	4.6	1.1	3.5	1.1
Quercus floribunda	*Trilong*	13.3	1.7	1.6	7.9
Quercus incana	*Banj*	15.4	2.1	13.3	7.2
Quercus semicarpifolia	*Kharsu*	14.3	2.1	12.2	3.6
Tectona grandis	*Sagwan*	2.9	1.3	1.6	0.3

Abbreviations: TPh: Total phenolics; NTPh: non-tannin phenolics; TTPh: total tannin phenolics; and CT: condensed tannins.

Source: Adapted from Dey et al. (2006).

There are two types of tannins viz. hydrolysable tannins (HTs), where gallic acid or ellagic acid are esterified with a carbohydrate moiety, and CT or PAs, where oligomers of the flavan-3-ols and related flavanol residues are connected by carbon-carbon bonds (Mangan, 1988). The molecular weight of CT varies with the degree of compression and varies from 1,000 to 20,000 Dalton (Mueller-Harvey & McAllan, 1992). Several factors viz. species, developmental stage and location determine the type and levels of tannins present in the feeds. High temperature stress and light intensity of tropical climate increases the tannins concentration in leaves (Makkar & Becker, 1998).

4.3 SAPONINS

The saponins are glycosides naturally present as surface-active phytogenic compounds. Although minor oceanic animals and some bacteria produce saponins, they are mostly produced in plants (Riguera, 1997; Yoshiki et al., 1998). The characteristic of stable, soap-like foam formation in aqueous solution has paid attention to human beings from prehistoric times. Chemically, saponins are groups of high molecular weight glycosides in which saccharide chain units (1–8) are linked to a triterpene (triterpene saponins) or steroidal (steroidal saponins) aglycone moiety (sapogenin). The C3 position is normally occupied by an oligosaccharide chain (monodesmosidic); however, an additional sugar moiety is attached in many saponins at C17 via C28 position (bidesmosidic). The variations in the aglycone structure, the property and positional attachment of the side chains to the aglycone moiety determine the complexity of saponins. The diverse properties of saponins have attracted researchers for their applications in livestock health and production, and enteric methane abatement by modulating rumen microbiome.

4.4 ESSENTIAL OILS (EOS)

Essential oils (EOs) are plant secondary metabolites which are volatile in nature and have plant-specific aroma and flavor. They have been used from ancient times as antimicrobial and preservative agents. Herbs and spices are the key sources of EOs. EOs are not 'essential' for animals and humans, and they are also not the traditional glycerol-based oils. The name 'essential oils' derives from the plant-specific aroma/smell of the compound with oily advent (Guenther, 1948). They were first established by Arabs through steam or hydro-distillation processes (Greathead, 2003) and used for their antiseptic and medicinal properties. EOs protect plants from herbivores with their strong smell and taste, producing aversion in predators. Many EOs also support pollination and the diffusion of seeds. Structurally, EOs are derivatives of terpenoids and phenylpropanoids with ester, alcohol, or aldehyde (Figure 4.1).

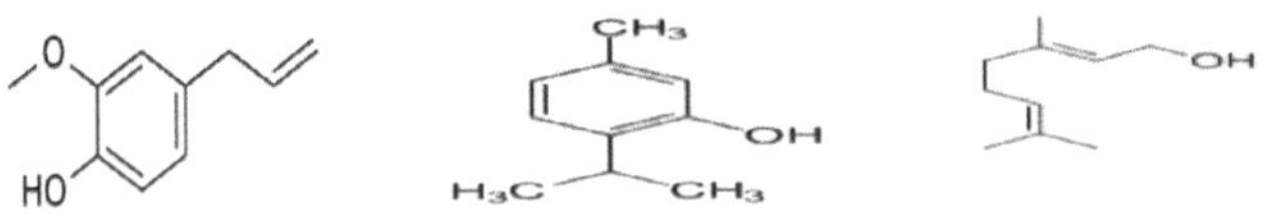

Eugenol Thymol Geraniol

FIGURE 4.1 Structure of some common essential oils.

Monoterpenes and sesquiterpenes under the terpenoids are abundant in the plant kingdom, produced from acetyl-CoA (Hart et al., 2008). These compounds are derivatives of the isoprene unit, a 5-carbon ($C5H8$) structure, and their complexity depends on the number of these units that exist. Phenylpropanoids are meager EOs related to terpenoids and produced via shikimate pathways from phenylalanine, an aromatic amino acid. Although EOs are found in every part of the plant (Table 4.2), viz. leaves, bark, roots, stems, flowers, etc., the growth stage and health of plants, along with environmental conditions, determine their concentration (Gershenzon et al., 2000).

TABLE 4.2 Some Essential Oils and Their Active Compounds

Essential Oils	Name of the Plants	Major Bioactive Compounds
Cinnamon	*Cinnamomum cassia*	Cinnamaldehyde, eugenol
Eucalyptus	*Eucalyptus citriodora*	Citronellal, citronellol
Capsicum	*Capsicum annum*	Capsaicin
Oregano	*Origanum vulgare*	Carvacrol, thymol
Clove	*Syzygium aromaticum*	Eugenol
Thyme	*Thymus vulgaris*	Thymol, carvacrol
Lemon grass	*Cymbopogon citratus*	Limonene, geraniol, citral, citronellal
Garlic	*Allium sativum*	Allicin, diallyl sulfite
Ginger	*Zingiber officinale*	Zingiberene, zingerone

4.5 PLANT BIO-ACTIVES IN LIVESTOCK HEALTH IMPROVEMENT

Recently, much emphasis has been given to the application of phytogenic compounds for the improvement of livestock health. Phenolic compounds, saponins, and EOs have been reported to modulate the immune system and improve the antioxidant status of animals, resulting in the improvement of livestock health.

4.5.1 ANTIOXIDANT STATUS

Oxidative stress leads to reduced activity of biomolecules viz. proteins, enzymes, or lipids, and is produced owing to excessive production of reactive oxygen and/or nitrogen species beyond the capacity of the body to overcome by endogenous system. Oxidative stress is the prime factor behind many degenerative diseases like cancer, coronary heart disease or aging (Ames et al., 1993). Studies have revealed the powerful antioxidant capacity of phenolics (Ho et al., 1999; Dey et al., 2015) than vitamins and carotenoids (Rice-Evans et al., 1995). The phenolic compounds facilitate antioxidant status either by scavenging free radical species or suppressing free radical formation through direct inhibition of enzymes involved or chelation of trace minerals. The phenolics also improve the antioxidant defense of the body (Cotelle, 2001). Antioxidant properties of PAs (condensed tannin (CT)) were demonstrated by many researchers (Ho et al., 1999; Hagerman et al., 1999; Lin et al., 2000) in various scattered in vitro experiments. Recent studies explored that the supplementation of CTs containing diets to crossbred cows (Dey & De, 2014) and lambs (Dey et al., 2015) improved the activity of antioxidant enzymes viz. glutathione (GSH), catalase, superoxide dismutase (SOD) and thiol groups in erythrocytes (Table 4.3). Lipid peroxidation (LPO) was also reduced significantly by feeding of tanniferous leaves. Feeding of a composite tannin rich leaf including Psidium guajava, Ficus infectoria and Ficus bengalensis was reported to enhance antioxidant status in goats (Dutta et al., 2012).

TABLE 4.3 Dietary Inclusion of Ficus infectoria Leaf Meal (FILM) on Erythrocytic Antioxidant Enzymes of Lambs

Attributes	Treatments				SEM
	CON	FILM-I	FILM-II	FILM-III	
Catalase (mK/g Hb)	9.5^a	10.1^b	10.7^b	10.4^b	0.12
SOD(mmol MTT formazan formed/gHb)	1.5^a	1.7^{ab}	2.0^b	1.9^b	0.11
LPO (nmol MDA/g Hb)	73.9^b	65.2^a	63.9^a	60.8^a	1.11
GSH (μmol/g Hb)	9.7^a	10.8^a	11.6^b	11.8^b	0.34
T-SH (μmol/ml PRBC)	40.1^a	46.0^b	46.1^b	48.3^b	0.82
NP-SH (μmol/ml PRBC)	4.4	4.4	4.6	4.6	0.12
P-SH (μmol/ml PRBC)	35.6^a	41.6^b	41.5^b	43.7^b	0.62

^{a,b}Mean bearing different superscript within a row differ significantly ($p < 0.05$).

Source: Adapted from Dey et al. (2015).

EOs improve health status by inhibition of inflammation and cellular production of free radicals (Harborne & Williams, 2000; Ri et al., 2017). Experiments on broiler birds with dietary inclusion of Salvia officinalis EOs (0.05%) which is rich in rosmarinic acid and phenolic diterpenes (Cuvelier et al., 1996) were demonstrated to enhance antioxidant enzymes and reduce peroxidation of body lipids (Ryzner et al., 2013). In a study with oregano (Oreganum vulgare) powder feeding (150 mg/kg) to broiler chicks, it was reported to improve body weight gain and antioxidant status of broiler chicks (Ri et al., 2017). The terpenoid compounds (carvacrol and thymol) present in oregano EOs are the key modulators of antioxidant enzymes of animals (Teixeira et al., 2013). Habibi et al. (2014) described enhanced SOD activity and reduced malondialdehyde (MDA) concentrations in the liver of broiler birds supplemented with ginger EOs (150 mg/kg). Dietary inclusion (1.5%) of a mixture of medicinal plants in the ration of Holstein calves was reported to lower the weaning age and enhance antioxidant status (Seifzadeh et al., 2016). Kumar et al. (2018) observed higher antioxidant status and reduced LPO in buffalo calves supplemented with a mixture of eucalyptus and poplar leaves.

4.5.2 IMMUNE-MODULATION

Phenolics can modulate the immune system either by a direct stimulatory effect or by sustaining cellular integrity through improving the antioxidant status of animals. Enhanced immunity was reported on feeding Ficus infectoria leaves CTs (1-2%) in lambs (Dey et al., 2015). Increased skin fold thickness (Figure 4.2) following intradermal injection of phytohaemagglutinin-P (PHA-P) to lambs supplemented with tannin-rich Ficus infectoria leaves suggested improved cell-mediated immune (CMI) response. An improved immunity was also demonstrated in goats fed a mixture of tanniferous leaf meal (Dutta et al., 2012). Kumar et al. (2018) observed enhanced CMI response as well as humoral immunity (Figure 4.3) in buffalo calves supplemented with EO-rich tree leaves. The study clearly demonstrated the critical role of EOs for enhancing the overall immune status of buffalo calves. Research on adjuvants has revealed that a low concentration of saponins in a mixture of adjuvants not only stimulated the immune system but also boosted antibody production (Oda et al., 2000). It was also reported to enhance the proliferation of immune cells by quillaja saponins or a crude mixture of saponin compounds (Plohmann et al., 1997; So et al., 1997; Lacaille-Dubois et al., 1999). Although the pathways of immune-stimulating action of saponins are not clear, it was reported to influence the production of cytokines, which is envisaged to stimulate the immune system (Jie et al., 1984; Kensil, 1996; Barr et al., 1998).

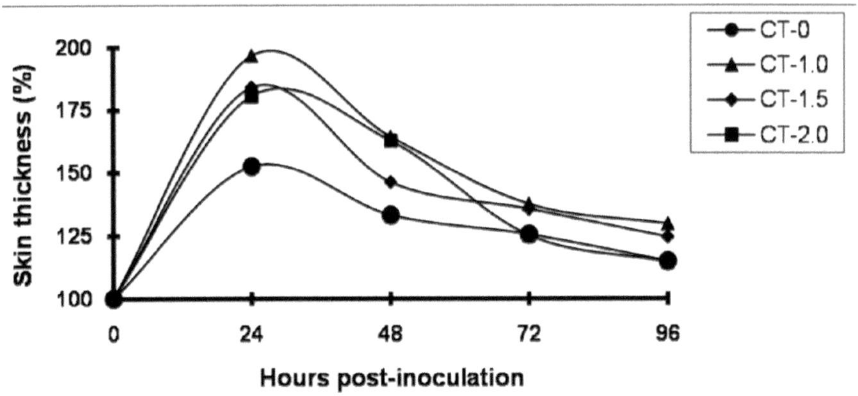

FIGURE 4.2 Effect of Ficus infectoria leaf meal supplementation on delayed type hypersensitivity response of lambs against PHA-P.

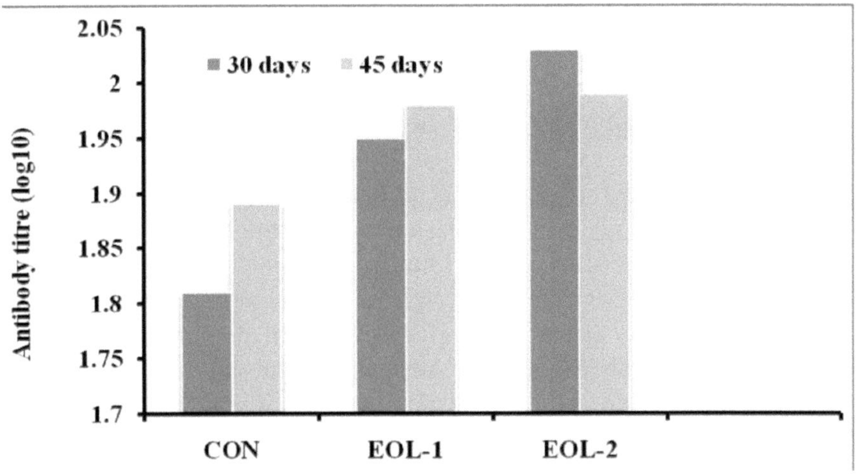

FIGURE 4.3 Effect of supplementing essential oils-rich feed additives on antibody titer (log10) against Pasteurella multocida in buffalo calves.

4.5.3 CANCER PREVENTION

Phenolic compounds are the most abundant plant metabolites and are a common part of the diet. They are involved in the modulation of the carcinogenesis process besides their role in antioxidant activity. Various berry extracts and mined polyphenols have been demonstrated to impede the growth of tumor cell lines (human oral, colon, breast, and prostate); however, the degree of inhibition depends on the concentration of extracts as well as the sensitivity of the cells (Zhang et al., 2008). Genistein, an isoflavone from soybean, has been shown to inhibit cancer cells of the prostate, breast, brain, and respiratory system (Sarkar & Li, 2002). Lala et al. (2006) reported that grape and berry extracts rich in anthocyanin declined the negative effects of the colon carcinogen, azoxymethane. The active ingredients present in the phenolic extract and bioavailability at the target tissue determine the effectiveness to inhibit tumor cells. Natural phenolics modulate carcinogenesis through improving body defense mechanisms and interfering with the functional activities of cells (Kampa et al., 2007).

4.5.4 *ALTERNATIVE ANTHELMINTICS*

Resistance of gastrointestinal parasites (GIP) to chemotherapeutic drugs has stimulated the search for the potential of plant secondary metabolites as an alternative GIP control strategy. Dietary supplementation of CTs through a mixture of dried leaves consisting of Psidium guajava, Ficus infectoria, and Ficus bengalensis was reported to reduce parasitic eggs in the feces (Figure 4.4) and improve production performance in goats (Dutta et al., 2012).

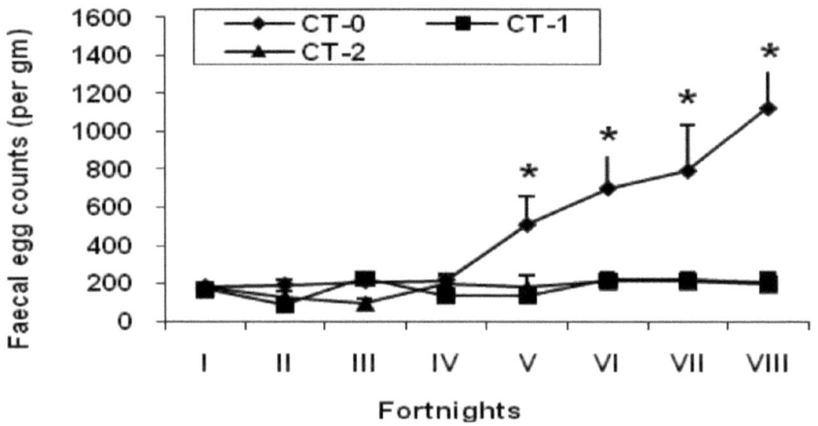

FIGURE 4.4 Effect of tanniferous leaf meal supplementation on fecal egg count in goats.

Various stages in the life cycle of hookworm (Haemonchus contortus) were inhibited and efficiency of nutrient utilization was improved in sheep with feeding of tanniferous leaf meal at 1.5% CTs in the dry matter (Pathak et al., 2013). Feeding of forages rich in CTs (Hedysarum coronarium/Lotus pedunculatus) to various categories of ruminants revealed increased growth of parasitized animals, decreased fecal egg counts, and direct inhibition of lungworm and gastrointestinal nematode larvae (Robertson et al., 1999; Athanasiadou et al., 2000).

Plant phenolics may directly interact with GI nematodes and can remarkably prevent infestation of numerous nematodes in sheep and goats by inhibiting infective larvae stage or egg hatching (Dey & Paul, 2014). CTs in the feeds increase the availability of rumen undegradable proteins by forming pH dependent tannin-protein complex thereby increasing absorption of essential amino acids including methionine and cystine in sheep (McNabb et al., 1993; Dey et al., 2008a). Improvement of host immunity due to feeding of CTs (Dey et al., 2007; Dutta et al., 2012) may also be important with selection for immunity to GIP. The aqueous defatted crude extracts of CT (freeze dried) from tanniferous leaves were described to have strong anthelmintic effect against Haemonchus contortus (Pathak et al., 2013). The experiment demonstrated very sluggish movement of H. contortus larvae after two hours of incubation with CT extracts, however, mortality started after four hours. The results after 12 hours of incubation showed the highest (85%) mortality of larvae with Psidium guajava extract. The adult H. contortus in the control group were seen active throughout the incubation period. The potential benefit of tannins on GIP also depends on the nature of CT, their source, and dose levels to animals. The type of parasites present in the GI tract; their habitat and degree of worm load also determine the efficacy of CTs to gastrointestinal parasitism.

4.6 PLANT BIO-ACTIVES IN ENVIRONMENT PROTECTION

Phytogenic bioactive compounds, especially EOs, tannins, and saponins, modulate the rumen microbiome depending on the concentrations and constituents due to their antimicrobial action. The diet of the animals also plays a significant role in this mechanism. They reduce nitrogen excretion to the environment by reducing protein degradation and inhibiting rumen bacteria responsible for a high rate of ammonia production (HAP) and enteric methane emission by inhibiting ruminal methanogens.

4.6.1 INHIBITION OF RUMINAL PROTEIN DEGRADATION AND NITROGEN EXCRETION

Proteins in the ruminants' diet are instantly solubilized and release more than half of the nitrogen in the rumen while chewing of feeds;

resulting in significant (~30%) losses of nitrogen through absorption of ammonia via rumen wall (Lapierre et al., 2005). The absorbed ammonia is converted into urea in the liver and a greater part is excreted through urine, resulting in lowered efficiency of protein utilization, and environmental pollution by ammonia emission (Castillo et al., 2001). An experiment at CIRB Hisar (Mery Chanu, 2018) on 16S rRNA sequences of enrichment culture of buffalo rumen content revealed Proteobacteria (61.1 to 68.2%) as the most predominant bacteria followed by Unclassified bacteria (24–29%), Firmicutes (2.8–6.9%), Bacteroidetes (0.9–2%), Actinobacteria (0.1–1.1%), Fusobacteria (0.04–0.7%) and Synergistetes (0.1–0.4%) at phylum level. At family and genus level, most of the sequences remained unclassified. Among sequences with valid family name Acidaminococcaceae was most predominant constituting 1.5 to 4.7% of sequences. Similarly, at genus level, among sequences with valid genus name Acidaminococcus was most predominant constituting 1.4–4.1% of total bacterial sequences.

Protein degradation by the rumen microbes is inhibited by EOs; subsequently the ammonia production is reduced. It is envisaged that amino acids deamination and ammonia formation is inhibited by EOs through reduction in the number of hyper ammonias producing bacteria viz. Clostridium sticklandii, Peptostreptococcus anaerobius (Wallace et al., 2002). Although higher doses (> 40 ppm) inhibited other bacterial species, resulting in less volatile fatty acid production, lower doses reduce the number of HAP bacteria and hyperactive starch and protein degrading bacteria, such as Ruminobacter amylophilus (Szumacher-Strabel & Cieślak, 2012). Although a very small number of HAP bacteria are present in the rumen (< 0.01% of the total ruminal bacteria), they produce a very high rate of ammonia that exceeds the capacity of microbial protein production (Russell & Strobel, 1988). Blended EOs supplementation in sheep was reported to decrease degradation of soybean meal but not the rapeseed meal (Newbold et al., 2004). Molero et al. (2004) demonstrated reduced ruminal degradation of sunflower meal in comparison to soybean or fish meal by blended EOs in cattle, suggesting substrate specificity of EOs on rumen microbes. EOs interfere with the attachment and colonization of rumen microbes with the plant materials (Wallace, 2004). Higher concentration of EOs exhibits general inhibition of rumen microbes, resulting reduced fermentation of feed in the rumen (Fernandez-Lopez et al., 2005; Dey et al., 2016).

Eucalyptus oil supplementation was shown to reduce in vitro ruminal ammonia production at a higher concentration (1 μl/ml of buffered rumen fluid); however, it was ineffective at a lower (0.5 μl/ml) dose (Singh et al., 2019). A dose-dependent reduction of ruminal ammonia production was also reported with lemongrass oil supplementation (Singh et al., 2018a). The extracts from poplar leaves, eucalyptus leaves, and clove buds were demonstrated to reduce ammonia production without affecting feed digestion under an in vitro fermentation system (Kumar et al., 2017). The blends of these extracts were shown to have positive associative effects in rumen fermentation pattern. Several plant bioactive compounds, including the organosulfur compounds present in garlic oil, have the capability to reduce the deamination process in the rumen through inhibition of microbes associated with protein degradation (Wallace et al., 2002). Graded levels of garlic oil supplementation were described to decrease ammonia-N production linearly with impeding digestibility of oats hay substrate (Dey et al., 2016). Cardozo et al. (2004) also observed a decrease in ammonia production with the addition of garlic oil. Thymol EO, garlic oil and saponins from soap bark tree (Quillaja saponaria) was observed to reduce protease activity and ammonia production (Figure 4.5) in the enrichment culture of mixed rumen bacteria of buffalo, suggesting inhibition of HAP (Mery Chanu, 2018).

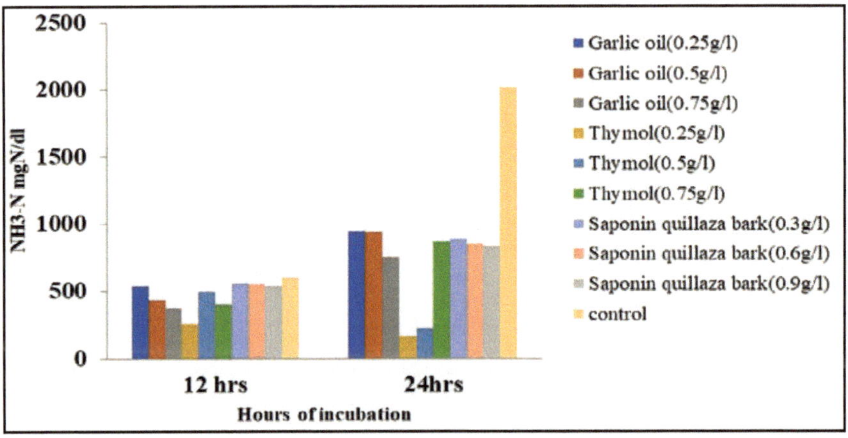

FIGURE 4.5 Effects of garlic oil, thymol, and saponin quillaja bark on ammonia production by mixed rumen microflora in in vitro enrichment culture.

CTs in the diet are suggested to be used as an organic protectant of proteins, for enhancing rumen by-pass of the protein fraction and reducing loss by microbial deamination (Dey & Dutta, 2014). Although lower levels of CT in the diet affect rumen function and animal performance favorably, however, a higher concentration reduces feed utilization and animal performance. An ideal concentration has been recommended to be 15 to 40 g/kg feed dry-matter intake (Dey et al., 2008b; Pathak et al., 2013). When CT-rich herbage is masticated, a pH-dependent tannin-protein complex is formed, which remains stable at a pH range of 3.5–7.0. The complex bonds are dissociated at a lower pH of the abomasum and a higher pH of the small intestine. Thus, proteins are protected from ruminal degradation and good quality amino acids are available for intestinal absorption. Supplementation of CT at graded levels (1–2%) from various tropical tree leaves in the substrate significantly reduced the nitrogen degradability (IVDN) of groundnut cake under in vitro system (Table 4.4). The leaves of Ficus glomerata and Ficus infectoria were demonstrated to reduce IVDN significantly (35–37% IVDN of GNC) (Dey et al., 2006). The improvement in N-retention by reducing urinary-N excretion (Table 4.5) was shown with dietary supplementation of CT (1–2%) in sheep (Dey et al., 2008a), suggesting better availability of protected protein for animal performance. Studies indicated that the rumen protective effect of CT was much higher for essential amino acids than non-essential amino acids, which is likely to account for at least part of higher growth rates and N retention in sheep given diets containing CTs (Dey et al., 2008a; Dubey et al., 2013). The CT alters nutrient partitioning for higher microbial protein production than the volatile fatty acids, thereby benefiting the host (Pathak et al., 2014).

TABLE 4.4 Various Levels of Condensed Tannins on Nitrogen Degradability (IVDN) of Groundnut Cake

Sources of CT	IVDN at different CT levels (%)						SEM
	0	0.5	1.0	1.5	2.0	2.5	
Artocarpus heterophyllus	0.92^{d}	0.91^{d}	0.85^{cC}	0.75^{bC}	0.68^{aC}	0.66^{aC}	0.05
Ficus bengalensis	0.92^{d}	0.91^{d}	0.79^{cB}	0.68^{bB}	0.61^{aB}	0.58^{aB}	0.07
Ficus glomerata	0.92^{d}	0.91^{d}	0.71^{cA}	0.61^{bA}	0.57^{aA}	0.55^{aA}	0.08
Ficus infectoria	0.92^{d}	0.92^{d}	0.80^{cB}	0.67^{bB}	0.59^{aB}	0.57^{aAB}	0.07

$^{abcd/ABC}$ means with different superscript within a row/column differ significantly ($P < 0.01$).

Source: Adapted from Dey et al. (2006).

TABLE 4.5 Effect of CT Supplementation on Nitrogen Balance in Lambs

Attributes	**Treatments**				**SEM**
	CT-0	**CT-1.0**	**CT-1.5**	**CT-2.0**	
	Nitrogen Balance (g per day)				
Intake	15.2[a]	15.9[ab]	17.0[b]	14.7[a]	0.37
Fecal loss	4.6[a]	6.0[ab]	6.6[b]	5.7[a]	0.15
Urinary loss	5.7[c]	5.0[bc]	4.7[b]	3.6[a]	0.21
	Nitrogen Excretion (% intake)				
Fecal	37.0	38.0	38.5	38.6	0.58
Urinary	37.4[c]	31.3[b]	27.5[ab]	25.5[a]	1.06
	Retention (%)				
N intake	25.7[a]	30.7[b]	30.2[b]	37.0[c]	1.29
N absorbed	40.7[a]	49.3[b]	54.6[bc]	60.1[c]	1.75

Abbreviations: CT-0: Control supplement; CT-1.0: 1% CT-containing supplement; CT-1.5: 1.5% CT-containing supplement; and CT-2.0: 2% CT-containing supplement.

Source: Adapted from Dey et al. (2008a).

The resistance and resilience to infections is increased in animals with improved protein utilization; however, it remains more evident in young ones and animals with pathophysiological changes of body tissue. Dietary saponins affect metabolism of nutrients in ruminants (Cheeke, 1996). Research on saponin-rich Y. schidigera plant extract revealed enhanced feed utilization efficiency, health status, and body weight gain in ruminants (Mader & Brumm, 1987). Quillaja saponins were reported to decrease protein degradation and improve microbial protein synthesis in vitro (Makkar & Becker, 1996; Mery Chanu, 2018). Intra-ruminal administration of lucerne saponins was demonstrated to reduce total protozoa count in sheep (Lu & Jorgensen, 1987). Saponins bind with sterols present in the surface membrane of protozoa only and are detrimental to the survival of the protozoa. Yucca saponins control rumen ammonia concentrations by binding and releasing ruminal ammonia, thereby supporting enhanced microbial protein production (Hussain & Cheeke, 1995; Makkar et al., 1999). Dietary supplementation of saponin-rich Sesbania sesban leaves was shown to improve ruminal flow of protein to the lower tract with inhibition of protozoal activity (Newbold et al., 1997). However, rumen bacteria are capable of degrading saponins and it was also reported that sheep saliva can degrade saponin if fed a saponin-containing diet for a longer time (Teferedegne, 2000). Wang et al. (2000a, b) demonstrated inhibition of cellulolytic bacteria without affecting amylolytic bacteria in ruminants fed a high-grain diet, thus playing a positive role by saponins in rumen functions. However, adaptation of rumen microbial population to saponins might be responsible for the absence of any effects on animal performance (Wu et al., 1994; Cheeke, 1996). These studies demonstrated the role of different plant bioactive compounds in enhancing protein utilization efficiency and abatement of environmental pollution through reduced ammonia emission by ruminants.

4.6.2 METHANE MITIGATION

Rumen fermentation of feeds produces methane by methanogenic archaea present in the rumen microbial ecosystem. Tannins reduce methane production both by direct inhibition of methanogenic archaeal population as well as by reducing feed digestion, resulting in less hydrogen available for methanogenesis (Tavendale et al., 2005). Hess et al. (2003) demonstrated that decreasing the number of protozoa associated with rumen methanogens by dietary tannins is responsible for reduced methane production. Various studies revealed a linear decrease in methane production with increasing (0.2–2% of DM) levels of CT from different plant sources in the ruminants' diet (Tan et al., 2011; Cieslak et al., 2012) without hampering rumen fermentation. However, many reports described the ineffectiveness of tannins on ruminal methanogenesis (Beauchemin et al., 2007; Oliveira et al., 2007). It is suggested that the effects of tannins on rumen microbial consortium, thereby methane production are inconsistent owing to source and type of tannins and levels of dietary inclusion. Studies with various flavonoids described that naringin and quercetin suppress methanogenesis by reducing total protozoa and methanogens, without adversely affecting rumen microbial fermentation (Oskoueian et al., 2013).

Addition of Ficus bengalensis leaves extract containing CTs was observed to reduce methane production under in vitro fermentation system; however, higher doses reduced feed fermentation (Singh et al., 2019). Pathak et al. (2017) reported that addition of CT from Psidium guajava and Ficus infectoria leaf meal mixture at 1-2% in diet considerably reduced the methane emission by lambs. Therefore, strategic use of tanniferous leaves in the diet of ruminants could decrease enteric methane emissions. EOs lessen methane production either by reduction of methanogenic archaeal number or inhibition of archaeal genes responsible for methane production (Ohene-Adjei et al., 2008). A reduction (0.93–89%) of methane production was evident with inclusion of various doses (1–5 mM) of cinnamaldehyde to hay-based substrate (Macheboeuf et al., 2008). Singh et al. (2019) demonstrated a decreased (82.60–85.30%) methane production with addition (2–3 ml/l of rumen fluid) of eucalyptus oils (ECOs). Similarly, a dose-dependent reduction in methane production (58%, 1.66 ml/l, Kumar et al., 2009 and 90.3%, 2 ml/l, Sallam et al., 2009) was reported with supplementation of eucalyptus oil to in vitro fermentation medium. Kumar et al. (2017) reported reduced methane production with inclusion of extracts of EO-rich eucalyptus leaves, clove buds, or poplar leaves to oat hay substrate. Beef cattle supplemented with lemongrass powder (100 g/d) were demonstrated to improve nutrient utilization and rumen microbial protein production (Wanapat et al., 2008). A dose-dependent inhibitory effect on methane production was reported (Singh et al., 2018a) on inclusion of lemongrass (Cymbopogon citrates) oil with oats hay substrate. However, reduced feed digestibility was reported at higher doses owing to overall antimicrobial action on ruminal microbes. The concentrations and types of active components in EOs are the determining factors for inhibiting methanogenesis. As the effects on methane inhibition and improving animal performance also depend on the animals, type of feed offered, methods of feeding, etc., long-duration animal experimentations are required to be conducted for recommendation as a feed additive for animals.

4.7 CONCLUSION

Phytogenic bioactive compounds from agroforestry-based medicinal plants have the potential to modulate gut microbial consortium, thereby reducing methane production and nitrogen excretion by ruminants. They improve livestock health by stimulating the immune system and improving antioxidant status, nutrient utilization efficiency, and reducing environmental pollution. The ideal concentration of bioactive compounds or plants and their suitable blends in relation to ration composition of livestock is advocated for their application as phytogenic feed additives in livestock ration.

KEYWORDS

- **antioxidant status**
- **farm animals**
- **immune response**
- **methane mitigation**
- **milk production**
- **nitrogen excretion**
- **plant-bioactive**
- **ruminal protein degradation**

REFERENCES

Ames, B. N., Shigenaga, M. K., & Hagen, T. M., (1993). *Proc. National Acad. Sci. U.S.A., 90*(17), 7915–7922.

Athanasiadou, S., Kyriazakis, I., Jackson, F., & Coop, R., (2000). *Int. J. Parasitol., 30*(9), 1025–1033.

Barr, I. G., Sjolander, A., & Cox, J. C., (1998). *Adv. Drug Deliv. Rev., 32*, 247–271.

Beauchemin, K., McGinn, S., Martinez, T., & McAllister, T., (2007). *J. Anim. Sci., 85*(8), 1990–1996.

Cardozo, P., Calsamiglia, S., Ferret, A., & Kamel, C., (2004). *J. Anim. Sci., 82*, 3230–3236.

Carulla, J., Kreuzer, M., Machmüller, A., & Hess, H., (2005). *Crop Pasture Sci., 56*(9), 961–970.

Castillo, A., Kebreab, E., Beever, D., Barbi, J., Sutton, J., Kirby, H., & France, J., (2001). *J. Anim. Sci., 79*, 247–253.

Chaieb, I., (2010). *Tunis J. Plant Prot., 5*, 39–45.

Cheeke, P. R., (1996). In: Waller, G. R., & Yamasaki, Y., (eds.), *Saponins Used in Food and Agriculture* (pp. 377–386). Plenum Press, New York.

Cieslak, A., Zmora, P., Pers-Kamczyc, E., & Szumacher-Strabel, M., (2012). *Anim. Feed Sci. Technol., 176*(1), 102–106.

Cotelle, N., (2001). *Curr. Top. Med. Chem., 1*(6), 569–590.

Cuvelier, M. E., Richard, H., & Berset, C., (1996). *J. American Oil Chem. Soc., 73*, 645–652.

D'Archivio, M., Filesi, C., Di Benedetto, R., Gargiulo, R., Giovannini, C., & Masella, R., (2007). *Annali-Istituto Superiore di Sanita., 43*(4), 348.

Dai, J., & Mumper, R. J., (2010). *Molecules, 15*, 7313–7352.

Dey, A., & De, P. S., (2014). *Asian Aust. J. Anim. Sci., 27*(3), 342–348.

Dey, A., & Dutta, N., (2014). *Univ. J. Pharm., 3*(2), 20–25.

Dey, A., & Paul, S. S., (2014). *Indian J. Anim. Nutri., 31*(2), 101–109.

Dey, A., Dutta, N., & Sharma, K., (2009b). *Indian Vet. J., 86*(11), 1156–1158.

Dey, A., Dutta, N., Pattanaik, A. K., & Sharma, K., (2015). *Jap. J. Vet. Res., 63*(1), 15–24.

Dey, A., Dutta, N., Sharma, K., & Pattanaik, A., (2006). *Anim. Nutri. Feed Tech., 6*(2), 215–222.

Dey, A., Dutta, N., Sharma, K., & Pattanaik, A., (2007). *J.Immunol. Immunopath., 8*, 130, 131.

Dey, A., Dutta, N., Sharma, K., & Pattanaik, A., (2008a). *Small Rumim. Res.*, *75*(2), 105–114.

Dey, A., Dutta, N., Sharma, K., & Pattanaik, A., (2008b). *Indian J. Anim. Sci.*, *78*(1), 102–106.

Dey, A., Dutta, N., Sharma, K., & Pattanaik, A., (2009a). *Indian J. Anim. Sci.*, *79*(1), 58–62.

Dey, A., Paul, S., Dahiya, S. S., & Punia, B. S., (2016). In: Sreekumar, D., Jacob, N., Mahender, M., & Rajanna, N., (eds.), *Int. Livest. Conf. (INDIGENOUS) and 23rd Annual Convention Indian Society of Animal Production and Management(ISAPM)* (p. 41). Prof. Jayashankar Telangana State Agricultural University, Hyderabad, Telangana.

Dubey, M., Dutta, N., Sharma, K., Banerjee, P. S., Pattanaik, A. K., & Singh, M., (2013). *Indian J. Anim. Sci.*, *83*(12), 64–70.

Dutta, N., Dubey, M., Banerjee, P., Pattanaik, A., Sharma, K., Kumar, P., & Narang, A., (2012). *Livest. Res. Rural Dev.*, *24*(2). http://www.lrrd.org/lrrd24/2/dutt24035.htm (accessed on 18 December 2023).

FAO, (2010). *FAO Statistical Databases*. FAO, Rome.

Fernandez-Lopez, J., Zhi, N., Aleson-Carbonell, L., Perez-Alvarez, J., & Kuri, V., (2005). *Meat Sci.*, *69*, 371–380.

Firkins, J. L., Yu, Z., & Morrison, M., (2007). Ruminal nitrogen metabolism: Perspectives for integration of microbiology and nutrition for dairy. *J. Dairy Sci., 90*. https://doi.org/10.3168/jds.2006-518.

Gano, J. M., (2013). *Amino Acid-Fermenting Bacteria from the Rumen of Dairy Cattle Enrichment, Isolation, Characterization, and Interaction with Entodinium Caudatum.* The Ohio State University.

Gershenzon, J., McConkey, M. E., & Croteau, R. B., (2000). *Plant Physiol.*, *122*, 205–214.

Greathead, H., (2003). *Proc. Nutri. Soc.*, *62*, 279–290.

Guenther, E., (1948). *The Essential Oils* (Vol. I). Van Nostrand Company Inc., New York.

Habibi, R., Sadeghi, G., & Karimi, A., (2014). *Brit. Poultry Sci.*, *55*, 228–237.

Hagerman, A. E., Riedl, K. M., & Rice, R. E., (1999). In: Gross, G. G., Hemingway, R. W., & Yoshida, T., (eds.), *Plant Polyphenols, 2. Chemistry, Biology, Pharmacology, Ecology* (pp. 495–505). Kluwer Academic Publishers/Plenum Press, Dordrecht/New York.

Harborne, J. B., & Williams, C. A., (2000). *Phytochem.*, *55*, 481–504.

Hart, K., Yanez-Ruiz, D., Duval, S., McEwan, N., & Newbold, C., (2008). *Anim. Feed Sci. Technol.*, *147*, 8–35.

Hess, H. D., Kreuzer, M., Dıaz, T., Lascano, C. E., Carulla, J. E., Soliva, C. R., & Machmüller, A., (2003). *Anim. Feed. Sci. Technol., 109*(1), 79–94.

Ho, K., Huang, J., Tsai, C., Lin, T., Hsu, Y., & Lin, C. *J*., (1999). *Pharm. Pharmacol.*, *51*(9), 1075–1078.

Hussain, I., & Cheeke, P. R., (1995). *Anim. Feed Sci. Technol.*, *51*, 231–242.

IPCC, (2019). *Inter-Governmental Panel for Climate Change 2019: Special Reports on Climate Change and Land Use*. IPCC, Geneva.

Jie, Y. H., Cammisuli, S., & Baggiolini, M., (1984). *Agents and Actions*, *15*, 386–391.

Kampa, M., Nifli, A. P., Notas, G., & Castanas, E., (2007). In: *Reviews of Physiology, Biochemistry and Pharmacology* (pp. 79–113). Springer.

Kensil, C. R., (1996). *Critical Rev. Therap. Drug Carrier Sys.*, *13*, 1–55.

Kumar, K., Dey, A., & Rose, M. K., (2017). In: *3rd International Conference of Bioresource and Stress Management* (p. 204) Jaipur, Rajasthan.

Kumar, K., Dey, A., & Rose, M. K., (2018). In: *IX Asian Buffalo Congress* (pp. 103, 104). Asian Buffalo Association, CIRB., Hisar, India.

Kumar, R., Kamra, D., Agarwal, N., & Chaudhary, L., (2009). *Anim. Nutri. Feed Technol., 9*, 237–243.
Lacaille-Dubois, M. A., Hanquet, B., Cui, Z. H., Lou, Z. C., & Wagner, H., (1999). *J. Natural Prod., 62*, 133–136.
Lala, G., Malik, M., Zhao, C., He, J., Kwon, Y., Giusti, M. M., & Magnuson, B. A., (2006). *Nutr. Cancer, 54*(1), 84–93.
Lapierre, H., Berthiaume, R., Raggio, G., Thivierge, M., Doepel, L., Pacheco, D., Dubreuil, P., & Lobley, G., (2005). *Anim. Sci., 80*, 11–22.
Lin, C. C., Hsu, Y. F., & Lin, T. C., (2000). *Anticancer Res., 21*(1A), 237–243.
Lu, C. D., & Jorgensen, N. A., (1987). *J. Nutr., 117*, 919–927.
Macheboeuf, D., Morgavi, D., Papon, Y., Mousset, J. L., & Arturo-Schaan, M., (2008). *Anim. Feed Sci. Technol., 145*, 335–350.
Mader, T. L., & Brumm, M. C., (1987). *J. Anim. Sci., 65*, 9–15.
Makkar, H. P. S., & Becker, K., (1996). In: Waller, G. R., & Yamasaki, Y., (eds.), *Saponins Used in Food and Agriculture* (pp. 377–386). Plenum Press, New York.
Makkar, H. P. S., & Becker, K., (1998). *Agrofor. Sys., 40*(1), 59–68.
Makkar, H. P. S., Aregheore, E. M., & Becker, K., (1999). *J. Agric. Sci., 132*, 313–321.
Mangan, J., (1988). *Nutri. Res. Rev., 1*(1), 209–231.
McNabb, W., Waghorn, G., Barry, T., & Shelton, I., (1993). *Brt. J. Nutri., 70*(2), 647–661.
Mery, C. Y., (2018). *PhD Dissertation.* NDRI (Deemed University), Karnal, Haryana, India. 1–131.
Min, B. R., Fernandez, J., Barry, T., McNabb, W., & Kemp, P., (2001). *Anim. Feed Sci. Technol.,92*(3), 185–202.
Molero, R., Ibars, M., Calsamiglia, S., Ferret, A., & Losa, R., (2004). *Anim. Feed Sci. Technol., 114*, 91–104.
Mueller-Harvey, I., & McAllan, A., (1992). *Adv. Plant Cell Biochem. Biotech., 1*, 151–217.
Newbold, C. J., El Hassan, S. M., Wang, J., Ortega, M. E., & Wallace, R. J., (1997). *Brt. J. Nutr., 78*, 237–249.
Newbold, C., McIntosh, F., Williams, P., Losa, R., & Wallace, R., (2004). *Anim Feed Sci Technol., 114*, 105–112.
Oda, K., Matsuda, H., Murakami, T., Katayama, S., Ohgitani, T., & Yoshikawa, M., (2000). *Biol. Chem., 381*, 67–74.
Ohene-Adjei, S., Chaves, A., McAllister, T., Benchaar, C., Teather, R., & Forster, R., (2008). *Microb. Ecol., 56*, 234–242.
Oliveira, S. G., Berchielli, T. T., Dos, S. P. M., Primavesi, O., Frighetto, R., & Lima, M. A., (2007). *Anim. Feed Sci. Technol., 135*(3), 236–248.
Oskoueian, E., Abdullah, N., & Oskoueian, A., (2013). *BioMed. Res. Int. 2013*, Article ID 349129, https://doi.org/10.1155/2013/349129.
Pathak, A. K., Dutta, N., & Sharma, K., (2014). *Int. J. Biol. Biologic. Sci., 2*(3), 209–224.
Pathak, A. K., Dutta, N., Banerjee, P., Pattanaik, A., & Sharma, K., (2013). *Asian Aust. J. Anim. Sci., 26*(10), 1446.
Pathak, A. K., Dutta, N., Pattanaik, A. K., Chaturvedi, V. B., & Sharma, K., (2017). *Asian Aust. J. Anim. Sci., 30*(12), 1702–1710.
Plohmann, B., Bader, G., Hiller, K., & Franz, G., (1997). *Pharmazie, 52*, 953–957.
Ramírez-Restrepo, C., Barry, T., Pomroy, W., López-Villalobos, N., McNabb, W., & Kemp, P., (2005). *Anim. Feed Sci. Technol., 122*(3), 197–217.

Ri, C. S., Jiang, X. R., Kim, M. H., Wang, J., Zhang, H. J., Wu, S. G., Bontempo, V., & Qi, G. H., (2017). *Italian J. Anim. Sci.*, *1*, 7.
Rice-Evans, C. A., Miller, N. J., Bolwell, P. G., Bramley, P. M., & Pridham, J. B., (1995). *Free Radic. Res.*, *22*(4), 375–383.
Riguera, R., (1997). *J. Marine Biotech.*, *5*, 187–193.
Robertson, H., Niezen, J., Waghorn, G., & Charleston, W., (1995). *Proc New Zealand Soc Anim Prod.*, *55*, 199–201.
Russell, J., & Strobel, H., (1988). *J. Anim. Sci.*, *66*, 552–558.
Ryzner, M., Takáčová, J., Čobanová, K., Plachá, I., Venglovská, K., & Faix, Š., (2013). *Acta Vet. Brno.*, *82*, 43–48.
Sallam, S., Bueno, I., Brigide, P., Godoy, P., Vitti, D., & Abdalla, A., (2009). *Options Medi.*, *85*, 267–272.
Sarkar, F. H., & Li, Y., (2002). *Cancer Metast. Rev.*, *21*(3, 4), 265–280.
Seifzadeh, S., Aghjeh-Gheshlagh, F. M., Abdi-Benemar, H., Seifdavati, J., & Navidshad, B., (2016). *Iranian J. Appl. Anim. Sci.*, *6*.
Singh, R. K., Dey, A., Paul, S. S., Singh, M., & Punia, B. S., (2018a). *Indian J. Anim. Nutri.*, *35*(2), 174–179.
Singh, R. K., Dey, A., Paul, S. S., Singh, M., Dahiya, S. S., & Punia, B. S., (2019). *Agrofor. Sys.*, 10.1007/s10457-019-00395-3.
Singh, R. K., Dey, A., Singh, M., Punia, B. S., & Paul, S. S., (2018b). *Int. J. Curr. Microbiol. Appl. Sci.*, *7*, 4666–4672.
So, H. S., Yoon, H. S., Kwoon, Y. S., Sung, J. H., Lee, T. G., Park, E. N., Cho, H. S., et al., (1997). *Mol. Biol. Cells*, *7*, 176–186.
Szumacher-Strabel, M., & Cieślak, A., (2012). In: *Dietary Phytochemicals and Microbes* (pp. 285–309). Springer.
Tan, H., Sieo, C., Abdullah, N., Liang, J., Huang, X., & Ho, Y., (2011). *Anim. Feed Sci. Technol.*, *169*(3), 185–193.
Tavendale, M. H., Meagher, L. P., Pacheco, D., Walker, N., Attwood, G. T., & Sivakumaran, S., (2005). *Anim. Feed Sci. Technol.*, *123*, 403–419.
Teferedegne, B., (2000). *Proc. Nutri. Soc.*, *59*, 209–214.
Teixeira, B., Marques, A., Ramos, C., Serrano, C., Matos, O., Neng, N. R., Nogueira, J. M., et al., (2013). *J. Sci. Food Agric.*, *93*, 2707–2714.
Wallace, R. J., (2004). *Proc. Nutri. Soc.*, *63*, 621–629.
Wallace, R. J., McEwan, N. R., McIntosh, F. M., Teferedegne, B., & Newbold, C. J., (2002). *Asian Aust. J. Anim. Sci.*, *15*, 1458–1468.
Wanapat, M., Cherdthong, A., Pakdee, P., & Wanapat, S., (2008). *J. Anim. Sci.*, *86*, 3497–3503.
Wang, Y. X., McAllister, T. A., Yanke, L. J., & Cheeke, P. R., (2000a). *J. Appl. Microbiol.*, *88*, 887–896.
Wang, Y. X., McAllister, T. A., Yanke, L. J., Xu, Z. J., Cheeke, P. R., & Cheng, K. J., (2000b). *J. Sci. Food Agric.*, *80*, 2114–2122.
Wu, Z., Sadik, M., & Sleiman, F. T., (1994). *J. Anim. Sci.*, *72*, 1038–1042.
Yoshiki, Y., Kudou, S., & Okubo, K., (1998). *Biosci. Biotech. Biochem.*, *62*, 2291–2299.
Zhang, Y., Seeram, N. P., Lee, R., Feng, L., & Heber, D., (2008). *J. Agric. Food Chem.*, *56*(3), 670–675.

CHAPTER 5

Ethnobotany: Benefits, Research, and Development

SUMIT CHAKRAVARTY,[1] SAJITHA SIRIL,[1] N. N. SHAHINA,[1] SAROJ BISWAKARMA,[1] MANENDRA SINGH,[1] SUBRATA DAS,[1] GOPAL SHUKLA,[1,2] VINEETA PANDEY,[3] JAHANGEER A. BHAT,[4] and RAINER W. BUSSMANN,[5]

[1]*Uttar Banga Krishi Viswavidyalaya, Pundibari, West Bengal, India*

[2]*Department of Forestry, North Eastern Hill University, Tura Campus, Tura, Meghalaya, India*

[3]*National Medicinal Plants Board, Ministry of Ayush, New Delhi, India*

[4]*Piramal Foundation for Education Leadership (State Transformation Program), Jammu and Kashmir, India*

[5]*Department of Ethnobotany, Institute of Botany, Ilia State University, Tbilisi, Georgia*

ABSTRACT

People living in forest fringe areas depend primarily on plant resources for their livelihoods. Orally transmitted traditional knowledge (TK) carries considerable information that helps future generations in problem-solving and is unique and specific to the socio-cultural context. In this scenario, ethnobotany evolved to demystify the importance of the relationship between plants and people. The studies and associated documentation helped to understand and preserve various plants, which have value as food, fodder, medicine, and drugs. So, with the evolving scientific research and technology, the ethnobotanical research pattern needs a change noteworthy at the policy level. Hence, the chapter discusses diverse aspects of ethnobotany

Bioprospecting Ethnomedicinal Plant Resources: Sustainable Utilization and Restoration.
Gopal Shukla, Jahangeer A. Bhat, A. P. Das, and Sumit Chakravarty (Eds.)

uses, conservation needs, research trends, threats, and recommendations. Furthermore, it highlights the need for systematic and modern technology-based research and development that conserves ethnobotanical knowledge.

5.1 INTRODUCTION

Forests are primary resources for the population living around its fringe areas. Rural households, especially in developing countries, depend on forests and other ecosystems for their daily needs, e.g., for supplemental food, complement crops, fuels for cooking and a wide range of traditional medicines, fodder, hygiene products, clothing, and construction products. Furthermore, plant resources provide income essential for sustaining livelihoods during adverse seasons (Reddy et al., 2007; Pandey, 2009). Naturally, these communities interact closely with forests, not only to obtain their economic benefits but often also for their cultural and spiritual identity (Byron & Arnold, 1999). Many people living in poverty globally depend directly or indirectly upon forests for their livelihoods (Raj et al., 2018; Vineeta et al., 2020, 2021). India has a forest-dwelling population of over 100 million, belonging to 550 communities of 227 ethnic groups, of which 60% live in forest areas and depend on forests (Yesodharan & Sujana, 2006). India is renowned for its rich floral diversity with more than 47,000 species and is counted among the 12 mega biodiversity countries of the world (Bain & Premi, 2019). In India, the traditional knowledge (TK) about a wide variety of ethnobotanical plants is distributed among diverse ethnic groups, and indigenous communities draw their sustenance predominantly from that knowledge (Siva, 2007; Pandey, 2009). The various ethnic groups and their culture are braided through forest biological resources, and only knowing the ethnobotanical wealth will unfold the truth (Gangwar & Ramakrishnan, 1990; Kala, 2005).

5.2 ETHNOBOTANY

Ethnobotany highlights the people-plants relationship, especially plant use, often by indigenous people, for food, medicine, pesticides, clothing, shelter, and other purposes (Ashebo, 2019). Its popularity is acknowledged in Asian countries because studies are numerous and well-acknowledged among the scientific community (Palit & Gurung, 2008; Sharma & Kumar, 2013). The diverse flora and different cultures, beliefs, and needs of the various traditional communities richly contribute to the folklore (Jain, 1981; Ashebo, 2019). The TK of any region represents a unique picture because it is location-specific in the socio-cultural context and varies among people and communities (Ashebo,

2019). Hence, proper documentation and research are mandatory to reveal and explore valuable information that might be beneficial worldwide. In addition, skills like taxonomic identification, comprehending cultural concepts and transcribing the local language are essential for successful documentation (Pandey & Tripathy, 2017). Ethnobotany is an interdisciplinary science which includes various subjects: ethnobryology, ethnopharmacology, ethnoecology, ethnomedicine, ethnoarchaeology, ethnonarcotics, and ethnoagriculture (Nolan & Turner, 2011; Pandey & Tripathy, 2017; Figure 5.1). Therefore, by accounting for their involvement in traditional plant use, ethnobotanical studies are breakthroughs for research from multiple scales.

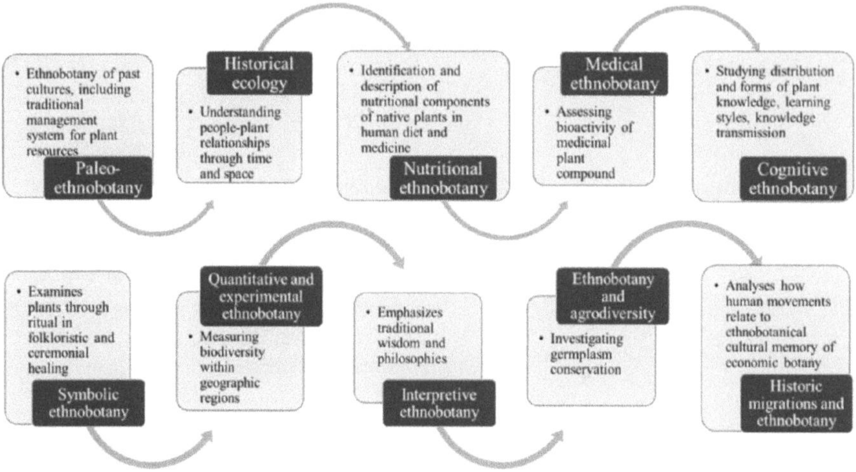

FIGURE 5.1 Some contemporary ethnobotanical research aspects.

Source: Adapted from: Nolan & Turner (2011).

5.3 ETHNOBOTANICAL USES

Researchers explored the ethnobotanical uses of numerous plants in various locations and communities in India (Table 5.1). Focusing on the present-day situation of the ever-increasing exploitation of plants and natural resources, the main reason for showing interest in ethnobotany is its vast outcome beneficial for every living being (Singh & Bharti, 2015). Apart from their traditional use as food (including fermented food and beverages), most wild plants have potential uses as medicine, veterinary use, drug development, fodder, timber, spices, fiber, cosmetics, fuel wood, aromatic, cultural, and spiritual uses (Reddy et al., 2007; Ajesh et al., 2012; Mathur & Joshi, 2013; Jennings et al., 2015).

TABLE 5.1 Location Wise Documentation and Reports of Ethnobotanical Uses

Sl. No.	Location	Number of Species	Primary Uses	References
1.	Bankura and Purulia, West Bengal	43–46	Medicinal	Ghosh et al. (2013); Sinhababu & Banerjee (2013)
2.	Bhotiya tribe, Niti valley, Central Himalaya, India	86	Medicine	Phondani et al. (2010)
3.	Bhoxa community, Dehra Dun, Uttarakhand	50	Treatment of dysentery and diarrhea.	Sharma et al. (2013a)
4.	Birbhum district, West Bengal and Dumka district, Jharkhand	28	Medicine	Mondal & Rahaman (2012)
5.	Boda and Kolli hills, Eastern Ghats	31	Medicine, traditional healing.	Xavier et al. (2011); Sathiyaraj et al. (2015); Anand & Deborah (2016)
6.	Cooch Behar, West Bengal	79–125	Food, medicine, fodder, furniture, fiber, cosmetics, fuelwood, hut construction, fencing, and agricultural equipment.	Bandyopadhyay & Mukherjee (2009); Shukla & Chakravarty (2012)
7.	Darjeeling and Sikkim Himalayas	24–281	Treat common ailments like cough and cold, fever, diarrhea, and dysentery, bone-related problems.	Sarkar et al. (2015)
8.	Darjeeling hills, West Bengal	48	Medicine, food, fodder, sacred/religious use.	Chettri et al. (2014)
9.	Eastern Himalayan Homegardens	260	Medicine, food.	Pala et al. (2019)
10.	Different parts of India	214	Remedies for jaundice.	Sharma et al. (2012)
11.	Dzongu, Sikkim	76	Vegetables, spices, fruits.	Kumar et al. (2012)
12.	Garhwal Himalaya hills	126	Cure fever, common cold, cough, diarrhea, dysentery, cuts, wound, burnt, blood-related problems, diabetes, liver, and bone-related problems.	Pala et al. (2012)
13.	Guna, Madhya Pradesh	32	Treating various human ailments.	Samar et al. (2015)
14.	Hill tribes and aborigines of the West Rarh region of West Bengal.	46	Treat various ailments.	Ghosh (2008)
15.	Jaintia Community, Meghalaya	40	Food and medicine.	Jaiswal (2010)
16.	Jalpaiguri, West Bengal	115	Food, medicine	Bose et al. (2015)

TABLE 5.1 *(Continued)*

Sl. No.	Location	Number of Species	Primary Uses	References
17.	Jammu and Kashmir	937	Treatment of dermatological problems, cold, cough, and throat related ailments, fever, joint and muscle-related ailments, gastrointestinal disorders, urogenital ailments, respiratory ailments, body pain and gynecological disorders.	Gairola et al. (2014)
18.	Kani tribals, Tirunelveli hills, Tamil Nadu	14	Treatment of skin problems.	Ayyanar & Ignacimuthu (2005)
19.	Kani tribes, Thodu hills, Kerala	35	Medicine	Xavier et al. (2014)
20.	Kanji Wildlife Sanctuary, Kargil, Northwest Himalaya	30	Fodder, fuel, vegetables, medicine, miscellaneous purposes.	Hamid & Raina (2014)
21.	Karbi Anglong, Assam	51	Cultural importance and dream interpretation use.	Kar et al. (2008); Terangpi et al. (2015)
22.	Kumaun, Uttarakhand	206	Medicinal, economic (aromatic, timber, spices, fuel, condiments, cosmetics, etc.), fodder, firewood, timber, food, and spiritual purposes.	Mathur & Joshi (2013)
23.	Kupwara district, Kashmir	48	Medicinal, aromatic, and cultural use.	Lone et al. (2012)
24.	Madurai, Tamil Nadu	60	Cure skin diseases, poison bites, wounds, cuts, burns, stomach aches and nervous disorders.	Ignacimuthu et al. (2006)
25.	Manipur	68	Nutritive and therapeutic purposes.	Konsam et al. (2016)
26.	Marathwada, India	40	Medicinal, against fever, asthma, bronchitis, kidney stones, dysentery, and skin diseases.	Kachare & Fatima (2012)
27.	Naxalbari, West Bengal	35	Food, medicine.	Biswakarma et al. (2015)
28.	Paniyaas and Malavettuvas in Kasargod, Malappuram, and Wayanad districts, Kerala	15	Fever, cough, stomach problems, kidney troubles, dysentery, wound healing during childbirth, burns, snake bites, psoriasis, ear infections, liver problems, circulatory system disorders, dermatological disorders, and skeletomuscular system disorders.	Jose et al. (2015)

TABLE 5.1 *(Continued)*

Sl. No.	Location	Number of Species	Primary Uses	References
29.	PedabayaluMandalam, Visakhapatnam, Andhra Pradesh	80	Treatment of dysentery, skin disease and fever.	Padalet al. (2013)
30.	Poshina forest range, Sabarkantha, Gujarat	35	Medicinal	Patel & Patel (2013)
31.	Purulia, West Bengal	57	Medicine	Chakraborty & Bhattacharjee (2006)
32.	Raigarh, Chhattisgarh	89	Medicine, food, fodder, furniture, fiber, and cosmetics.	Singh & Bharti (2015)
33.	Rava tribe, Jalpaiguri, West Bengal	41	Medicine	Bose (2011)
34.	Reang and Tripuri community, Tripura	33	Medicine	Das et al. (2009)
35.	Renukoot forest division, Sonbhadra, Uttar Pradesh	105	Medicine	Singh et al. (2012a)
36.	Sikkim	190	Fruit and nuts/seeds.	Chettri et al. (2005)
37.	Sonebhadra, Uttar Pradesh	143	Medicine	Singh & Dubey (2012)
38.	Southern Odisha	110	Food and medicine.	Misra & Misra (2016)
39.	Tharu community, Udham Singh Nagar, Uttarakhand	90	Treatment of skin diseases (wounds, boils, cuts, leprosy, etc.).	Sharma et al. (2011, 2014)
40.	The Achi Mayans of Guatemala	44	Food	Turreira-Garcíaet al. (2015)
41.	Tikamgarh, Bundelkhand, Central India	41	Medicine	Verma (2014)
42.	Tipturtaluk, Tumkur, Karnataka	127	Medicine	Achar et al. (2015)
43.	Tripura	61	Food, medicine, dye, and aroma.	Malla et al. (2012)
44.	Uraon tribe, Surguja, Chhattisgarh	15	Diabetes	Shrivastava & Kanungo (2013)
45.	Uttar Karnataka	31	Skin diseases	Harsha et al. (2003)
46.	Uttarakhand	64	Treatment of jaundice and epilepsy.	Sharma et al. (2012, 2013b)
47.	West Midnapore, West Bengal	21	Medicine	Mondal & Samanta (2014)
48.	Yimchunger-Naga tribes	56	Treat various diseases and relieve the pain.	Jamir et al. (2008)

5.3.1 EDIBLE USE

Ethnobotanical studies demystify the use of several plants that serve as alternatives to staple food. During food insufficiency, they provide valuable supplements for a nutritionally balanced diet (Sõukand et al., 2015). In addition, those edible plants ensure alternative sources of income for communities, especially the tribes residing in and around the forests (Deshmukh & Waghmode, 2011; Suresh et al., 2014). Minor fruits, nutrient-rich leaves, seeds, and other parts of plants play a vital role in maintaining people's health in areas of poor market accessibility. Rai et al. (2005) reported cheap and readily available tree fruits with vibrant taste appeal with medicinal, industrial, and nutritional values from Sikkim Himalayas. Besides their nutritional value, easy availability and inexpensiveness make the wild food plants a significant cultural and socio-economic entity (Hazarika et al., 2012; Narzary et al., 2013). In India, around 800 wild species are consumable and are utilized primarily by ethnic communities (Singh & Arora, 1978).

5.3.2 FERMENTED FOOD AND BEVERAGES

Globally, fermented food and beverages are widely recognized for their role in domestic healthcare strategies and therapeutic values (Quave & Pieroni, 2014, 2015). Sõukand et al. (2015) documented 116 species from 37 families with abandoned use as fermented food in eastern European countries. The availability of fermented foods contributes to food security and sovereignty, especially during the long winter periods when fresh produce is unavailable in isolated communities (Quave & Pieroni, 2015). Fermented products are potentially probiotics and have a remarkable role in human health and preventative medicine (Kumar et al., 2013; Singh & Bunger, 2014).

5.3.3 ETHNOMEDICINAL USE

Since ancient times, medicinal plants that treat various diseases are included in human civilization (Van der Merwe et al., 2001; Lewis & Elwin, 2003; Singh & Lahiri, 2010) and are of immense importance (Singh et al., 1998). Medicinal properties of plants mainly were discerned through trial and error and influenced by people's belief systems, often intertwined with spiritual and mythic traditions (Mathias et al., 1996). Among traditional medicines, around 85% are plant-derived (Farnsworth et al., 1985). Traditional medicinal use of plants is strongly related to the physiological and pharmacological activity of active plant ingredients (Fourie et al., 1992). The importance of medicinal plants is increasingly being recognized today, from ecological, social, and economic perspectives (Larsen et al., 2000; Olsen, 2005; Pala et al., 2012, 2019; Tripathi et al., 2013; Bhutia et al., 2018; Biswakarma et al., 2018; Raj et al., 2018; Vineeta et al., 2018a, b, 2020, 2021; Lepcha et al., 2019; Dinesha et al., 2021a, b; Manohar et al., 2021). Specific uses of ethnomedical plants include anticarcinogenic, antioxidants, antimicrobial, and anti-inflammatory, cure disorders in lung, stomach, and liver, cough, asthma, cardiovascular diseases, kidney stone, piles, diuretic, diarrhea, dysentery, eczema, malaria, epilepsy, gastric problem, blood-related problem, diabetes, blood pressure, gingivitis, diabetes, blood sugar, arthritis, anemia, and glucosuria (Ranjithakani et al., 1992; Subitha et al., 2011; Xavier et al., 2011; Biswakarma et al., 2015, 2017a, b, 2018; Sarkar et al., 2015).

The review by Uprety et al. (2016) reported 739 species of Non-Timber Forest Plants (NTFPs) with 24 different uses employed by the local people of Kanchenjunga landscape.

Of these, more species were documented from India (377 species), Nepal (363) and Bhutan (245). According to the study, local people utilized the medicinal value of plants more, which cured about 27 serious ailments, mainly gastrointestinal disorders. About 85% of India's rural population utilized wild plants for the treatment of various ailments (Farnsworth, 1994; Jain, 1992). Suresh et al. (2013) reported that specific plant parts treat complicated disease ailments.

The plant parts such as bark, leaves, roots, branches, stems, fruits, and seeds are utilized often considerably (Shukla & Chakravarty, 2012; Mondal & Samanta, 2014). Ethnobotanists documented many unique possibilities for using medicinal plants in various ailments while applications of some plants were indicated in Table 5.2 from the study published in diverse literature.

5.3.4 ETHNOVETERINARY MEDICINE (EVM)

The application of plants in animal disease treatment has developed in parallel with human disease treatments (Van der Merwe et al., 2001). Since time immemorial, livestock owners in India have been using traditional medication based on plant formulations. This knowledge has been passed on verbally, by example and sometimes through writing (Mathias et al., 1996). Ethno veterinary medicine (EVM) is essential in animal healthcare in developing countries (Cunningham & Zondi, 1991). It has become a recognized

TABLE 5.2 Some Important Ethnobotanically Utilized Species and Their Medicinal Application

Species Name	Medicinal Application	References
Adhatoda vasica	Cough, cold, and asthma.	Gao et al. (2008)
Aegle marmelos	Digestive disorders	Kesari et al. (2006); Mishra et al. (2010)
Annona squamosa leaves and seeds.	Leaves have antimicrobial effects in cuts and wounds; seeds have abortifacient activity.	Gupta et al. (2005); Kumar et al. (2007); Chavan et al. (2010)
Asteracantha longifolia	Anemia	Pawar et al. (2010)
Asparagus racemosus	Antitussive action.	Bopana & Saxena (2007)
Butea monosperma	Convulsions, gynecological disorder, and antimicrobial activities.	Kasture et al. (2002); Somani et al. (2006); Vidyasagar & Prashantkumar (2007); Shahavi & Desai (2008)
Azadirachta indica	Fever and inflammation.	Kosalge & Fursule (2009); Namsa et al. (2009)
Dalbergia sissoo	Gonorrhea	Hajare et al. (2001); Rana et al. (2009)
Moringa oleifera	Inflammation	Fayazuddin et al. (2013)

field of research that includes traditional veterinary theory, medicines, surgical methods, diagnostic procedures, and animal husbandry practices (Mathias et al., 1996). EVM is important in animal healthcare worldwide (Malla & Chhetri, 2012; Sheikh et al., 2013; Sanghi, 2014) because of its accessibility, ease of preparation and administration, at little or no cost at all to the farmers. This age-old practice covers every area of veterinary specialization and all livestock species and varies from place to place.

5.4 ETHNOBOTANY FOR DRUG DEVELOPMENT

Ethnobotanical studies have discovered multiple plants with influential healing properties, which were previously unexplored by Western science (Schultes, 1960, 1962). Plants play a prominent role in many pharmacological investigations aimed at developing valuable drugs. They either directly contribute to the medicinally vital plant-derived constituents or trigger the formation of drugs (Mukherjee, 2003). Plants, as a conventional treatment (CT) source, are beneficial for the development of new pharmaceuticals (Cox, 1994; Kinghorn & Balandrin, 1993; Iwu, 1994). Compounds derived from plants are used in approximately 25% of allopathic drugs in clinical use (Kinghorn & Balandrin,

1993). Active constituents were found in most tested plants selected based on their healing properties, while only a few medicinal plants tested have potentially harmful toxic effects (Fourie et al., 1992).

Ethnobotanical studies lead to the scientific assessment of the traditional medicines that provide vital information on drug development (Rout et al., 2009; Ghosh et al., 2013). Plant-oriented drug development approach shows an increasing trend with new foci. International and national organizations, NGOs (non-government organizations), and their agendas highlight the need to systematically estimate indigenous pharmacopeias to contribute to improved healthcare in rural areas (Heinrich, 2000). Previous years were a milestone in applying TK to develop drugs for several deadly diseases including cancer (Heinrich & Bremner, 2006). Sheng-Ji (2001) highlights the role of indigenous knowledge in the production of drugs, especially in China and Himalayan countries.

Ethnopharmacology helped to introduce various drugs into international markets, and about 25% of prescriptions incorporate functional principles emanated from or modeled after compounds from higher plants. Treatments for cancer, diabetes, hypertension, epilepsy, malaria, and arthritis highlight the potential of plants for drug development. The essential drugs developed through ethnomedical investigations are reserpine (hypertension treatment), podophyllotoxin (anticancer drug component), vinblastine (certain cancers treatment), and prostratin (human immunodeficiency virus treatment) (Pandey & Tripathi, 2017). A list of ethnobotanical plants used for drug development with their active ingredient is included in Table 5.3. Apart from designing new drugs, ethnobotany develops improved therapeutic options (Heinrich, 2000): (i) Preservation of indigenous knowledge; (ii) The selection of valuable species and the exclusion of toxic or ineffective ones; and (iii) The dissemination of the ethnobotanical and phytochemical pharmacological results.

5.5 NEED FOR CONSERVING TRADITIONAL KNOWLEDGE (TK)

Traditional knowledge (TK) can be adaptive and resilient (Ashebo, 2019; Reyes-García et al., 2014; Kujawska & Pardo-de-Santayana, 2015). The medicinal practices among tribal people and villagers in India today follow the same pattern they did 2,000 years ago, and there is hardly any change (Anita et al., 2013). Information on plant use for therapeutic purposes has been passed from generation to generation orally (Delacour, 2021; Kala et al., 2006). This knowledge, however, started to decline due to the lack of proper documentation and interest loss in younger generations (Kala,

TABLE 5.3 Drugs Derived from Plants Through Ethnobotanical Approaches

Active Ingredient	Action	Plant Source
Acetyldigoxin	Cardiotonic	*Digitalis lanata* (Grecian foxglove, woolly foxglove).
Adoniside	Cardiotonic	*Adonis vernalis* (Pheasant's eye, red chamomile).
Aescin	Anti-inflammatory	*Aesculus hippocastanum* (Horse chestnut).
Aesculetin	Antidysentery	*Frazinusrhychophyll*
Agrimophol	Anthelmintic	*Agrimonia supatoria*
Ajmalicine, serpentine	Treatment for circulatory disorders.	*Rauvolfia serpentina*
Allyl isothiocyanate	Rubefacient	*Brassica nigra* (Back mustard).
Andrographolide	Treatment for bacillary dysentery, hepatoprotective.	*Andrographis paniculata*
Anisodamine	Anticholinergic	*Anisodustangulicus*
Anisodine	Anticholinergic	*Anisodustangulicus*
Arecoline	Anthelmintic	*Areca catechu* (Betel nut palm).
Asiaticoside	Vulnerary	*Centella asiatica* (*Gotu cola*)
Atropine	Anticholinergic	*Atropa belladonna* (Deadly nightshade).
Berberine	Treatment for bacillary dysentery.	*Berberis vulgaris* (Common barberry).
Bergenin	Antitussive	*Ardisia japonica* (Marl berry).
Betulinic acid	Anticancer	*Betula alba* (Common birch).
Bromelain	Anti-inflammatory, proteolytic.	*Ananus comosus* (Pineapple).
Caffeine	CNS stimulant.	*Camelia sinensis* (Tea, also coffee, cocoa, and other plants).
(+)-Catechin	Hemostatic	*Potentilla fragarioideres*
Chymopapain	Proteolytic, mucolytic.	*Carica papaya* (Papaya).
Cocaine	Local anesthetic	*Erythroxylum coca* (Coca plant).
Codeine	Analgesic, antitussive	*Papaver somniferum* (Poppy).
Colchicine	Antitumor, antigout	*Colchicum autumnale* (Autumn crocus).
Convallatoxin	Cardiotonic	*Convallaria majalis* (Lily-of-the-valley).
Curcumin	Choleretic	*Curcuma longa* (Turmeric).
Cynarin	Choleretic	*Cynara scolymus* (Artichoke).
Danthron	Laxative	*Cassia species*

TABLE 5.3 *(Continued)*

Active Ingredient	Action	Plant Source
Deserpidine	Antihypertensive, tranquilizer	*Rauvolfia ceneseens*
Deslanoside	Cardiotonic	*Digitalis lanata* (Grecian foxglove, Woolly foxglove).
Digitalin	Cardiotonic	*Digitalis purpurea* (Purple foxglove).
Digitoxin	Cardiotonic	*Digitalis purpurea* (Purple foxglove).
Digoxin	Cardiotonic	*Digitalis purpurea* (Purple or common foxglove).
Emetine	Amoebicide, emetic	*Cephaelis ipecacuanha*
Ephedrine	Sympathomimetic, antihistamine	*Ephedra sinica* (Ephedra ma Huang).
Etoposide	Antitumor agent	*Podophyllum peltatum* (May apple).
Gitalin	Cardiotonic	*Digitalis purpurea* (Purple or common foxglove).
Glaucarubin	Amoebicide	*Simarouba glausa* (Paradise tree).
Glycymhizin	Sweetener, treatment for Addison's disease.	*Glysyrrhiza glabra* (Licorice)
Gossypol	Male contraceptive	*Gossypium* species (Cotton).
Hemsleyadin	Treatment for bacillary dysentery.	*Hemsleya amabilis*
Hydrastine	Hemostatic, astringent	*Hydratis Canadensis* (Goldenseal).
Hyoscyamine	Anticholinergic	*Hyoscyamus niger* (Black henbane, stinking nightshade, henpin).
Irinotecan	Anticancer, antitumor agent.	*Camptotheca acuminata*
Kainic acid	Ascaricide	*Digenea simplex* (Wire weed).
Kawain	Tranquilizer	*Piper methysticum* (Kava kava).
Khellin	Bronchodilator	*Ammi visnaga*
Lanatosides A, B, C	Cardiotonic	*Digitalis lanata*(Grecian foxglove, Woolly foxglove).
Lapachol	Anticancer, antitumor	*Tabebuia species* (Trumpet tree)
a-Lobeline	Smoking deterrent, respiratory stimulant.	*Lobelia inflate* (Indian tobacco).
Monocrotaline	Topical antitumor agent.	*Crotalaria sessiliflora*
Morphine	Analgesic	*Papaver somniferum* (Poppy).
Andrographolide	Treatment of dysentery	*Andrographis paniculate*
Noscapine	Antitussive	*Papaver somniferum*
Ouabain	Cardiotonic	*Strophanthus gratus* (Ouabain tree).
Papain	Proteolytic, mucolytic	*Carica papaya* (Papaya)

TABLE 5.3 *(Continued)*

Active Ingredient	Action	Plant Source
Phyllodulcin	Sweetener	*Hydrangea macrophylla* (Big leaf hydrangea, French hydrangea)
Physostigmine	Cholinesterase inhibitor	*Physostigma venenosum* (Calabar bean)
Picrotoxin	Analeptic	*Anamitra cocculus* (Fish berry)
Pilocarpine	Parasympathomimetic	*Pilocarpus jaborandi* (*Jabarandi*, Indian hemp)
Podophyllotoxin	Antitumor, anticancer agent	*Podophyllum peltatum* (May apple)
Protoveratrines A, B	Antihypertensives	*Veratrum album* (White false hellebore)
Pseudoephedrine	Sympathomimetic	*Ephedra sinica* (Ephedra, Ma Huang)
Nor-pseudoephedrine	Sympathomimetic	*Ephedra sinica* (Ephedra, Ma Huang)
Quinine	Antimalarial, antipyretic	*Cinchona ledgeriana* (Quinine tree)
Qulsqualic acid	Anthelmintic	*Quisqualis indica* (Rangoon creeper, Drunken sailor)
Rescinnamine	Antihypertensive, tranquilizer	*Rauvolfia serpentina*
Reserpine	Antihypertensive, tranquilizer	*Rauvolfia serpentina*
Rhomitoxin	Antihypertensive, tranquilizer	*Rhododendron mole* (Rhododendron)
Rorifone	Antitussive	*Rorippa indica*
Rotenone	Piscicide, insecticide	*Lonchocarpus nicou*
Rotundine	Analgesic, sedative, tranquilizer	*Stephania sinica*
Salicin	Painkiller	*Salix alba* (White willow)
Santonin	Ascaricide	*Artemisia matitma* (Wormwood)
Scillarin A	Cardiotonic	*Urginea maritime* (Squill)
Scopolamine	Sedative	Datura species (e.g., Jimsonweed)
Sennosides A, B	Laxative	Cassia species (Cinnamon)
Sylimarin	Antihepatotoxic	*Sylibum marianum* (Milk thistle)
Stevioside	Sweetener	*Stevia rebaudiana* (Stevia)
Stychnine	CNS stimulant	*Strychnos mux-vomica* (Poison nut tree)
Teniposide	Antitumor agent	*Podophyllam peltatum* (May apple or mandrake)
Levo-Tetrahydropalmatine	Analgesic, sedative, tranquilizer	*Carydalis ambigua*
Theobromine	Diureic, vasodilator	*Theobroma cacao* (Cocoa)
Theophyline	Diuretic, brochodilator	*Theobroma cacao* and others (Cocoa, Tea)

TABLE 5.3 *(Continued)*

Active Ingredient	Action	Plant Source
Topotecan	Antitumor, anticancer agent	*Comptotheca acuminata*
Trichosanthin	Abortifacient	*Trichosanthes kirilowii* (Snake gourd)
Tubocurarine	Skeletal music relaxant	*Chondodendron tomentosum* (Curare vine)
Valepotriates	Sedative	*Valeriana officinalis* (Valerian)
Yohimbine	Aphrodisiac	*Pausinystalia yohimbe* (Yohimbe)
Yuanhuacine	Abortifacient	*Daphne genkwa* (Lilac)
Yuanhuadine	Abortifacient	*Daphne genkwa* (Lilac)

Source: Adapted from: Farnsworth et al. (1985).

2000; Kala et al., 2006; Ashebo, 2019). Not only the knowledge but also the medicinal plants face threats because of their narrow geographic ranges, low population densities, slow growth rates, and lack of proper conservation (Kala, 1998; Nautiyal et al., 2002; Jan et al., 2019). Some of the plants became extinct (Jablonski, 2004).

People can conserve plant resources by engaging multiple stakeholders and local communities in sustainable management practices (Badola & Aitken, 2003; Hamid & Raina, 2014). Some of the endangered (EN) pharmaceutically important plants as a consequence of overexploitation from the wild are recognized by ethnobotanists and took measures to conserve them *in vitro* (Verma et al., 2012).

- *Elaeocarpus spharicus*;
- *Indigofera tinctoria*;
- *Rheum emodi*;
- *Picrorrhiza kurroa*;
- *Bergenia ciliata*;
- *Lavandula officinalis*;
- *Valeriana wallichii*;
- *Coleus forskohlii*;
- *Gentiana kurroo*;
- *Saussurea lappa*;
- *Stevia rebaudiana*;
- *Acorus calamus*;
- *Pyrethrum cinerariaefolium*;
- *Aloe vera*;
- *Bacopa monnieri*;

- *Salvia sclarea*;
- *Glycyrrhiza glabra*;
- *Swertia cordata*;
- *Psoralea corylifolia*;
- *Jurineamollis*;
- *Ocimum sanctum*;
- *Paris polyphylla*;
- *Papaver somniferum*.

These plants can be further tested for phytochemical investigation on the scientific scale and henceforth can improve the region's economy (Biswakarma et al., 2015, 2018).

Ethnobotanical studies, conducted among tribal communities, among tea garden workers and in homegardens, reported multiple values that uplift the community (Rai & Bhujel, 2002; Ghosh & Das, 2007). Homegardens serve as a refuge for legacy species and promote knowledge transmission and conservation at the family level (Kujawska & Pardo-de-Santayana, 2015). There is a need to domesticate the plants that possess ethnobotanical use value or to re-evaluate these species in local projects aimed at fostering endogenous strategies of food security as well as re-evaluating cultural heritage and sustaining small-scale food market circuits for protecting the natural resource to improve the socio-economic status and livelihoods of indigenous communities (Hong et al., 2015; Ahmad & Pieroni, 2016). A re-visitation of traditional, religious principles and approaches where contemporary conservation agendas could blend indigenous knowledge systems of local communities for conserving natural resources is needed (Eneji et al., 2012).

5.6 RESEARCH TRENDS AND RECOMMENDATION IN ETHNOBOTANY

Traditional medicine and ethnobotanical information play an essential role in scientific research, particularly when literature and fieldwork data have been adequately evaluated (Awadh et al., 2004). There has been a rapid expansion of allopathic medicinal treatment during the past century. Natural products, primarily plant products, are still widely used as medicine, particularly in inaccessible areas lacking modern healthcare systems (Namahata & Mukherjee, 1988, 1989; Mallick & Behera, 2009; Acharya & Mukherjee, 2010; Mallick & Mallick, 2012; Mondal & Biswas, 2012; Mallick et al., 2012). Unfortunately, information relating to medicinally valuable species, their uses, and associated traditional knowledge (TK) and practices is still very fragmentary (Yadav et al., 2012). Ethnobotany evolved to document such diverse uses and associated TK.

However, Gaoue et al. (2017) demonstrated how ethnobotanical studies overemphasize quantitative ethnobotany indices and underemphasize their theoretical foundation. The study listed 17 crucial theories and hypotheses that are important in the advancement of understanding plant-people relations.

Over the decades, ethnobotanical studies have helped to conserve, and document ethnopharmacological knowledge: medicinal uses of herbs, ointments, oils, animal products, and minerals.

Ethnobotanical publications are growing exponentially in recent years, according to a study by Malik et al. (2021) from 1974 to 2019. It found that the compound annual growth rate of global publications on ethnobotany was approximately 11%, which means the number of publications is doubling every six years. The review by Pathak & Bharati (2020) identified 2,123 research papers from India (also involving collaboration with other countries) on ethnobotany with a drastic increase in publications recently. The review documented 105 tribal groups of India where ethnobotanical analysis was carried out. Ethnobotanical research shows a clear positive trend, and people have become aware of declining TK that needs conservation. Ethnobotany is ideally suited for this task, as it includes multiple fields involving the interaction between humans and the environment and approaches to community stakeholders and policymakers (McClatchey & Gollin, 2005; McClatchey, 2005). Because ethnobotany is an interdisciplinary science, it requires collaboration with other fields, widening the scope of ethnobotanical research. Figure 5.2 shows the multidisciplinary aspects and significant contributions of ethnobotany.

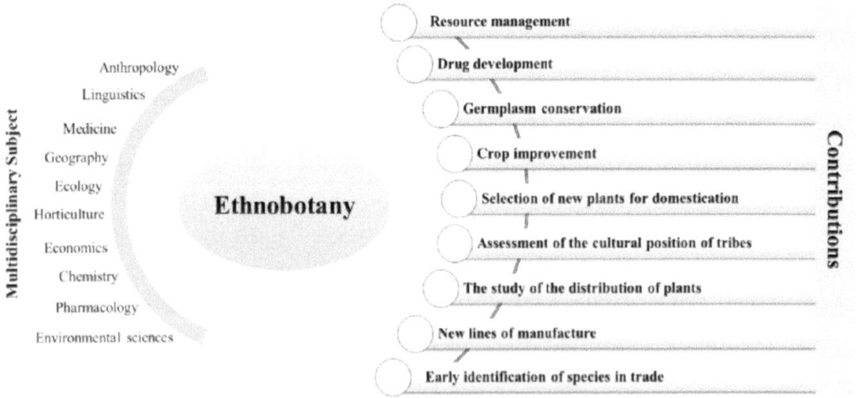

FIGURE 5.2 Ethnobotany is a multidisciplinary subject that contributes to diverse benefits.

Source: Adapted from: Pathak & Bharati (2020).

Even though ethnobotany is still a relatively new scientific discipline, its systematization and formalization have greatly advanced in the last decades. It is widely accepted worldwide because it helps to preserve indigenous, and TK. Oliveira et al. (2010) considered ethnobotany as ancient in practice but young in theory. Ethnobotanical education is still training-oriented, potentially creating an environment to learn about herbal medicine, psychoactive plants, or the natural connection to land and organisms (Gaoue et al., 2017). Ethnobotanical knowledge contributes to sustainable development goals (SDGs). The review by Kumar et al. (2021) suggested that out of the 17 SDGs, at least seven have a strong association with traditional ethnobotanical knowledge. Ethnobotanical knowledge contributes to poverty alleviation, health, and wellbeing, sustainable production and consumption and safe life on land free from climate change (Figure 5.3).

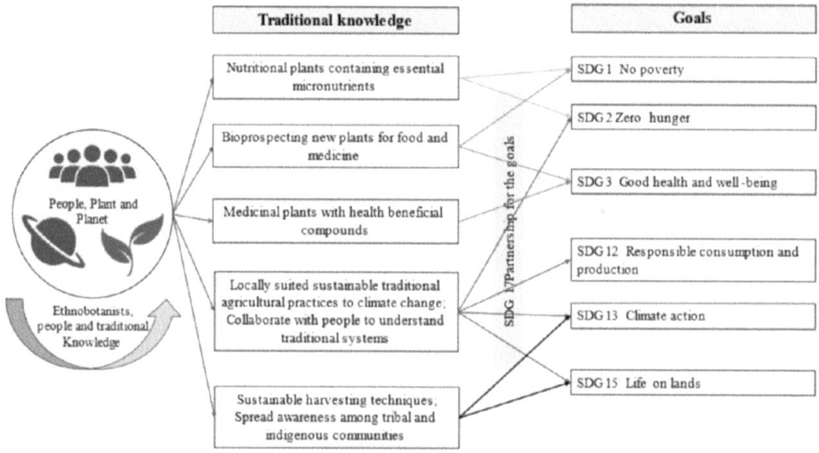

FIGURE 5.3 Schematic representations highlighting the potential role of ethnobotanists and ethnobotanical knowledge towards achieving seven sustainable development goals and targets.

Source: Adapted from: Kumar et al. (2021).

Currently, the exploration of ethnobotanical information has considerable importance and scope. However, ethnobotanical data need further systematic research. One of the significant concerns regarding herbal medicines is their safe usage. Herbal medicines have, e.g., sometimes been found to be contaminated with heavy metals like arsenic, lead, and mercury, leading to health hazards (Ernst, 2002; Chan, 2003). Traditional herbal medicines need to be safe and efficient in use. Thus, issues regarding quality standards of herbal formulations and their safety must be addressed. The proper pharmacological activity of the medicinal plants needs to be substantiated through extensive and elaborate experimental studies to ensure the safe, efficacious, and rational utilization of plants.

Studies also highlight the overexploitation of the endangered medicinal plants by the residents who are unaware of their importance in the ecosystem (Bopana & Saxena, 2007). There is a great and urgent necessity to educate the local population and healers to adopt conservation measures as necessary so that over-collection of such species can be avoided and will not lead to their extinction in their territory, which signifies the loss of their source medicinal material (Tripathi et al., 2013).

Another concern is the lack of proper documentation and the associated loss of biodiversity, culture, and TK. Last, the possibility of intellectual theft of indigenous knowledge has been raised, although the ratification of the "Nagoya Protocol on Access to Genetic Resources and Fair and Equitable Participation in the Benefits from their Use in the Convention on Biological Diversity (CBD)" (https://www.cbd.int/abs/about/) has brought a boost to the recognition of the privileges of indigenous and regional communities, their participation in ethnobotanical research often is still fragmentary. In this scenario, local participants should finally get the role they deserve – to not only be participants, but investigators, and co-authors, fully participating in data analysis and publication, and application of the results, as well as sharing the benefits (Sheng-Ji, 2001; Pathak & Bharati, 2020). Nevertheless, awareness and extension activities must go in tandem at local and regional levels with research-based policy changes. It may help prevent the uncontrolled transboundary trade of medicinal plants. Furthermore, it will encourage the local people to share information on the value of plants with outsiders. Kumar et al. (2021) put forward some valuable recommendations for the ethnobotany research development, represented in Figure 5.4.

Marketing of useful plants is another research thrust, where ethically and sustainably produced products that can sustain both biological and cultural systems (Flaster, 1998). The lack of markets and proper market channels often discourages the sale and availability of ethnobotanically valuable plants.

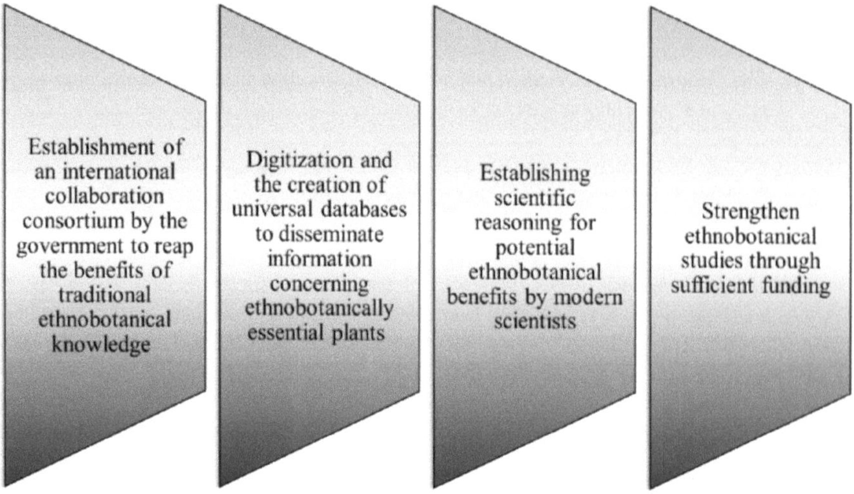

FIGURE 5.4 Recommendations for ethnobotany research improvement.

Source: Adapted from: Kumar et al. (2021).

5.7 CONCLUSION

Ethnobotany research is of significant importance nowadays because it highlights the cultural, social, and economic importance of plant resources. Even though numerous studies highlight ethnobotanical uses through documentation, their exploration as an interdisciplinary science with policy-level implications is limited so far.

KEYWORDS

- **drug development**
- **ethnobotany**
- **ethnoveterinary medicine**
- **forest**
- **medicinal plant**
- **overexploitation**
- **policy-level implications**
- **traditional knowledge**

REFERENCES

Achar, K. G., Boosanur, V. S., &Shivanna, M. B., (2015). Ethno-medico-botanical knowledge of Tiptur taluk in Tumkur district of Karnataka India. *Indian Journal of Traditional Knowledge, 1*, 147–154.

Acharya, J., & Mukherjee, A., (2010). Herbal therapy for urinary ailments as documented from Bankura district West Bengal. *Indian Journal of Science Research, 1*, 67–69.

Ahmad, K., & Pieroni, A., (2016). Folk knowledge of wild food plants among the tribal communities of Thakht-e-Sulaiman Hills North-West Pakistan. *Journal of Ethnobiology and Ethnomedicine, 12*, 17–31.

Ajesh, T. P., Abdulla, N. S. A., &Kumuthakalavalli, R., (2012). Ethnobotanical documentation of wild edible fruits used by Muthuvan tribes of Idukki Kerala- India. *International Journal of Pharma and Bio Science, 3*, 479–487.

Anand, S. P., & Deborah, S., (2016). Enumeration of wild edible fruits from Boda hills and Kolli hills. *International Journal of Applied Biology and Pharmaceutical Technology, 7*, 96–102.

Anita, R., Samuel, R., & Rajendra, Y., (2013). Ethnomedicinal plants used by the people of Darjeeling Hills in the eastern Himalaya of India. *Universal Journal of Pharmacy, 2*, 122–134.

Ashebo, T., (2019). A review on the qualitative method of the study of people-plants' relationship in their environment. *International Journal of Environmental Science and Natural Research, 22*(1), 556078.

Awadh, A., Ali, N., Al-Rrahwi, K., & Lindequist, U., (2004). Some medicinal plants used in Yemeni herbal medicine to treat Malaria. *African Journal of Traditional Complementary and Alternative Medicines, 1*, 72–76.

Ayyanar, M., &Ignacimuthu, S., (2005). Ethnomedicinal plants used by the tribals of Tirunelveli hills to treat poisonous bites and skin diseases. *Indian Journal of Traditional Knowledge, 4*, 229–236.

Badola, H. K., & Aitken, S., (2003). The Himalayas of India: A treasury of medicinal plants under siege. *Biodiversity, 4*, 3–13.

Bain, B., & Premi, J. K., (2019). An investigation on the ethnogynecological medicinal knowledge of the Birhor tribe of Chhattisgarh, India. *Research Journal of Pharmacy and Technology, 12*(11), 5138–5150.

Bandyopadhyay, S., & Mukherjee, S. K., (2009). Wild edible plants of Koch Bihar district West Bengal India. *Natural Product Radiance, 8*, 64–72.

Bhutia, K. D., Suresh, C. P., Pala, N. A., Shukla, G., & Chakravarty, S., (2018). Nutraceutical potential of some wild edible fruits of Sikkim Himalaya, India. *Studies onEthno-Medicine, 12*(1, 2), 106–112.

Biswakarma, S., Pala, N. A., Shukla, G., Vineeta, & Chakravarty, S., (2017a). Plants for liver and jaundice treatment: A case study from forest fringe communities in North Bengal, India. *Forestry Ideas, 23*, 145–151.

Biswakarma, S., Pala, N. A., Shukla, G., Vineeta, & Chakravarty, S., (2017b). Ethnomedicinal plants used to cure stomach disorders in forest fringe communities in northern part of West Bengal. *Indian Journal of Natural Products and Resources (Natural Product Radiance), 8*, 370–380.

Biswakarma, S., Pala, N. A., Shukla, G., Vineeta, Pradhan, K., & Chakravarty, S., (2018). Influence of socio-economic factors on attitude of ethno-botanical users among forest fringe communities in sub-Himalayan region of West Bengal, India. *Indian Journal of Agroforestry*, 27–35.

Biswakarma, S., Sarkar, B. C., Shukla, G., Pala, N. A., & Chakravarty, S., (2015). Traditional application of ethno-medicinal plants in Naxalbari area of West Bengal, India. *International Journal of Forest Usufructs Management*, *16*(1), 36–42.

Bopana, N., & Saxena, S., (2007). *Asparagus racemosus*- ethnopharmacological evaluation and conservation needs. *Journal of Ethnopharmacology*, *110*, 1–15.

Bose, D., (2011). An ethno-medicobotanical investigation among Rava tribe of Jalpaiguri district. *NBU Journal of Plant Sciences*, *5*, 61–65.

Bose, D., Roy, J. G., Mahapatra, S. D., Datta, T., Mahapatra, S. D., & Biswas, H., (2015). Medicinal plants used by tribals in Jalpaiguri district West Bengal India. *Journal of Medicinal Plants Studies*, *3*, 15–21.

Byron, N., & Arnold, J. E. M., (1999). What futures for the people of the tropical forests? *World Development*, *27*, 789–805.

Chakraborty, M. K., & Bhattacharjee, A., (2006). Some common ethnomedicinal uses for various diseases in Purulia district West Bengal. *Indian Journal of Traditional Knowledge*, *5*, 554–558.

Chan, K., (2003). Some aspects of toxic contaminants in herbal medicines. *Chemosphere*, *52*, 1361–1371.

Chavan, M. J., Wakte, P. S., & Shinde, D. B., (2010). Analgesic and anti-inflammatory activity of caryophyllene oxide from *Annona squamosa* L. bark. *Phytomedicine*, *17*, 149–151.

Chettri, D., Moktan, S., & Das, A. P., (2014). Ethnobotanical studies on the tea garden workers of Darjeeling Hills. *Pleione*, *8*, 124–132.

Chettri, N., Sharma, E., & Lama, S. D., (2005). Non timber forest produces utilization distribution and status in a trekking corridor of Sikkim India. *Lyonia*, *8*, 89–101.

Cox, P. A., (1994). The ethnobotanical approach to drug discovery: Strengths and limitations. In: Ethnobotany and the search for new drugs. In: Chadwick, J., & Marsh, J., (eds.), *Ciba Foundation Symposium* (Vol. 185, pp. 25–41). Wiley Chichester.

Cunningham, A. B., & Zondi, A. S., (1991). *Cattle Owners and Traditional Medicines Used for Livestock.* Investigational Report No. 69. Institute of Natural Resources University of Natal Pietermaritzburg.

Das, H. B., Majumdar, K., Datta, B. K., & Ray, D., (2009). Ethnobotanical uses of some plants by Tripuri and Reang tribes of Tripura. *Natural Product Radiance, 8*, 172–180.

Delacour, F., (2021). The importance of biodiversity in therapeutically promising plant species native to North East India. *Journal of Genetics and Molecular Biology, 5*, 5.

Deshmukh, B. S., &Waghmode, A., (2011). Role of wild edible fruits as a food resource: Traditional knowledge. *International Journal of Pharmacy and Sciences*, *2*, 919–924.

Dinesha, S., Shukla, G., Mondal, H. A., & Chakravarty, S., (2021a). Macro- proliferation cup technique for mass multiplication of *Woodfordiafruticosa* (L.) Kurz (*Dhawai*) in sub-humid foothills of eastern Himalaya. *Indian Journal of Ecology*, *48*(5), 1305–1312.

Dinesha, S., Vineeta, Shukla, G., Rakesh, S., Roy, B., & Chakravarty, S., (2021b). Effects of growing media and growth hormones on the sprouting, rooting and field establishment of *Woodfordiafruticosa* (L.) Kurz in sub humid foothills of eastern Himalaya. *Journal of Plant Development Sciences*, *13*, 257–264.

Eneji, C. V. O., Ntamu, G. U., Unwanade, C. C., Godwin, A., Bassey, J. E., Willaims, J. J., & Joseph, I., (2012). Traditional African religion in natural resources conservation and management in cross river State Nigeria. *Environment and Natural Resources Research, 2*, 45–53.

Ernst, E., (2002). Toxic heavy metals and undeclared drugs in Asian herbal medicines. *Trends in Pharmacological Sciences, 23.*

Farnsworth, N. R., (1994). Ethnopharmacology and drug development. *Ciba Foundation Symposium 145*, 42–51.

Farnsworth, N. R., Akerele, O., Bingel, A. S., Soejarto, D. D., & Guo, Z., (1985). Medicinal plants in therapy. *Bulletin of the World Health Organization, 63*(6), 965.

Fayazuddin, M., Ahmad, F., Kumar, A., & Yunus, S. M., (2013). An experimental evaluation of anti-inflammatory activity of *Moringa oleifera* seeds. *International Journal of Pharmacy and Pharmaceutical Sciences, 5*, 717–721.

Flaster, T., (1998). Assessing and developing new markets for ethnobotanicals and general Medicinals. *North American Conference on Enterprise Development Through Agroforestry: Farming the Agroforest for Specialty Products* (pp. 69–73).

Fourie, T. G., Swart, I., &Snyckers, F. O., (1992). Folk medicine: A viable starting point for pharmaceutical research. *South African Journal of Science, 88*, 190–192.

Gairola, S., Sharma, S., & Bedi, Y. S., (2014). A cross-cultural analysis of Jammu Kashmir and Ladakh (India) medicinal plant use. *Journal of Ethnopharmacology, 155*, 925–986.

Gangwar, A. K., & Ramakrishnan, P. S., (1990). Ethnobiological notes on some tribes of Arunachal Pradesh northeastern India. *Economic. Botany, 44*, 94–31.

Gao, H., Huang, Y. N., Gao, B., Li, P., Inagaki, C., & Kawabata, J., (2008). Inhibitory effect on a-glucosidase by *Adhatodavasica*Nees. *Food Chemistry, 108*, 965–972.

Gaoue, O. G., Coe, M. A., Bond, M., Hart, G., Seyler, B. C., &McMmillen, H., (2017). Theories and major hypotheses in ethnobotany. *Economic Botany, 71*(3), 269–287.

Ghosh, A., (2008). Ethnomedicinal plants used in West Rahrr region of West Bengal. *Natural Product Radiance, 75*, 461–465.

Ghosh, C., & Das, A. P., (2007). Plants of ethnobotanical significance among the tea garden workers of Darjeeling Hills in West Bengal. In: Das, A. P., & Pandey, A. K., (eds.), *Advances in Ethnobotany* (pp. 133–147). CBS Publishers New Delhi.

Ghosh, S. K., Guria, N., Sarkar, A., & Ghosh, A., (2013). Traditional herbal remedies for various ailments within the rural communities in the district of Bankura and Purulia West Bengal India. *International Journal of Pharmacy and Pharmaceutical Sciences, 5*, 195–198.

Gupta, R. K., Kesari, A. N., Murthy, P. S., Chandra, R., Tandon, V., & Watal, G., (2005). Hypoglycemic and antidiabetic effect of ethanolic extract of leaves of *Annona squamosa* L. in experimental animals. *Journal of Ethnopharmacology, 99*, 75–81.

Hajare, S. W., Chandra, S., Sharma, J., Tandan, S. K., Lal, J., & Telang, A. G., (2001). Anti-inflammatory activity of *Dalbergia sissoo* leaves. *Fitoterapia, 72*, 131–139.

Hamid, A., & Raina, A. K., (2014). Ethnobotanical uses of plants in and around Kanji Wildlife Sanctuary North West Himalaya. *International Journal of Science and Research (IJSR), 3*, 538–545.

Harsha, V. H., Hebber, S. S., Shripathi, V., & Hedge, G. R., (2003). Ethnomedicobotany of Uttar Kannada District in Karnataka India- plants in treatment of skin diseases. *Journal of Ethnopharmacology, 84*, 37–40.

Hazarika, T. K., Lalramchuana, & Nautiyal, B. P., (2012). Studies on wild edible fruits of Mizoram India used as ethno-medicine. *Genetic Resource and Crop Evolution, 59*, 1767–1776.

Heinrich, M., & Bremner, P., (2006). Ethnobotany and ethnopharmacy – their role for anti-cancer drug development. *Current Drug Targets*, *7*, 239–245.

Heinrich, M., (2000). Ethnobotany and its role in drug development. *Phytotherapy Research*, *14*, 479–488.

Hong, H., Zhuo, J., Lei, Q., Zhou, J., Ahmed, S., Wang, C., Long, Y., Li, F., & Long, C., (2015). Ethnobotany of wild plants used for starting fermented beverages in Shui communities of southwest China. *Journal of Ethnobiology and Ethnomedicine*, *11*, 42–62.

Ignacimuthu, S., Ayyanar, M., & Sivaraman, K. S., (2006). Ethnobotanical investigations among Tribes in Madurai District of Tamil Nadu (India). *Journal of Ethnobiology and Ethnomedicine, 2*, 25–31.

Iwu, M. M., (1994). African medicinal plants in the search for new drugs based on ethnobotanical leads. In: Chadwick, D. J., & Marsh, (eds.), *Ethnobotany and the Search for New Drugs:Journal of Ciba Foundation Symposium* (Vol. 185, pp. 116–129).

Jablonski, D., (1994). Extinction: Past and present. *Nature*, 427.

Jain, S. K., (1981). *Glimpses of Indian Ethnobotany*. Oxford and IBH Publishing Co. New Delhi.

Jain, S. K., (1992). Ethnopharmacology and drug development. In: Chadwick, D. J., & March, U., (eds.), *Ethnobotany and Search for New Drugs, Ciba Foundation Symposium* (p. 153). Wiley Chichester.

Jaiswal, V., (2010). Culture and ethnobotany of Jaintia tribal community of Meghalaya Northeast India: A mini review. *Indian Journal of Traditional Knowledge*, *9*, 38–44.

Jamir, N. S., Jungdan, S., & Madhabi, S., (2008). Traditional knowledge of medicinal plants used by the Yimchunger-Naga tribes in Nagaland. *Pleione*, *2*, 223–228.

Jan, M., Singh, S., Maqbool, F., &Nawchoo, I. A., (2019). Effect of explant source and different hormonal combinations on *in vitro* regeneration of *Heracleum candicans* wall: An important medicinal herb. *African Journal of Biotechnology*, *18*(28), 707–712.

Jennings, H. M., Merrell, J., Thompson, J. L., & Heinrich, M., (2015). Food or medicine? The food–medicine interface in households in Sylhet. *Journal of Ethnopharmacology*, *167*, 97–104.

Jose, T., Sebastian, A., & Antony, V. T., (2015). Ethnobotanical study of traditional medicinal plants used by indigenous people in North Kerala. *Indian Journal of Applied Research*, *5*, 184–186.

Kachare, S. V., & Fatima, S., (2012). Ethnobotanical investigations in Marathwada India. *InternationalJournal of Current Research*, *4*, 36–39.

Kala, C. P., (1998). *Ethnobotanical Survey and Propagation of Rare Medicinal Herbs for Small Farmers in the Buffer Zone of the Valley of Flowers National Park Garhwal Himalaya*. International Centre for Integrated Mountain Development Kathmandu Nepal.

Kala, C. P., (2000). Status and conservation of rare and endangered medicinal plants in the Indian trans-Himalaya. *Biological Conservation*, *93*, 371–379.

Kala, C. P., (2005). Current status of medicinal plants used by traditional Vaidyas in Uttaranchal State of India. *Ethnobotany Research and Application*, *3*, 267–278.

Kala, C. P., Dhyani, P. P., &Sajwan, B. S., (2006). Developing the medicinal plants sector in northern India: Challenges and opportunities. *Journal of Ethnobiology and Ethnomedicine*, *2*, 32.

Kar, A., Borkataki, S., & Borthakur, S. K., (2008). Wild edible fruits of the Karbi's of Karbi Anglong district of Assam India. *Pleione*, *2*, 175–181.

Kasture, V. S., Kasture, S. B., &Chopde, C. T., (2002). Anticonvulsive activity of *Butea monosperma* flowers in laboratory animals. *Pharmacology Biochemistry and Behavior*, *72*, 965–972.

Kesari, A. N., Gupta, R. K., Singh, S. K., Diwakar, S., & Watal, G., (2006). Hypoglycemic and antihyperglycemic activity of *Aegle marmelos* seed extract in normal and diabetic rats. *Journal of Ethnopharmacology*, *107*, 374–379.

Kinghorn, A. D., &Balandrin, M. F., (1993). Human medicinal agents from plants. *ACS Symposium Series*, 534. American Chemical Society Washington.

Konsam, S., Thongam, B., &Handique, A. K., (2016). Assessment of wild leafy vegetables traditionally consumed by the ethnic communities of Manipur Northeast India. *Journal of Ethnobiology and Ethnomedicine*, *12*, 9–23.

Kosalge, S. B., &Fursule, R. A., (2009). Investigation of ethnomedicinal claims of some plants used by tribals of Satpuda Hills in India. *Journal of Ethnopharmacology*, *121*, 456–461.

Kujawska, M., & Pardo-de-Santayana, M., (2015). Management of medicinally useful plants by European migrants in South America. *Journal of Ethnopharmacology*, *172*, 347–355.

Kumar, A., Avasthe, R. K., Shukla, G., & Pradhan, Y., (2012). Ethnobotanical edible plant biodiversity of Lepcha tribes. *Indian Forester*, *138*, 798–803.

Kumar, A., Kumar, S., Komal, Ramchiary, R. N., & Singh, P., (2021). Role of traditional ethnobotanical knowledge and indigenous communities in achieving sustainable development goals. *Sustainability*, *13*, 3062.

Kumar, B., Vijayakumar, M., Govindarajan, R., &Pushpangadan, P., (2007). Ethnopharmacological approaches to wound healing- exploring medicinal plants of India. *Journal of Ethnopharmacology*, *114*, 103–113.

Kumar, R. S., Kanmani, P., Yuvaraj, N., Paari, K. A., Pattukumar, V., & Arul, V., (2013). Traditional Indian fermented foods: A rich source of lactic acid bacteria. *International Journal of Food Science and Nutrition*, *64*, 415–428.

Larsen, H. O., Olsen, C. S., & Boon, T. E., (2000). The non-timber forest policy process in Nepal: Actors objectives and power. *Forest Policy and Economics*, *1*, 267–281.

Lepcha, L. D., Shukla, G., Vineeta, Sarkar, B. C., Tamang, M., Chettri, R., Manohar, A. K., et al., (2019). Ethnobotanical plant diversity in home gardens-based agroforestry system in Kanchendzonga Biosphere Reserve, Sikkim, India. *Journal of Tree Sciences, 38*, 59–69.

Lewis, W. H., Elwin, & Lewis, M. P., (2003). *Medical Botany Plants Affecting Human Health* (p. 812). John Wiley and sons New York.

Lone, F., Lone, S., Aziz, M. A., & Malla, M. A., (2012). Ethnobotanical studies in the tribal areas of district Kupwara Kashmir India. *International Journal of Pharma and Bio Sciences*, *3*, 399–411.

Malik, B. A., Malik, Z. A., Naushad, A. P. M., & Bussmann, R. W., (2021). Detecting seminal research contributions to the development of ethnobotany by reference publication year spectroscopy (RPYS), *Nordic Journal of Botany*, 03102.

Malla, B., & Chhetri, R. B., (2012). Ethnoveterinary practices of some plant species by ethnic people of Parbat district Nepal. *Kathmandu University Journal of Science Engineering and Technology*, *8*, 44–50.

Malla, S., Shukla, G., & Chakravarty, S., (2012). Utilization and conservation of wild plants by the tribal communities of Tripura. *Indian Forester*, *138*, 1002–1007.

Mallick, H., & Mallick, S. K., (2012). Medicinal plants used by the tribals of Natungram village district Bankura West Bengal. *International Journal of Basic and Applied Sciences*, *1*, 131–133.

Mallick, S. K., & Behera, N., (2009). Phytosociological analysis of ethnomedicinal tree species in Goaltore forest area of Medinipur West Bengal. *Non-Timber Forest Products, 16*, 187–190.

Mallick, S. K., Banerjee, P., & Sand, S. A., (2012). Medicinal plants used by the tribals of Ratanpur village of Bankura West Bengal. *International Journal of Life Sciences, 1*, 82–86.

Manohar, A. K., Shukla, G., & Chakravarty, S., (2021). The spice tree of India: *Cinnamomum tamala*. *Journal of Tree Sciences, 40*, 92–101.

Mathias, E., McCorkle, C. M., &Schillhorn, V. V. T. W., (1996). Introduction: Ethnoveterinary research and development. In: McCorkle, C. M., Mathias, E., &Schillhorn, V. V. T. W., (eds.), *Ethnoveterinary Research and Development*.

Mathur, A., & Joshi, H., (2013). Ethnobotanical studies of the terai region of Kumaun Uttarakhand India. *Ethnobotany Research and Applications, 11*, 175–203.

McClatchey, W., & Gollin, L. X., (2005). An ethnobotany research training workshop in Madagascar. *Ethnobotany Research and Applications, 3*, 309–327.

McClatchey, W., (2005). Medicinal bioprospecting and ethnobotany research. *Ethnobotany Research and Applications, 3*, 189, 190.

Mishra, B. B., Singh, D. D., Kishore, N., Tiwari, V. K., & Tripathi, V., (2010). Antifungal constituents isolated from the seeds of *Aegle marmelos*. *Phytochemistry, 71*, 230–234.

Misra, S., & Misra, M. K., (2016). Ethnobotanical and nutritional evaluation of some edible fruit plants of southern Odisha India. *International Journal of Advances in Agricultural Science and Technology, 3*, 1–30.

Mondal, S., & Rahaman, C. H., (2012). Medicinal plants used by tribal people of Birbhum district of West Bengal and Dumka district of Jharkhand in India. *Indian Journal of. Traditional. Knowledge, 11*, 674–679.

Mondal, T., & Biswas, S., (2012). Ethnoveterinary uses of some medicinal plants of Bankura district West Bengal. *Life Sciences Leaflets, 5*, 47–49.

Mondal, T., & Samanta, S., (2014). An ethnobotanical survey on medicinal plants of Ghatal blocks West Midnapur District West Bengal India. *International Journal of Current Research in Bioscience and Plant Biology, 1*, 35–37.

Mukherjee, P. K., (2003). *GMP for Indian Systems of Medicine* (pp. 99–112). Business Horizons New Delhi.

Namahata, D., & Mukherjee, A., (1988). Ethnomedicine in Bankura district West Bengal. *IndianJournal of Applied and Pure Biology, 3*, 53–55.

Namahata, D., & Mukherjee, A., (1989). Some common practices of herbal medicines in Bankura district West Bengal. *Indian Journal of Forestry, 12*, 318–321.

Namsa, N. D., Tag, H., Mandal, M., Kalita, P., & Das, A. K., (2009). An ethnobotanical study of traditional anti-inflammatory plants used by the Lohit community of Arunachal Pradesh India. *Journal of Ethnopharmacology, 125*, 234–245.

Narzary, H., Brahma, S., & Basumatary, S., (2013). Wild edible vegetables consumed by Bodo tribe of Kokrajhar District (Assam) North-East India. *Archives of Applied Science Research, 5*, 182–190.

Nautiyal, S., Rao, K. S., Maikhuri, R. K., Negi, K. S., & Kala, C. P., (2002). Status of medicinal plants on way to Vashuki Tal in Mandakini Valley Garhwal Uttaranchal. *Journal of Non-Timber Forest Product, 9*, 124–131.

Nolan, J. M., & Turner, N. J., (2011). Ethnobotany: The study of people–plant relationships. In: *Ethnobiology* (pp. 133–147). John Wiley & Sons, Inc.

Oliveira, F. C., Albuquerque, U. P., Fonseca-Kruel, V. S., &Hanazaki, N., (2010). Advances in ethnobotany research in Brazil. *Recent Developments and Case Studies in Ethnobotany*, 153–185.

Olsen, C. S., (2005). Valuation of commercial central Himalayan medicinal plants. *Ambio, 34*, 607–610.

Padal, S. B., Chandrasekhar, P., & Satyavathi, K., (2013). Ethnomedicinal investigation of medicinal plants used by the tribes of PedabayaluMandalam Visakhapatnam District Andhra Pradesh India. *International Journal of Computational Engineering Research, 3*, 8–13.

Pala, N. A., Negi, A. K., Gokhale, Y., Razvi, R., &Todaria, N. P., (2012). Medicinal plant resources in sacred forests of Garhwal Himalaya. *Non Wood Forest Products Journal, 19*, 291–296.

Pala, N. A., Sarkar, B. C., Shukla, G., Chettri, N., Deb, S., Bhat, J. A., & Chakravarty, S., (2019). Floristic composition and utilization of ethnomedicinal plant species in home gardens of the Eastern Himalaya. *Journal of Ethnobiology and Ethnomedicine,15*, 14.

Palit, D., & Gurung, S., (2008). Some phytoremedies used traditionally by Gurungs in Darjeeling West Bengal India. *Pleione, 2*, 175–181.

Pandey, A. K., & Tripathi, Y. C., (2017). Ethnobotany and its relevance in contemporary research. *Journal of Medicinal Plants Studies, 5*(3), 123–129.

Pandey, R., (2009). Forest resources utilization by tribal community of Jaunsar Uttarakhand. *Indian. Forester., 135*, 655–662.

Patel, H. R., & Patel, R. S., (2013). Ethnobotanical plants used by the tribes of RDF Poshina Forest range of Sabarkanta district north Gujarat. *International Journal of Scientific and Research Publications, 3*.

Pathak, M., & Bharati, K. A., (2020). Mapping ethnobotany research in India. *Ethnobotany Research and Applications, 20*, 1–12.

Pawar, R. S., Jain, A. P., Lodhi, S., &Singhai, A. K., (2010). Erythropoietic activity of *Asteracantha longifolia* (Nees.) in rats. *Journal of Ethnopharmacology, 129*, 280–282.

Phondani, P. C., Maikhuri, R. K., Rawat, L. S., Farooquee, N. A., Kala, C. P., Vishvakarma, S. C. R., Rao, K. S., & Saxena, K. G., (2010). Ethnobotanical uses of plants among the Bhotiya tribal communities of Niti valley in Central Himalaya India. *Ethnobotany Research and Applications, 8*, 233–244.

Quave, C. L., & Pieroni, A., (2014). Fermented foods for food security and food sovereignty in the Balkans: A case study of the Gorani people of North-eastern Albania. *Journal of Ethnobiology, 34*, 28–43.

Quave, C. L., & Pieroni, A., (2015). A reservoir of ethnobotanical knowledge informs resilient food security and health strategies in the Balkans. *Natural Plants, 1*, 14021.

Rai, A. K., Sharma, R. M., & Tamang, J. P., (2005). Food value of common edible wild plants of Sikkim. *Journal of Hill Research, 18*, 99–103.

Rai, S. K., & Bhujel, R. B., (2002). Ethnic uses of some monocotyledonous plants in the Darjeeling Himalayan region. In: Das, A. P., & Bishen, S. B., & Mahendra, P. S., (eds.), *Perspectives of Plant Biodiversity* (635–644). M. P. Dehra Dun.

Raj, A. J., Biswakarma, S., Pala, N. A., Shukla, G., Vineeta, Kumar, M., Chakravarty, S., & Bussman, R. W., (2018). Indigenous uses of ethnomedicinal plants among forest-dependent communities of northern Bengal, India. *Journal of Ethnobiology and Ethnomedicine, 14*, 8.

Rana, V., Kumar, V., & Soni, P. L., (2009). Structure of the oligosaccharides isolated from Dalbergia sissoo Roxb. leaf polysaccharide. *Carbohydrate Polymers, 78*, 520–525.

Ranjithakani, P., Geetha, S., Lakshmi, G., & Murugan, S., (1992). Preliminary survey of wild edibles of Kolli hills of Salem. *Ancient Science of Life, 11*, 133–136.

Reddy, K. N., Pattanaik, C., Reddy, C. S., & Raju, V. S., (2007). Traditional knowledge on wild food plants in Andhra Pradesh. *Indian Journal of Traditional Knowledge, 6*, 223–229.

Reyes-García, V., Aceituno-Mata, L., Calvet-Mir, L., Garnatje, T., Gómez-Baggethun, E., Lastra, J. J., Ontillera, R., et al., (2014). Resilience of traditional knowledge systems: The case of agricultural knowledge in home gardens of the Iberian Peninsula. *Global Environment Change, 24*, 223–231.

Rout, S. M., Choudary, K. A., Kar, D. M., Das, L., & Jain, A., (2009). Plants in traditional medicinal system- future source of new drugs. *International Journal of Pharmacy and Pharmaceutical Sciences, 1*, 1–23.

Samar, R., Shrivastava, P. N., & Jain, M., (2015). Ethnobotanical study of traditional medicinal plants used by tribe of Guna District Madhya Pradesh India. *International Journal of Current Microbiology and Applied Sciences, 4*, 466–471.

Sanghi, S. B., (2014). Study of some ethno-veterinary medicinal plants of Tendukheda District Narsinghpur Madhya Pradesh. *Life Sciences Leaflets, 55*, 1–5.

Sarkar, B. C., Biswakarma, S., Shukla, G., Pala, N. A., & Chakravarty, S., (2015). Documentation and utilization pattern of ethnomedicinal plants in Darjeeling Himalayas, India. *International Journal of Forest Usufructs Management, 16*(2), 3–11.

Sathiyaraj, R., Sarvalingam, A. A., Balachandran, A., & Reddy, R. K., (2015). Diversity of ethnomedicinal plants in Bodamalai hills Eastern Ghats Namakkal District Tamil Nadu. *Journal of Plant Sciences, 3*, 77–84.

Schultes, R. E., (1960). Taping our heritage of ethnobotanical ore. *Economic Botany, 14*, 257–262.

Schultes, R. E., (1962). The role of the ethnobotanist in the search of new medicinal plants. *Llioydia, 25*, 257.

Shahavi, V. M., & Desai, S. K., (2008). Anti-inflammatory activity of Butea monosperma flowers. *Fitoterapia, 79*, 82–85.

Sharma, J., Gairola, S., Gaur, R. D., &Painuli, R. M., (2012). The treatment of jaundice with medicinal plants in indigenous communities of the Sub-Himalayan region of Uttarakhand India. *Journal of Ethnopharmacology, 143*, 262–291.

Sharma, J., Gairola, S., Gaur, R. D., Painuli, R. M., & Siddiqi, T. O., (2013b). Ethnomedicinal plants used for treating epilepsy by indigenous communities of sub-Himalayan region. *Journal of Ethnopharmacology, 150*(1), 353–370.

Sharma, J., Gairola, S., Gaur, R. D., Siddiqi, T. O., &Painuli, R. M., (2013a). Plants used for treatment of dysentery and diarrhoea by the Bhoxa community of district Dehra Dun Uttarakhand India. *Journal of Ethnopharmacology,150*, 969–1006.

Sharma, J., Gairola, S., Gaur, R., &Painuli, R. M., (2011). Medicinal plants used for primary healthcare by Tharu tribe of Udham Singh Nagar Uttarakhand, India. *International Journal of Medicinal and Aromatic Plants, 1*, 228–233.

Sharma, J., Gairola, S., Sharma, Y. P., & Gaur, R. D., (2014). Ethnomedicinal plants used to treat skin diseases by Tharu community of district Udham Singh Nagar, Uttarakhand, India. *Journal of Ethnopharmacology, 2*(158), 140–206.

Sharma, M., & Kumar, A., (2013). Ethnobotanical uses of medicinal plants: A review. *International Journal of Life Science and Pharma Research, 3*, 52–57.

Sheikh, M., Ahmad, A., Haq, E., Ahmad, B., Khan, Z., & Rizvi, R., (2013). Survey of ethnoveterinary medicine amongst ethnopractitioners of western Uttar-Pradesh province of India. *Journal of Medicinal Plants Research, 7*, 509–516.

Sheng-Ji, P., (2001). Ethnobotanical approaches of traditional medicine studies: Some experiences from Asia, *Pharmaceutical Biology, 39*(1), 74–79.

Shrivastava, S., & Kanungo, V. K., (2013). Ethnobotanical survey of Surguja district with special reference to plants used by Uraon tribe in treatment of diabetes. *International Journal of Herbal Medicine, 1*, 127–130.

Shukla, G., & Chakravarty, S., (2012). Ethnobotanical plant use of Chilapatta reserved forest in West Bengal. *Indian Forester, 138*, 1116–1124.

Singh, A., & Dubey, N. K., (2012). An ethnobotanical study of medicinal plants in Sonebhadra district of Uttar Pradesh India with reference to their infection by foliar fungi. *Journal of Medicinal Plants Research, 6*, 2727–2746.

Singh, A., Singh, G. S., & Singh, P. K., (2012a) Medico-ethnobotanical inventory of Renukoot forest division of district Sonbhadra Uttar Pradesh India. *Indian Journal of Natural Products and Resources, 3*, 448–457.

Singh, H. B., & Arora, R. K., (1978). Raishan (*Digitaria* sp.)– A minor millet of Khasi Hills India. *Economic Botany, 26*, 376–380.

Singh, K. K., Saha, S., & Maheshwari, J. K., (1998). Ethnobotanical use of some fern amongst the tribal area of Utter Pradesh. *Indian Fern Journal, 6*, 66, 67.

Singh, U., & Bharti, A. K., (2015). Ethnobotanical study of plants of Raigarh area Chhattisgarh India. *International Research Journal of Biological Sciences, 4*, 36–43.

Singh, U., & Lahiri, N., (2010). *Ancient India: New Research.* Oxford University Press, India.

Singh, V. A., & Bunger, R., (2014). Probiotics and gut health. *Journal International Medicinal Science Academy, 27*, 41–43.

Sinhababu, A., & Banerjee, A., (2013). Ethno-botanical study of medicinal plants used by tribals of Bankura district West Bengal India. *Journal of Medicinal Plants Studies, 1*, 98–104.

Siva, R., (2007). Status of natural dyes and dye-yielding plants in India. *Current Science, 92*, 916–925.

Somani, R., Kasture, S., &Singhai, A. K., (2006). Antidiabetic potential of *Butea monosperma* in rats. *Fitoterapia, 77*, 86–90.

Sõukand, R., Pieroni, A., Biró, M., Dénes, A., Dogan, Y., Hajdari, A., Kalle, R., et al., (2015). An ethnobotanical perspective on traditional fermented plant foods and beverages in Eastern Europe. *Journal of Ethnopharmacology, 170*, 284–296.

Subitha, T. K., Ayyanar, M., Udayakumar, M., & Sekar, T., (2011). Ethnomedicinal plants used by Kanitribals in Pechiparai Forests of Southern Western Ghats Tamil Nadu India. *International Research Journal of Plant Science, 2*, 349–354.

Suresh, C. P., Bhutia, K. D., Shukla, G., Pradhan, K., & Chakravarty, S., (2013). Free list of wild edible fruit species of Sikkim Himalayas and their uses. In: *Proc. Second International Symposium on Minor Fruits and Medicinal Plants* (pp. 17–37).

Suresh, C. P., Bhutia, K. D., Shukla, G., Pradhan, K., & Chakravarty, S., (2014). Wild edible tree fruits of Sikkim Himalayas. *Journal of Tree Sciences, 33*, 43–48.

Terangpi, R., Phangchopi, U., & Teron, R., (2015). Ethnobotany of dreams and dream interpretations: A study among the Karbis of India. *Ethnobotany Research and Applications, 14*, 111–121.

Tripathi, S., Ray, S., Mondal, A. K., & Verma, N. K., (2013). Rare ethnomedicinal plants of south West Bengal India with their different medicinal uses needs conservation. *International Journal of Life Sciences Biotechnology and Pharma Research, 2*, 114–122.

Uprety, Y., Poudel, R. C., Gurung, J., Chettri, N., & Chaudhary, R. P., (2016). Traditional use and management of NTFPs in Kanchenjunga landscape: Implications for conservation and livelihoods. *Journal of Ethnobiology and Ethnomedicine, 12*, 19–77.

Van, D. M. D., Swana, G. E., & Botha, C. J., (2001). Use of ethnoveterinary medicinal plants in cattle by Setswana-speaking people in the Madikwe area of the North West Province of South Africa. *Journal of South African Veterinary Association, 72*, 189–196.

Verma, P., Mathur, A. K., Jain, S. P., & Mathur, A., (2012). *In vitro* conservation of twenty-three over exploited medicinal plants belonging to the Indian Sub-Continent. *The Scientific World Journal*, 12.

Verma, R. K., (2014). An ethnobotanical study of plants used for the treatment of livestock diseases in Tikamgarh District of Bundelkhand Central India. *Asian Pacific Journal of Tropical Biomedicine, 4*, 460–467.

Vidyasagar, G. M., &Prashantkumar, P., (2007). Traditional herbal remedies for gynaecological disorders in women of Bidar district Karnataka India. *Fitoterapia, 78*, 48–51.

Vineeta, Manohar, A. K., Shukla, G., Sarkar, B. C., & Chakravarty, S., (2020). Use of Ethnomedicinal plants by ethnic people for the treatment of dermatological problems. *Journal of Plant Development Sciences, 12*, 7–15.

Vineeta, Pala, N. A., Shukla, G., & Chakravarty, S., (2018a). Traditionally used medicinal plants for treatment of stomach disorders in West Bengal, India: A scrutiny and analysis from secondary literature. *Studies on Ethno-Medicine, 12*(3), 163–183.

Vineeta, Shukla, G., Bhat, J. A., & Chakravarty, S., (2021). Species richness and folk therapeutic uses of ethnomedicinal plants in West Bengal, India – a meta-analysis. *Phytomedicine Plus, 2*(1), 100158.

Vineeta, Shukla, G., Pala, N. A., Dobhal, S., & Chakravarty, S., (2018b). Influence of seed priming on germination and seedling growth of *Adeantherapavonina* in sub humid region of West Bengal, India. *Indian Journal of Tropical Biodiversity,26*, 87–91.

Xavier, T. F., Kannan, M., Lija, L., Auxillia, A., Freeda, R. A. K., & Kumar, S. S., (2014). Ethnobotanical study of Kani tribes in Thodu hills of Kerala South India. *Journal of Ethnopharmacology, 152*, 78–90.

Xavier, T. F., Rose, A. F., & Dhivyaa, M., (2011). Ethnomedicinal survey of Malayali tribes in Kolli hills of Eastern Ghats of Tamil Nadu India. *Indian Journal of Traditional Knowledge, 10*, 559–562.

Yadav, M., Khan, K. K., & Beg, M. Z., (2012). Ethnobotanical plants used for curing skin diseases by tribals of Rewa district (Madhya Pradesh). *Indian Journal of Life Science, 1*, 123–126.

Yesodharan, K. K., & Sujana, K. A., (2007). Wild edible plants traditionally used by the tribes in the Parambikulam wildlife sanctuary Kerala India. *Natural Product Radiance, 6*, 74–80.

CHAPTER 6

Problems and Prospects of Medicinal and Aromatic Plants with Special Reference to Coordinated Research in India Under ICAR

SATYAJIT ROY, K. A. GEETHA, and R. NAGARAJA REDDY

ICAR–Directorate of Medicinal and Aromatic Plants Research, Boriavi, Anand, Gujarat, India

ABSTRACT

Traditional healthcare systems are largely in favor of people across the globe after the COVID-19 outbreak. Medicinal and aromatic plants (MAPs) offer tremendous opportunity to generate income, employment, and foreign exchange in India and the world. As consumption of herbal medicines has increased globally, India can shine in global exports of MAPs. To address the emerging research needs in MAP, the Indian Council of Agricultural Research (ICAR) has institutionalized certain project instruments viz. All-India Co-ordinated Research Project on Medicinal and Aromatic Plants and Betelvine (AICRP-MAP&B) and National Research Center for Medicinal and Aromatic Plants (NRC-MAP) renamed as ICAR–Directorate of Medicinal and Aromatic Plants Research (ICAR-DMAPR) to carry out basic, strategic, and applied research on the identified MAPs to meet the present-day research needs of MAPs, to convert the research achievements into deliverables for the benefit of all the stakeholders with a focus on farmers and to address future national and international research and development needs and challenges. These collaborative platforms carry out multi-disciplinary and multi-location research that is very well essential for the promotion of herbal plants in the country and the globe. This chapter envisions the various

Bioprospecting Ethnomedicinal Plant Resources: Sustainable Utilization and Restoration.
Gopal Shukla, Jahangeer A. Bhat, A. P. Das, and Sumit Chakravarty (Eds.)

research platforms and their prospects and problems in MAPs cultivation. Conservation, plant introduction, developing good agricultural practices (GAPs), production of quality raw materials of medicinal plants as per the demand are the major mandate of these programs. The research in MAPs in coordinated mode is needed to meet the growing global demand and thus to bring in sustainability in the supply of raw drugs to the user industry.

6.1 INTRODUCTION

Medicinal and aromatic plants (MAP) provide vital medicines to alternative systems of medicine such as AYUSH and also provide livelihood to millions of people across the globe. The last two decades have seen renewed interest in AYUSH systems of medicine owing to their advantages that they are natural products, safe, effective, and accessible to folklores. MAP offers alternative therapies with a growing opportunity to generate income, employment, and foreign exchange in India and the world. The consumption of herbal medicines has increased in recent years as compared to modern medicines to be around US $62 billion globally and is projected to reach US $5 trillion by 2050 (Anon., 2006). India, with its rich traditional knowledge (TK) related to the herbal sector, employs an abysmally large number of plants in AYUSH systems of medicine. It is estimated that around 2,000 species are in use in Ayurveda, 1121 in Siddha, 751 in Unani, and 337 species in Tibetan as raw drugs. The estimated trade value of the domestic herbal industry has grown from Rs. 627.9 crores in 2004–2005 to Rs. 1,950 crores in 2014–2015 (three times). A substantially large number of plant-based medicines are also exported to various countries with a trade value of Rs. 3,211 crores annually (2014–2015) (Goraya & Ved, 2017). Thus, India is emerging as a major player after China in the global trade of plant-based medicines.

Promotion of MAP is possible only with the development and validation of appropriate good agricultural practices (GAPs) and varieties. To cater to the research needs, ICAR started the All-India Co-ordinated Research Project on Medicinal and Aromatic Plants (AICRP-MAP) in 1972 followed by the establishment of the National Research Center on Medicinal and Aromatic Plants (NRC-MAP) in 1992, which was upgraded to the ICAR-DMAPR in 2010 headquartered in Anand, Gujarat, India. These institutions are entrusted to carry out basic, strategic, and applied research on some of the highly demanded MAPs in the country. All-India Co-ordinated Research Project (AICRP) is a collaborative platform for multi-disciplinary and multi-location research in agriculture. The research in MAPs is in the formative stage, and lots of research works in coordinated mode are needed to meet the growing national and international demand. There are a few national research institutes working on MAPs, the AICRP-MAP&B generate technology for the farming communities and for better and safe healthcare, which has a unique advantage.

6.2 HISTORY OF AICRP-MAP&B

The AICRP-MAP project was started during the Fourth Five Year Plan with five centers, at Indian Agricultural Research Institute (IARI), New Delhi and Indian Institute of Horticultural Research (IIHR), Bengaluru which are ICAR research institutes and three in State Agricultural Universities such as Gujarat Agricultural University, Anand, Dr. Y. S. Parmar University of Horticulture and Forestry, Solan, and Tamil Nadu Agricultural University (TNAU) at Kodaikanal. The project started its operation from the Plant Introduction Division (now National Bureau of Plant Genetic Resources (NBPGR)), IARI, New Delhi. Later, the project was shifted to NRCMAP, Anand, Gujarat. Later the AICRP-MAP was further strengthened with the merger of AICRP on Betelvine by the ICAR in the 10th plan. Under the umbrella of AICRP-MAP&B there are 26 centers working across different agro-climatic zones of India (Figure 6.1) on 47 important MAPs (Table 6.1).

6.3 MANDATE

- Conservation and plant introduction studies with a view to develop GAPs and making available, in a decentralized manner, quality raw materials of many medicinal plants as per the therapeutic requirements of the community based PHCs located in various agro-climatic zones of the country.
- Developing high-yielding and high-quality varieties of selected medicinal and aromatic plants having market potential - both domestic and export.
- Establishment of medicinal plants gardens.
- Development of appropriate GAP and post-harvest technology for economically important medicinal and aromatic plants for use in their commercial cultivation under different agro-climatic conditions.
- Maintenance of Betelvine germplasm and its utilization in selected centers.

FIGURE 6.1 AICRP-MAP&B centers across different agro-climatic zones of India.

Source: Adapted from ICAR. https://dmapr.icar.gov.in/aicrp%20on%20map/center.html

TABLE 6.1 MAPs Under Study by Various AICRP-MAP&B

Sl. No.	EN/VN	SN
1.	Aloe	*Aloe barbadensis*
2.	*Ashoka*	*Saracaasoca*
3.	*Asalio*	*Lepidium sativum*
4.	*Ashwagandha*	*Withania somnifera*
5.	Basil	*Ocimum basilicum*
6.	*Ban kakdi*	*Podophyllym hexandrum*
7.	*Babchi*	*Psoralea corylifolia*
8.	Betel vine	*Piper betel*
9.	Bitter snake gourd	*Tricosanthus cucumarina*
10.	*Brahmi*	*Bacopa monnieri*
11.	*Chitrak*	*Plumbago* spp.

TABLE 6.1 *(Continued)*

Sl. No.	EN/VN	SN
12.	*Chirayita*	*Swertia chirayita*
13.	*Dodi*	*Leptadenia reticulata*
14.	Dwarf marigold	*Tegetusminuta*
15.	*Giloe*	*Tinospora cordifolia*
16.	*Guggal*	*Commiphora wightii*
17.	Glory lily	*Gloriosa superba*
18.	*Harde*	*Terminalia chebula*
19.	Indian valerian	*Valeriana jatamansi*
20.	*Isabgol*	*Plantago ovata*
21.	*Jeevak*	*Malaxis acuminata*
22.	*Kalmegh*	*Andrographis panicualata*
23.	KhurasaniAjmo	*Hyoscyanus niger*
24.	Lemongrass	*Cymbopogon* spp.
25.	Long pepper	*Piper longum*
26.	Lotus	*Nelumbo nucifera*
27.	Liquorice	*Glycyrrhiza glabra*
28.	Makoi	*Solanum nigrum*
29.	Madhunasini	*Gymnena sylvester*
30.	Mandukaparni	*Centella asiatica*
31.	Mentha	*Mentha* spp.
32.	Mucuna	*Mucuna pruriens*
33.	Neel	*Indigofera tinctorea*
34.	Opium poppy	*Papaver somniferum*
35.	Palmarosa	*Cymbopogon martinii*
36.	Safed musli	*Chlorophytum borivilianum*
37.	Sarpgandha	*Rauvolfia serpentina*
38.	Salparni	*Desmodium gangeticum*
39.	Shatavari	*Asparagus racemosus*
40.	Shankapusphi	*Convolvulus pluricaulis*
41.	Senna	*Cassia angustifolia*
42.	St. John's Wort	*Hypericum perforatum*
43.	Silybum	*Silybum marianum*
44.	Sugandhmantri	*Homalomena aromatica*
45.	Tulsi	*Ocimum sanctum*
46.	Vach	*Acorus calamus*
47.	Vetiver	*Vetiveria zizaniodes*

Abbreviations: Sn: Serial number; EN: English name; VN: Vernacular name; and SN: Scientific name.

6.4 MULTI-DISCIPLINARY RESEARCH

Multi-disciplinary research has been undertaken in AICRP-MAP&B project: (i) crop improvement discipline focuses on conservation of plant genetic resources, varietal development and quality planting material (QPM) production; (ii) crop production discipline mandates on development of region specific agrotechnology development, refinement of agrotechnologies developed and technologies for organic cultivation; (iii) crop protection discipline aims to develop integrated pest and disease management technologies and forecast of pests and diseases in MAPs; and (iv) phytochemistry and post-harvest management discipline is engaged in development of standard operating protocols (SOPs) for quality analysis, development of efficient extraction technologies and value added products based on high value compounds.

6.5 RESEARCHABLE ISSUES IN MAPS

- Increasing demand for MAPs has raised concerns about the conservation of their genetic diversity, habitat loss, quality in terms of cleanliness (free from physical and chemical contaminants/adulterants), consistency (dependable production and bioactivity levels), and certified (identifiable for origin and history) products.
- Inconsistent product quality caused by genotypic variation within the plant species, environmental effects that alter chemical constituent's level and distribution, and increased use of wild-simulated species is also a matter of concern in the use of medicinal and aromatic plants.
- In many instances, switching to the cultivation of MAPs has raised concerns of duplicating the demanding environmental conditions for a particular species in the cultivated fields that may affect the quality of produce.
- Climate change is another area of concern as it might cause adverse effect on food production in general and MAPs in particular, besides creating newer health concerns.
- Marketing of MAPs is informal, secretive, and opportunistic. As a result, the raw material supply is shaky and unsustainable. This results in depletion of resource base and non-availability of quality herbal drugs for domestic consumption as well as for export.
- Despite the availability of plenty of resources (biological, human, and financial), the MAPs sector is facing challenges like inadequate

quality standards for finished products, claim support studies, standard operating procedures of various products, scientific documentation, trade regulation, and intellectual property rights (IPRs) implementation.

- Several constraints also exist due to inadequate awareness, insufficient investments in research and development, manufacturer-exporter dissonance, lack of quality and standardization norms, insufficient marketing, and trade information.
- The integrated efforts of the scientists of various disciplines and farmers have contributed significantly to increasing the total production and productivity of MAPs. As the frequency of extreme climate events has increased, it is important to evaluate the genetic materials for their resilience against fragile environments as well as sustaining yields and soil fertility.

6.6 SOME SIGNIFICANT CONTRIBUTIONS

- Varieties developed in medicinal and aromatic plants: 34 varieties in 18 MAPs viz., Vallabh Kalmegh-1 and Anand Kalmegh-1 of Kalmegh, Anand Kuvarpata Aloe-1 of Aloe, Anand Safed Musli-1 of Safed musli, GI2, Vallabh Isabgol-1, Vallabh Isabgol-2 and Vallabh Isabgol-3 of Isabgol, GAB-1 for Basil, Gujarat Anand Senna-1 of Senna and AWS-1 of Ashwagandha at AAU, Anand, HI-5 of Isabgol, HM-1 of Mulhati, Prabhat (Selection-1) of Periwinkle, RH-49 of Palmarosa and HLS-4 of Asalioat CSSHAU, Hissar, Viswam of Long pepper and Aswani of *Ashoka* at KAU, Trichur, Chetak Aphim of Opium Poppy and Prapap Babchi-1 of Babchi at MPUAT, Udaipur, Kirtiman, NOP-1 and Ghazipur Local in Opium, NLG-84 in lemon grass, IC-111226 in Babchi, NDAS-4 in sataver and Vallabh Medha in Mandook Parni at NDUAT, Faizabad, Jawahar Aphim-16 (JA-16), Jawahar Opium Poppy-539, Jawahar Opium Poppy-540 of opium, Jawahar Ashwagandha-134 (JA-134), Jawahar Ashwagandha-20 (JA-20), Raj Vijay Ashwagandha-100 (RVA-100) of *Ashwagandha*, Jawahar Isabgol-4 of Isabgol, Jawahar *Safed musli*-405 (JSM-405) and Raj Vijay *Safed musli*-414 (RVSM-414) of *Safed Musli* and Raj Vijay Asalio-1007 (RVA-1007) of Asalio at RVSKVV, Mandsaur, CO 1 of Coleus (*Coleus forskohlii*) at TNAU, Coimbatore, and Swarna swara of Bach at YSRHU, Venkataramannagudem, high yielding varieties were developed.

- Varieties developed in Betel vine: Five varieties were developed viz., Bidhan Pan 1 at BCKV, Kalyani, Godi Bangla and Swarna Kapoori suitable for growing under Areca nut support at IIHR, Bangalore, Krishna Pan (Dpb-6) at MPKV, Rahuri, Utkal Sudam at OUAT, Bhubaneswar. Besides, nine promising Betel vine hybrids from IIHR, Bangalore entered in MLT evaluation.
- Registration of germplasm: Five trait-specific germplasm viz., heterostyle stigma (IC 561246 of Isabgol) of AAU, Anand, High valepotriates yielding (> 4%) strain (INGR 11036) of *Valeriana jatamansi*, and Three accessions (differing on the basis of morphological and hypericin content) of *Hypericum perforatum* were identified, multiplied, and registered with NBPGR under National Identity No. INGR 10074, INGR 10075, and INGR 10076. Germplasm accession INGR 10075 was identified as the best in terms of biomass yield and hypericin content. Hypericin extracted from its flowers and leaves showed slow degradation upon storage at YSPUHF, Solan were registered.
- Regions specific agro-technologies or GAPs for Isabgol, *Guggal*, *Kalmegh*, *Asalio*, *Safed musli* and *Shankapusphi* developed at AAU, Anand, Satawar, *Ashwagandha*, *Mulhati* have been developed at CCSHAU, Hissar, GAP for Asalio, Aloe, *Ashwagandha*, *Palmarosa*, *Sarpagandha*, and *Patchouli* was developed at IGKV, Raipur, GAP for *Plumbago rosea*, *Brahmi*, Sida, *Indigofera tinctoria* and Ashoka was developed at KAU, Trichur, Gap for *Opium poppy*, *Isabgol*, *Babchi*, *Safed musli*, *Ashwagandha*, and *Kalmegh* was developed at MPUAT, Udaipur, GAP for Opium poppy, Lemon grass, Java citronella, Palmarosa, Rose Vetiver, Tuberose, Matricaria, *Tulsi*, Mentha, Sweet basil, Hanebane, *Babchi*, *Isabgol*, *Kalmegh*, *Sataver*, *Ashwagandha*, *Sarpgandha*, Aloe, *Safed musli*, *Asalio* was developed at NDUAT, Faizabad, GAP for *Sarpagandha* and Long pepper developed was developed at OUAT, Bhubaneswar, *Ashwagandha*, *Kalmegh*, *Kawach* at PDKV, Akola, GAP for opium poppy, *Ashwagandha*, *Safed musli*, *Tulsi*, *Isabgol*, and *Kalmegh* was developed at RVSKVV, Mandsaur, GAP for *Swertia chirayita*, *Aconitum heterophyllum* and *Rheum austra* was developed at YSPUHF, Solan, *Mokai* (*Solanum nigrum*), *Gloriosa superb* and *Coleus forskohlii* was developed at TNAU, Coimbatore, GAP for *Mokai*, Bach, Senna, and Coleus was developed at YSRHU, Venkataramanagudam, GAP for *Valeriana jatamansi* and *Swertia chirayita* was developed at UUHF,

Bharsar, and GAP for *Swertia chirayita* and *Valeriana jatamansi* was developed at UBKV, Kalimpong were developed and being refined.

- Region specific agro technology for cultivation for Betel vine was developed at BCKV, Kalyani for North-eastern region, BAU, Islampur, and RAU Pusa for UP and Bihar, OUAT, Bhubaneswar for Orissa and YSRHU, Venkataramanagudam for Andhra Pradesh. Developed technologies against the wilt disease of Betel vine at MPKV, Rahuri. Integrated Crop Management module was developed for Betel vine at YSRHU, Venkataramanagudam.
- Organic farming practices for maximum yield in *Solanum nigrum* and *Coleus forskohlii* at YSRHU, Venkataramanagudam.
- Developed integrated pest management (IPM) module for management of major diseases of Opium poppy, Ashwagandha, Isabgol, and Safed musli at RVSKVV, Mandsaur.
- Developed a disease forecasting system for downy mildew of opium poppy based on weather data and yield contributing parameters at RVSKVV, Mandsaur.
- SOP were developed for estimation of glycyrrhizin content in *Glycyrrhiza glabra,* Ashwagandha (*Withania somnifera*) roots and Aloin of *Aloe vera* at CCSHAU, Hissar, Catechol, Pyrocatechol, Resorcinol, Guaicol, and Phthalic acid and the presence of fatty acids like myristic, aleic, and arachidic acid in Asoka at KAU, Trichur, morphine in Opium poppy, sapogenin in Safed mulsi at MPUAT, Udaipur, alkaloids of Ashwagandha root powder and leaves, fiber content and swelling factor of Isabgol, alkaloid present in flower petals of opium poppy at RVSKVV, Mandsaur, Picroside-I (0.35% in roots; 0.999% in rhizomes and 5.53% in leaves) and Picroside-II at YSPUHF, Solan, and iridoid glycosides Picroside-I and Picroside-II through HPLC at YSPUHF, Solan.
- Quick detection of adulteration in the raw drug market samples of Asoka, Tree Turmeric and Sida by TLC/HPLC fingerprint methods at KAU, Trichur.
- Post-harvest management technologies were developed for Safed musli at AAU, Anand, *Satawar* (*Asparagus racemosus*) tubers, *Ashwagandha* (*Withania somnifera*) roots, *Guar Patha* (*Aloe vera/ Aloe barbadensis*) and *Mulhati* (*Glycyrrhiza glabra*) at CCSHAU, Hissar, *Safed musli* at MPUAT, Udaipur, *Safed musli* and *Shatavari* roots were developed at PDKV, Akola, and *Ashwagandha*, *Safed musli*, *Asalio* at RVSKVV, Mandsaur.

- Value added products using herbs were developed, including herbal soft drink (Naruneendi Syrup), hair growth tonic (Kesaraksha oil), Dandhapala skin oil, Rasnadi, and Adalodakachoornam for cold and cough, and cosmetic products like herbal kajal and herbal soap. Turmeric and arrowroot powder at KAU, Trichur.
- Basic information on floral biology and standardized hybridization techniques in Betel vine at IIHR, Bangaluru.
- Standardized hybrid seedling technology and identified promising Betel vine hybrids for the first time in Betel vine at IIHR, Bangalore.
- Inter-specific Betel vine hybrids between Piper betle and Piper colubrinum developed for the first time at IIHR, Bangalore.
- Developed sex-specific SCAR marker and validated in 60 germplasm lines of Betel vine at IIHR, Bangalore.
- Molecular profiles of 82 germplasm lines of Betel vine were generated using 28 ISSR primers at IIHR, Bangalore.
- A total of 3,783 germplasm accessions of 39 MAP&B crops are being maintained at various centers.

6.7 IMPACT

- Preference of plant-based drugs over conventional medicines in the national and international market, which still contains about 25% plant-derived compounds, is growing and will further grow with the rising global population from the present 6.8 billion to 9.1 billion by the year 2050. Besides, their role in prevention as well as curing of human health problems and the global market, MAPs are also increasingly being recognized as a source of significant livelihood opportunities for many rural communities, especially primitive forest-dwellers, landless poor, and marginalized farmers. In spite of the scope, importance, and awareness, and several scientific innovations in the field of MAPs, still more than half of the global population does not have access to primary healthcare facilities. Medicinal plants and their by-products constitute an important part of foreign trade, showing continuous growth in export over import and thus signifying their importance outside the boundaries of the territory.
- The AICRP, a multi-location research project, has released 34 high yielding varieties in 18 MAPs and five high yielding varieties of

Betel vine which are popular among the farmers. Location specific agro-technologies for 40 MAPs and Betel vine have been developed. In crops like *Isabgol*, *Ashwagandha*, *opium poppy*, etc., the varieties and technologies developed through AICRP have made a good impact in the national and international markets by augmenting exports besides increasing domestic consumption. The technologies developed from AICP-MAP&B have made a significant impact on the livelihood of poor farmers in the country besides ensuring health security. An increase in the area of several MAPs has been witnessed, viz., *Senna*, *Isabgol*, *Ashwagandha*, *Kalmegh*, *Tulsi*, and *Sataver*, due to the availability of location-specific technologies and high yielding varieties. There is a need to develop improved agro technologies, varietal development, post-harvest management, processing, and marketing to enhance the cultivation area of MAPs. There is also a need for low-cost processing equipment for the greater benefit of the farming community. The changing climate presents new challenges for scientists to work in cropping systems wherein MAPs can be grown easily with less care. Its cultivation can be a good substitution for the farmers in fetching additional income and better utilization of land.

6.8 CONCLUSION

Recent trends indicate a preference for plant-based drugs over conventional medicines in the national and international market, which is growing and will continue to grow with the rising global population. Institutionalization can help in developing new varieties and reaching out to the public. Varieties and technologies developed through such organizations have made a significant impact in the national and international markets by increasing exports and domestic consumption. Adverse climate events, pollution, improper harvesting, etc., have deteriorated the quality of plants, leading to the extinction of many. To ensure sustained yield and quality, it is important to evaluate the genetic materials for their resilience against fragile environments as well. Problems in the marketing of MAPs, which are unstable and unsustainable, can be addressed with proper institutional mechanisms. AICRP-MAP&B, ICAR-DMAPR, NMPB are working progressively to achieve these goals.

KEYWORDS

- **agro-climatic zones**
- **All-India Coordinated Research Project**
- **conventional medicines**
- **farming community**
- **genetic materials**
- **hybridization**
- **India**
- **medicinal plant**

REFERENCES

Anonymous, (2006). *Report of the Working Group on "Access to Health Systems Including AYUSH."* Planning commission, Government of India.

Goraya, G. S., & Ved, D. K., (2017). *Medicinal Plants in India: An Assessment of Their Demand and Supply*. National Medicinal Plants Board, Ministry of AYUSH, Government of India, New Delhi and Indian Council of Forestry Research & Education, Dehra Dun.

CHAPTER 7

Important Medicinal Plants of Bangladesh: Uses and Prospects

MD. AZIZUL HOQUE and MD. MASUD KARIM PRODHAN

Department of Horticulture, Bangabandhu Sheikh Mujibur Rahman Agricultural University, Salna, Gazipur, Bangladesh

ABSTRACT

From ancient times, medicinal plants have been used as medication for different diseases. Bangladesh is a country where approximately 500 species of plants are used for remedial purposes. Of them, several species are widely used, and the demand trend of those plants is increasing severely. Besides, there is an export opportunity for different medicinal plants to different countries. Over-exploitation and habitat destruction have made the medicinal plants endangered. Although medicinal plants have a tremendous prospect in Bangladesh, some plants are becoming unavailable due to a lack of proper conservation and awareness. Therefore, it is time to identify and save the stock of medicinal plants for future use and availability. A number of authors have worked on different medicinal plant species of Bangladesh. In this chapter, important medicinal plants of Bangladesh and their uses have been highlighted from the work of different authors, which will be helpful for different stakeholders concerned with medicinal plants. Moreover, academicians will be able to learn different aspects of important medicinal plants available in Bangladesh.

7.1 INTRODUCTION

Medicinal plants provide important healing agents and are also valuable raw materials for preparing many traditional and recent medicines. The use of

Bioprospecting Ethnomedicinal Plant Resources: Sustainable Utilization and Restoration.
Gopal Shukla, Jahangeer A. Bhat, A. P. Das, and Sumit Chakravarty (Eds.)

medicinal plants worldwide is increasing day by day. It is estimated from a study that 70–80% of individuals worldwide rely mainly on traditional, mostly herbal medication, to fulfill the requirements of their primary healthcare (Mukul et al., 2007). In fact, these natural products are readily obtainable, inexpensive, environmentally friendly, and have no side effects. Bangladesh, due to its location and favorable climate, is exceptionally gifted with a large number of vegetation, including medicinal plants. There are about 5,700 angiosperms in the country, of which about 500 are thought to have medicinal characteristics. These are thus prescribed for the remedy of several diseases in traditional drug (Yusuf et al., 1994; Akter et al., 2013). The majority of people in Bangladesh live in rural regions. That is why they are unable to use modern healthcare facilities due to insufficient road transport, lack of allopathic physicians and hospitals or health centers, non-affordability to purchase allopathic medicines, and long-term dependency on outdated local medical practitioners (Haque et al., 2014). Even nowadays, particularly in remote rural populations and poorly accessible regions, the practice of using medicinal plants in primary healthcare systems is very important. Therefore, the long practice of traditional herbal medication systems by the people based on a variety of native plants is considered an important element of the primary healthcare structure. Currently, different traditional medicinal systems coexist within Bangladesh, namely the Homeopathic, Unani, Ayurvedic, and the traditional system. The traditional system is accomplished by village or old-fashioned medical practitioners, who are locally referred to as *Kavirajes* or traditional medical practitioners (Ghani, 2003). There are more than 87,000 villages in Bangladesh, and in most villages, there are one or two *Kavirajes* who practice with medicinal plants. For further scientific studies, the long experiences of the *Kavirajes* about the use of different medicinal plant species of Bangladesh can be an important source for having better drugs with lesser side effects from those plants used (Rahman et al., 2013). The government of Bangladesh has included the system of Unani and Ayurvedic Medicine in the National Drug Policy of Bangladesh in 1982 to ensure availability, commercial preparation, and promotion of better-quality medicines (Ahsan et al., 1997). Besides, the government of Bangladesh has already appointed Ayurvedic and Unani doctors at district and *upazilla* (sub-division) public hospitals (Hoque, 2019).

Like other countries of the world, the demand for medicinal plants as raw material in the medicine industries of Bangladesh is increasing gradually. Besides the medicine industries, these plants are also used a lot in beauty parlors and the manufacturing of different cosmetics. But due to uncontrolled trade and over-exploitation of plants, the stock of medicinal plants in the world is under threat. This is now a major concern for conservationists and also for resource consumers. The other reasons for such threats are destructive techniques of harvesting, ignorance of the people about the medicinal value, and lack of conservation of important medicinal plants by the concerned people or executives. Similarly, it is a common phenomenon in Bangladesh to excessively and freely exploit most wild medicinal plants, which endangers their availability in the future. Unplanned and indiscriminate ways of collecting medicinal plants from wild sources by the local suppliers of traditional manufacturing companies have been noticed. Sometimes, they solely collect a particular plant species in such an extensive and exhaustive way that the plant cannot further regenerate. In this way, many plants that were once easily available and endemic are diminishing day by day, impacting the total medicinal plant population of the country. As a result, many widely used and important medicinal plant species are also on the brink of extinction. Many studies have documented medicinal plants of Bangladesh but they are scattered (Ghani, 2003; Rahman et al., 2012, 2013; Bhowmik et al., 2014; Haque et al., 2014; Hasan et al., 2014; Kabir & Saha, 2014; Rahman & Keya, 2015; Ahmmed et al., 2017; Bari et al., 2017; Khatun et al., 2018) and now is the time to bring together all the important medicinal plants with their uses. Therefore, an attempt has been undertaken in the present chapter to compile the information of important and extensively used medicinal plants of Bangladesh along with endangered species to raise awareness among the concerned stakeholders for protecting them from extinction.

7.2 SOME ENDANGERED MEDICINAL PLANTS OF BANGLADESH

Endangered medicinal plants will have to be identified first for their proper conservation. Several authors identified the endangered medicinal plant species of Bangladesh. Some of them are presented below. A total of 19 medicinal plant species have been identified as endangered in Barind Tract of Bangladesh where the rating of occupancy by the respondents was 91–100% (Siddique et al., 2005). They listed endangered species were *Withania somnifera* (*Ashwagandha*), *Aristolochia indica* (*Iswaramul*), *Hemidesmus indicus* (*Anantamul*), *Mimosa pudica* (*Lojjaboti*), *Curculigo orchioides* (*Talamulika*), *Ipomoea mauritiana* (*Bhumikumra*), *Leea macrophylla* (*Hathikana*), *Gloriosa superba* (Glory lily), *Scindapsus officinalis* (*Gajapipal*), *Cassia alata* (*Daadmordon*), *Tinospora cordifolia* (*Guloncho*),

Plumbago zeylanica (*Chitra*), *Vitex negundo* (*Nishinda*), *Asparagus racemosus* (*Shatamuli*), *Andrographis paniculata* (*Kalomegh*), *Rauvolfia serpentina* (*Shorpogonda*), *Rauvolfia tetraphylla* (*Gandhanakuli*), *Smilax zeylanica* (*Kumari lota*) *and Abrus precatorius* (*Lal kunch*). Similarly, Rahman & Fakir (2015) also listed 14 endangered medicinal plant species in part of their distribution ranges due to over harvesting and habitat destruction. These were *Aglaia roxburghiana* (*Pironga*), *Abrus precatorius* (*Roti/ Lal kunch*), *Caesalpinia bonduc* (*Nata*), *Bixa orellana* (*Doigota*), *Ipomoea digita* (*Bhuikumra*), *Mallotus philippinensis* (*Sinduri*), *Piper chaba* (*Choi*), *Vetiveria zizanioides* (*Ghandhabena*), *Terminalia chebula* (*Haritaki*), *Terminalia belerica* (*Bohera*), *Terminalia arjuna* (*Arjune*), *Rauvolfia serpentina* (*Sarpagandha*), *Holarrhena antidysenterica* (*Kurchi*) and *Litseamonopetela* (*Kharajora*). The study recommended for immediate action to conserve these medicinal and aromatic plant (MAP) genetic resources in threatened areas and restore and reproduce them in farmers'/users' arena to enrich biodiversity and to sustain and maintain the century's old customary medicinal culture of Bangladesh. Another study listed 28 threatened plants including nine medicinal plant species (Rashid et al., 2014). The medicinal plant species described in this study are presented in Table 7.1.

7.3 CONSERVATION OF ENDANGERED MEDICINAL PLANT SPECIES

Conservation of endangered medicinal plants is very much essential to ensure primary healthcare of rural as well as other mass communities. Otherwise, the country has to spend lots of foreign currency for import. Several steps are suggested below for conserving endangered medicinal plants of Bangladesh.

7.3.1 INVESTIGATION, EXPLORATION, AND DOCUMENTATION

Ethnobiological study should be conducted in the tribal zones, rural groves, and systematic records of all forest areas need to be carried out to accurately know about the medicinal plants. Furthermore, data should also be collected about the traditional databases from the medicinal plant users (i.e., *Kaviraj*), Unani, and Ayurvedic industries of the country to gather knowledge about the resource bases and possible needs. To achieve sustainable conservation and equitable progress of the medicinal plant segment, a "Medicinal Plants Board" similar to the one present in each of the states of India should also be established in Bangladesh.

TABLE 7.1 Enumeration of Threatened Medicinal Plant Species

Sl. No.	Local Name	Botanical Name	Habitat	Occurrence in Bangladesh	Status of Occurrence	Causes of Threat
1.	*Chhatim*	*Alstonianeriifolia*	Moist evergreen forests.	Chittagong, Sylhet, and all over Bangladesh.	Endangered	Habitat destruction and firewood collection.
2.	*Sarpagandha*	*Rauvolfia serpentina*	Mixed forests, plains, and foothills.	Chittagong, Rangpur, Mymensingh, and Sylhet region.	Endangered	Habitat destruction and over exploitation.
3.	*Lata Aam, Lati Aam*	*Willoughbeia edulis*	Evergreen forests.	Chittagong and Sylhet region.	Vulnerable	Habitat destruction.
4.	*Dudhi, Khirra, Dudhkoraiya, Bara kurch*	*Wrightia arborea*	Evergreen forests	Chittagong and Mymensingh region.	Endangered	Deforestation and habitat destruction.
5.	*Gwali-lata, Goali-Kata*	*Cayratiapedata*	Rainy forest areas.	Chittagong, Dhaka, and Sylhet region.	Near threatened.	Habitat destruction.
6.	Fox grape	*Cayratia tenuifolia*	Rain forests.	Chittagong region.	Near threatened.	Habitat destruction.
7.	*Sona-tola*	*Cissus pentagona*	Climbing over bushes and small trees in hilly forest areas.	Chittagong and Sylhet region.	Near threatened.	Habitat destruction.
8.	Princess vine	*Cissus verticillata*	Primary forests	Chittagong	Extinct	Habitat destruction, over harvest.
9.	*Kuanria*	*Tetrastigma dubium*	Hilly forests.	Chittagong and Sylhet region.	Endangered	Habitat destruction.

7.3.2 IN-SITU AND EX-SITU MAINTENANCE

Protected zones, recognized through statutory law, can be very beneficial for the preservation of medicinal plants. There are various forms (national parks, forest reserves, strict nature reserves, etc.), generally serving several purposes in addition to biodiversity conservation, and with different rules applying to the collection and conservation of medicinal plants. Usually, medicinal plant issues have been shown to be low among the priorities of the responsible organizations. In India, for instance, the Forest Department is much more concerned with timber than NTFPs and within the wide category of NTFPs, often in other types of products separately from medicinal plants, particularly those considered as being financially more lucrative (Jain, 2000).

"Medicinal plant village" should be established for *ex-situ* conservation of endangered and other medicinal plants. For implementation of establishing "Medicinal plant villages," concerned government organizations like the Department of Forest, Department of Agricultural Extension, and other organizations should take initiatives. Besides, different NGOs might be involved in this. Initiative should also be taken for planting medicinal plants in the homestead areas or roadsides as *ex-situ* conservation of medicinal plants. The country may be divided into different zones for growing medicinal plants where they can be grown successfully. Furthermore, existing unproductive lands may be brought under medicinal plant farming.

7.3.3 HIGH QUALITY PLANTING MATERIALS (QPMS)

To raise the stock of medicinal plants, a regular supply of quality planting materials (QPMs) is very much essential and hence, nurserymen of existing nurseries need to be trained up. Besides, the government should provide incentives for establishing new nurseries emphasizing on medicinal plant species at local levels where interested growers can get QPMs at a reasonable price.

7.3.4 FARM LEVEL CULTIVATION

The respective department of the government should encourage the farmers for cultivating medicinal plants. For this, demonstration plots on different medicinal plants should be set up and field day should be arranged for dissemination among the growers.

7.3.5 *STRENGTHENING OF RESEARCH AND DEVELOPMENT ACTIVITIES*

Research activities should be strengthened on medicinal plant species. Developing high yielding varieties of different medicinal plants through an organized research program, standardization of improved production technologies, and supply of QPMs as well as training to the farmers is essential for conservation of different medicinal plant species. Like other crop-based research organizations, a separate research organization specialized in medicinal plants should be established.

7.3.6 *AWARENESS OF BENEFICIARIES AND CAPACITY BUILDING*

Awareness should be raised among the beneficiaries of medicinal plants. Unplanned harvesting should be avoided strictly. While harvesting, the harvester will have to plant at least the same number of plants in that place or other suitable places. For creating awareness, help may be taken from different mass media like radio, television, internet, and Facebook. Involvement of mobile phones, internet, and other information technologies may also be used for raising awareness among the community. Publishing booklets, leaflets, banners, and posters, staging folk drama, group discussion, installing billboards, arrangement of exhibition, free distribution of seedlings in different occasions, organizing 'Nature Club,' observation of special day, etc., may also be helpful for raising awareness among people.

"Capacity building' in short is the investment in an organization's success and sustainability. Capacity may be built up in individual, institutional, and systemic levels. Individual capacity may be built up through training and mentoring of existing workers (*Kaviraj*, Ayurvedic, and Unani personnel) at home and abroad. Institutional capacity may be built up through different activities like fundraising, designing, and implementing need-based research programs, developing partnerships with similar other national and international organizations. At present, many organizations in this region (Indian sub-continent) are working on various issues of medicinal plants. They are maintaining different medicinal plants suitable for cultivation in Bangladesh. Therefore, making a liaison with those organizations may be helpful for the introduction, conservation, and utilization of different medicinal plants. Capacity building at the systemic level is much more difficult. Lobbying elected persons at different levels and lawmakers can be effective for changing the existing system. Therefore, involving public representatives in different inaugural session ceremonies related to medicinal plants may be effective for capacity building.

7.3.7 INSTITUTIONAL SUPPORT

For the expansion of any technology or idea, institutional support is very much essential. Likewise, for conserving the endangered medicinal plants of the country, concerned institutions should take initiatives. For encouraging commercial scale production of medicinal plants, the government should provide loans to the producers in simple terms and conditions through the banks or other financial organizations. The Department of Agricultural Extension can arrange training on medicinal plants and encourage people to plant medicinal plants. Organizations can distribute seeds or seedlings of different medicinal plants. The health departments can raise awareness among the masses about the safe use of medicinal plants. The research organizations may strengthen research on medicinal plants such as the development of locally adopted new varieties, standardization of cultivation techniques, conservation of endangered medicinal plants, etc. The NGOs that are working in the country may also contribute greatly to this aspect.

7.3.8 MARKETING SUPPORT

Presently, there are no markets where medicinal plants can be sold. As a result, the growers are getting a minimum price for their products. Hence, marketing of produce is very much essential for increasing cultivation of medicinal plants. If a producer fails to get the appropriate price for his produce, then he will be deprived and will not go for further cultivation of that crop. Therefore, farmers' cooperatives should be organized involving government personnel for proper marketing and ensuring an appropriate price for the growers who are growing medicinal plants. Concerned officials should also take initiative for making a linkage between the grower and the traders. Moreover, export markets should be searched by the concerned government officials.

7.3.9 PRIORITY RESEARCH FUNDING OF MEDICINAL PLANT-BASED PROJECTS

At present, different national and international donor agencies in Bangladesh like Ministry of Science and Technology, Ministry of Forest, Ministry of

Education, BARC, UGC, KGF, NATP, BAS, USAID, CARE, and HKI (Helen Keller International) are providing funds for conducting crop-based research activities. The donor agencies will have to provide funds on a priority basis for research projects regarding medicinal plants. Moreover, the researchers should be inspired to design medicinal plant-based projects. Different organizations in Bangladesh have all sorts of facilities to carry out research on medicinal plants. They have skilled personnel to run medicinal plant-based project activities. Several organizations are conducting research on medicinal plants on a limited scale.

Presently, BFRI (Bangladesh Forest Research Institute) in Chittagong has some ongoing research on medicinal plants under the program areas 'Biodiversity and Conservation,' 'Forest Pests and Diseases,' 'Bamboo and Non-timber Economic Crops,' 'Post Harvest Utilization-Chemical Processing,' 'Breeding and Tree Improvement,' 'Plantation technique and forest management.' Germplasm conservation and management practices of different medicinal plants are a major area of study. A total of 18,000 seedlings of medicinal tree species were raised and planted in the coastal areas in 2015 (Anon., 2019a). Besides, they have the 'BFRI Herbarium' and 'Nursery' as well. The institute publishes peer-reviewed journals, newsletters, booklets, leaflets, and other publications where different aspects of research on medicinal plants are published. For the extension of medicinal plants, the institute arranges training for farmers, extension workers, and other relevant personnel. They also distribute seedlings/saplings to the beneficiaries.

The National Botanic Garden of Bangladesh and the Bangladesh National Herbarium under the Department of Forestry have made the biggest plant conservation center in Bangladesh on around 84 hectares (210 acres). It is situated at Mirpur in Dhaka, beside the National Zoo and was established in 1961. It is one of the largest botanic gardens of Bangladesh, an information center for nature lovers and botanists, and also a tourist target. The herbarium has a collection of around 100,000 preserved samples of plants. A total of 150 different medicinal plants are conserved *in-situ* in this garden (Anon., 2014). The Forest Department upholds a forest nursery in each upazila. The nursery is managed and maintained by Social Forestry Plantation Centers under the control of Social Forestry Nursery and Training Centers at the district level. The nurseries produce good quality seedlings of different forest plant species, including medicinal plants, for plantation, distribution, and sale.

The Bangladesh Agricultural Research Institute (BARI), the largest national research institute in Bangladesh, has some research programs on various medicinal plants and has already developed several high-yielding varieties. The Bangabandhu Sheikh Mujibur Rahman Agricultural University (BSMRAU), Gazipur, Bangladesh, has included different aspects of medicinal plants in its agricultural and horticulture graduate programs. In addition, several theses and dissertations have already been published on different medicinal plants like mint, moringa, aloe, agor, etc., from this university. A quarterly journal is regularly published from the university named 'Annals of Bangladesh Agriculture' (web: https://bsmrau.edu.bd/aba/). A high-yielding variety of black cumin named 'BU kalojira-1' has been released from the Department of Horticulture (Anon., 2019b). There is a digital herbarium of this university, where images and databases of different crop plants comprising field, horticultural, forest, and medicinal plants are being digitized, and a virtual herbarium is developed (web: www.dhcrop.bsmrau.net). Besides, BSMRAU has an *ex-situ* gene bank where about 547 germplasm of different species, including MAPs, are maintained. Several other universities in the country are conducting research on a limited scale and have research programs on medicinal plants. In addition to these universities and institutions, Hamdard University Bangladesh has established an International Research Center for Natural Medicine, an International Convention Center, a Scholar House, and other required facilities (Anon., 2019c).

7.3.10 INTRODUCTION OF MEDICINAL PLANTS IN COURSE CURRICULA FROM SCHOOL TO UNIVERSITY EDUCATION

The government should introduce different aspects of medicinal plants in the curricula from secondary levels to university levels. Presently, a specialized university named "Hamdard University" is established and introducing a degree on Unani and Ayurveda. The Department of Botany in different universities and several other private universities have a limited syllabus on medicinal plants. Therefore, a separate course should be included in the curricula of public and private universities. Moreover, the present 27 government and private educational institutes that are offering degrees on Unani and Ayurveda should be strengthened and modernized. The curricula of those institutes should be reorganized by expert personnel.

7.4 IMPORTANT MEDICINAL PLANTS OF BANGLADESH AND UTILIZATION

Medicinal plants are used by different persons in different ways in different countries for various diseases. Parts of medicinal plants such as leaf, roots, stems, flowers, or fruits are used as dried powder, pill, poultice, infusion, decoction, or steam. These also depend on external or internal disorders of the body. Important medicinal plants of Bangladesh with their detailed information as well as their reported uses are listed in Table 7.2.

7.5 PROSPECTS OF MEDICINAL PLANTS IN BANGLADESH

Medicinal plants play a vital role in the everyday life of rural as well as urban people of Bangladesh. It not only keeps the human health sound but also has a countless impact on the economy of the country. Therefore, there is great prospect of medicinal plants in Bangladesh.

7.5.1 INCREASED NATIONAL AS WELL AS GLOBAL DEMAND

Nowadays, people are very much aware of their health and beauty. Therefore, the demand for herbal healthcare and beauty products is increasing day by day in the national and global market as these are cost-effective, available, and without side effects. So, there is a need for increased production of medicinal raw materials. The domestic market of Bangladesh for herbal medicine is Tk. 1,000 crores, which is growing by about 15% per year. On the other hand, the world trade for medicinal plant products is valued at about US $60 billion with the European market being valued at US $10 billion, which is growing about 10% annually (Anon., 2009).

7.5.2 INCOME SOURCE OF RURAL POOR

Presently, the rural people collect and dry medicinal plants from different places and transport them to local markets for sale. This offers an income and employment opportunities for the rural poor to improve their livelihood. Alternatively, it also increases the employment opportunities for jobless rural youths and female members of the family towards their empowerment. Besides, the local '*Kaviraj*' are also earning money from using different

TABLE 7.2 Important Medicinal Plants of Bangladesh with Their Possible Usages

Sl. No.	Vernacular Name	English Name	Botanical Name	Family	Uses	References
1.	*Ulotkombol*	Abroma/Devil's cotton	*Abroma augusta*(L.f.)	Malvaceae	It is used to treat various diseases like diabetes mellitus, as uterine stimulant in emmenagogue, dysmenorrhea, amenorrhea, sterility, and menstrual disarrays, rheumatic pains of joints and headache from sinusitis.	Yusuf et al. (1994); Dixie et al. (2003); Ghani (2003); Rahmatullah et al. (2009); Rashid &Eakram (2009); Rahman et al. (2013); Tuhin et al. (2013); Bhowmik et al. (2014); Haque et al. (2014); Kabir &Saha (2014); Rahman & Keya (2015); Bari et al. (2017); Khatun et al. (2018); Chowdhury et al. (2019); Das & Islam (2019)
2.	*Roti*, *Gunjika*, *Kaichguta*	Jequirity bean or rosary pea.	*Abrus precatorius*L.	Fabaceae	Leaves and young stems are consumed directly used for treatment of cough, dry discharge, hopping cough, asthma, constipation, and malnutrition.	Yusuf et al. (1994); Ahsan et al. (1997); Ghani (2003); Mukul et al. (2007); Chowdhurya et al. (2009); Rahmatullah et al. (2009); Rahman et al. (2013); Bhowmik et al. (2014); Rahman & Keya (2015); Ahmmed et al. (2017); Bari et al. (2017); Khatun et al. (2018)
3.	*Khayar*	Black catechu; catechu; cutch tree.	*Acacia catechu* (L.f.)/ *Senegalia catechu*	Fabaceae	Paste and powder made from bark are used for abdominal pain, skin ailments, cough, and sore throat. The bark is effective against diarrhea and the seeds are supposed to have an antibacterial action.	Yusuf et al. (1994); Ahsan et al. (1997); Mukul et al. (2007); Chowdhurya et al. (2009); Rahman et al. (2013); Bari et al. (2017); Khatun et al. (2018)
4.	*Muktajhuri*	Indian Copperleaf, Indian Mercury	*Acalypha indica* Linn.	Euphorbiaceae	Plant extract is used as cough syrup, laxative, and diuretic. Root infusion is used in bronchitis and to treat asthma. Plant or plant parts are used for treating many other diseases like gonorrhea, dysentery, chest pain, earache, piles, injuries, epilepsy, etc.	Ghani (2003); Rashid &Eakram (2009); Rahman et al. (2013); Tuhin et al. (2013); Hasan et al. (2014); Kabir & Saha (2014); Rahman & Jamila (2016); Khatun et al. (2018)
5.	*Apang*	Chaff-flower	*Achyranthes aspera* L.	Amaranthaceae	Root, stem, and leaf extract is used to facilitating delivery, and to treat bronchitis, cold, cough, indigestion, weakness, dropsy, dysentery, dog bite and skin diseases, etc.	Ghani (2003); Khan & Rashid (2006); Rashid &Eakram (2009); Rahman et al. (2013); Tuhin et al. (2013); Bhowmik et al. (2014); Hasan (2014); Rahman & Keya (2015); Rahman & Jamila (2016); Bari et al. (2017)

TABLE 7.2 *(Continued)*

Sl. No.	Vernacular Name	English Name	Botanical Name	Family	Uses	References
6.	*Bel*	Wood apple	*Aegle marmelos*L.	Rutaceae	The whole plant or plant parts are used in diabetes, liver toxicity, mycological contagion, contagious infection, irritation, pyrexia, and to release pain.	Ahsan et al. (1997); Chowdhurya et al. (2009); Ghani (2003); Mukul et al. (2007); Rahmatullah et al. (2009); Rashid &Eakram (2009); Rahman et al. (2013); Tuhin et al. (2013); Bhowmik et al. (2014); Haque et al. (2014); Hasan et al. (2014); Kabir & Saha (2014); Rahman & Keya (2015); Rahman & Jamila (2016); Ahmmed et al. (2017); Bari et al. (2017); Khatun et al. (2018)
7.	*Rokto-keshob*	Balipoovu	*Aerva sanguinolenta*L.	Amaranthaceae	It shows multiple biological activities such as antioxidant, antibacterial, antilithiatic, hepatoprotective, anti-diabetic, anticancer, anti-hyperlipidemic, and nephroprotective potential.	Ghani (2003); Rashid &Eakram (2009); Bhowmik et al. (2014); Rahman et al. (2013); Tuhin et al. (2013); Hasan et al. (2014); Kabir & Saha (2014); Rahman & Keya (2015); Rahman & Jamila (2016);Bari et al. (2017);Musaddiqa et al. (2018)
8.	Atichada	Chick weed	*Ageratum conyzoides*L.	Asteraceae	Leaf is used in diarrhea, boils, skin diseases, joint pain. Also used in gynecological diseases, leprosy, and other skin diseases.	Ghani (2003); Mukul et al. (2007); Rashid &Eakram (2009); Tuhin et al. (2013); Haque et al. (2014); Hasan et al. (2014); Kabir & Saha (2014); Rahman & Jamila (2016)
9.	*Alamonda*	Alamanda	*Allamanda cathartica*	Apocynaceae	For the preparation of therapeutic, leaves, flowers, and roots of almonds are used. The milky latex is known to have antibacterial and possibly anticancer properties. Small doses of leaf decoction are used as remedy for poisoning.	Ahsan et al. (1997); Ghani (2003); Tuhin et al. (2013); Hasan et al. (2014); Kabir & Saha (2014); Rahman & Keya (2015); Rahman & Jamila (2016); Ahmmed et al. (2017); Bari et al. (2017); Khatun et al. (2018)
10.	*Piyaaz*	Onion	*Allium cepa* L	Liliaceae	Cut pieces of bulbs are boiled with water and the releasing vapor is slowly breathed which is helpful for mucus congestion in cold.	Dixie et al. (2003); Ghani (2003); Chowdhurya et al. (2009); Rahman et al. (2013); Tuhin et al. (2013); Haque et al. (2014); Hasan et al. (2014); Kabir & Saha (2014); Rahman & Keya (2015); Rahman & Jamila (2016)

TABLE 7.2 *(Continued)*

Sl. No.	Vernacular Name	English Name	Botanical Name	Family	Uses	References
11.	*Roshun*	Garlic	*Allium sativum* L.	Liliaceae	Cloves of garlic are used to control heart disease, in lowering cholesterol level, blood sugar and troubles in urinating. It is also used to regulate blood pressure and in bronchitis.	Yusuf et al. (1994); Dixie et al. (2003); Chowdhurya et al. (2009); Rahmatullah & Mollik (2009); Rahman et al. (2013); Tuhin et al. (2013); Haque et al. (2014); Kabir & Saha (2014); Rahman & Keya (2015); Rahman & Jamila (2016); Khatun et al. (2018)
12.	Fenkochu, Mankochu	Giant taro	*Alocasia macrorrhizos*(L.) G.Don	Araceae	Stem, root, and rhizome of giant taro are used against rheumatism and diabetes. Rhizome is also used to treat furuncles as a poultice.	Yusuf et al. (1994); Ghani (2003); Mukul et al. (2007); Chowdhurya et al. (2009); Rahmatullah & Mollik (2009); Rahman et al. (2013); Tuhin et al. (2013); Bhowmik et al. (2014); Hasan et al. (2014); Kabir & Saha (2014); Rahman & Keya (2015); Rahman & Jamila (2016); Bari et al. (2017)
13.	*Ghritokumari*	Burn plant, aloe	*Aloe barbadensis*Mill.	Liliaceae	Leaf gel along with water is helpful for the treatment of gastric ulcer, constipation, and skin dullness. Leaf is used for stomach disarrays, lung infection and skin burns.	Ahsan et al. (1997); Ghani (2003); Khan & Rashid (2006); Mukul et al. (2007); Chowdhurya et al. (2009); Rahmatullah & Mollik (2009); Rahman et al. (2013); Tuhin et al. (2013); Bhowmik et al. (2014); Haque et al. (2014); Hasan et al. (2014); Kabir & Saha (2014); Rahman & Keya (2015); Rahman & Jamila (2016); Ahmmed et al. (2017); Bari et al. (2017); Khatun et al. (2018)
14.	*Chatim*	Devil's tree	*Alstoniascholaris* (L.)	Apocynaceae	Leaves are used for treatment in fever. Bark is used in skin diseases and as alterative and tonic. Exudates from bark and stem are used in diabetes and ulcer.	Mukul et al. (2007); Rahmatullah & Mollik (2009); Rashid &Eakram (2009); Rahman et al. (2013); Tuhin et al. (2013); Haque et al. (2014); Hasan et al. (2014); Rahman & Keya (2015); Rahman & Jamila (2016); Bari et al. (2017)

TABLE 7.2 *(Continued)*

Sl. No.	Vernacular Name	English Name	Botanical Name	Family	Uses	References
15.	*Haicha*	Creeping chaffweed	*Alternanthera sessilis*(L.)	Amaranthaceae	Leaf is used for the treatment of leucorrhea and gonorrhea. Plant is used to treat hepatitis and other lung troubles. Also used as a hair tonic and to stop bleeding.	Mukul et al. (2007); Chowdhurya et al. (2009); Rahman (2013); Bhowmik et al. (2014); Haque et al. (2014)
16.	*Katanatay*	Pigweed, soldier-weed, spiny Amaranthus	*Amaranthus spinosus* L.	Amaranthaceae	Leaves are used against constipation, piles, liver trouble and other urinary problems. Regular consumption of fried leaves increases milk production of lactating mother. Roots are used in dysentery and as diuretic. Poultice of the plant is used to cure abscess.	Ghani (2003); Rahmatullah & Mollik (2009); Rahman et al. (2013); Bhowmik et al. (2014); Haque et al. (2014); Hasan et al. (2014); Rahman & Keya (2015); Rahman & Jamila (2016); Bari et al. (2017); Khatun et al. (2018)
17.	*Anarosh*	Pineapple	*Ananas comosus* (L.) Merr.	Bromeliaceae	Pineapple lowers heart disease risk by lowering blood cholesterol, good for colds and coughs, strengthens bones, relief joint and muscle pain, increase male fertility and cure digestive problems. It also induces abortion, induce menstruation, expel worms, and kill parasitic amoebas.	Ghani (2003); Mukul et al. (2007); Rahmatullah & Mollik (2009); Tuhin et al. (2013); Bhowmik et al. (2014); Haque et al. (2014); Hasan et al. (2014); Kabir & Saha (2014); Rahman & Jamila (2016); Bari et al. (2017)
18.	*Kalomegh*	Creat, Green Chirayta, King of bitters	*Andrographis paniculate*	Acanthaceae	Used as blood purifier, liver tonic, releases worm, Leaves are used to treat fever, blood dysentery, scabies, common cold, indiges-tion, and respiratory tract infections.	Ghani (2003); Khan & Rashid (2006); Mukul et al. (2007); Chowdhurya et al. (2009); Tuhin et al. (2013); Bhowmik et al. (2014); Haque et al. (2014); Hasan et al. (2014); Kabir & Saha (2014); Rahman &Jamila (2016); Khatun et al. (2018)
19.	*Ata, Sarifa*	Sugar-apple/ custard apple	*Anona squamosa* Linn.	Annonaceae	Leaves and fruits are used against tumors problem. In treating boils, internal and external wounds crushed leaves are used. Crushed seeds are used against internal and external parasites, worms, and head lice. Leaf, bark, and roots are considered to have sedative action. Also used in ulcer treatment and as a nervine tonic.	Yusuf et al. (1994); Ghani (2003); Mukul et al. (2007); Rahmatullah & Mollik (2009); Rahman et al. (2013); Tuhin et al. (2013); Hasan et al. (2014); Kabir & Saha (2014); Rahman & Keya (2015)

TABLE 7.2 *(Continued)*

Sl. No.	Vernacular Name	English Name	Botanical Name	Family	Uses	References
20.	*Pitraj*	Pithraj tree	*Aphanamixispolystachya* (Wall.)	Meliaceae	Stem, leaf, bark is used for the treatment of cancer and oil as insecticide. Bark of the plant is used against colds and chest pain. Also used in the treatment of tumors, abdominal complaints, spleen, and liver diseases. Oil from seed is used against rheumatism.	Ghani (2003); Mukul et al. (2007); Rashid &Eakram (2009); Rahman et al. (2012, 2013); Bhowmik et al. (2014); Haque et al. (2014); Hasan et al. (2014); Kabir & Saha (2014)
21.	*Aptenia*	Baby sun-rose/ Dew plant	*Aptenia cordifolia*	Aizoaceae	Whole plant or fruits are used for inflammation, deodorant, headache poultice treatment. Leaves and burnt stems are sometime applied to joints to get relief from pain.	Ghani (2003); Mukul et al. (2007); Chowdhurya et al. (2009); Rahman et al. (2013); Bhowmik et al. (2014); Haque et al. (2014); Hasan et al. (2014); Kabir & Saha (2014); Khatun et al. (2018)
22.	*Supari*	Arecanut/betel nut palm	*Areca catechu* L	Arecaceae	Areca nut is proved to have anti-inflammatory, anti-hypertensive, antioxidant, anti-carcinogenic, and anti-allergic properties. It helps in digestion; stimulate appetite and central nervous system. Also used for prevention of dental cavities and toothache.	Yusuf et al. (1994); Ghani (2003); Mukul et al. (2007); Chowdhurya et al. (2009); Rahmatullah & Mollik (2009); Rahman et al. (2012, 2013); Tuhin et al. (2013); Bhowmik et al. (2014); Haque et al. (2014); Hasan et al. (2014); Kabir & Saha (2014); Khatun et al. (2018); Rahman & Keya (2015)
23.	*Kathal*	Jackfruit	*Artocarpus heterophyllus* Lam	Moraceae	Fruits and seeds are consumed for ulcer, constipation, and diarrhea. Gum is used to dry hemorrhoids and cracked heels. Decoction of root is used to reduce fever and treatment of diarrhea, asthma, and other skin diseases.	Ghani (2003); Mukul et al. (2007); Chowdhurya et al. (2009); Rahmatullah & Mollik (2009); Rahman et al. (2013); Bhowmik et al. (2014); Haque et al. (2014); Hasan et al. (2014); Kabir & Saha (2014); Rahman & Keya (2015); Rahman & Jamila (2016); Ahmmed et al. (2017); Bari et al. (2017); Khatun et al. (2018)

TABLE 7.2 *(Continued)*

Sl. No.	Vernacular Name	English Name	Botanical Name	Family	Uses	References
24.	*Deowa*	Monkey jack	*Artocarpus lakoocha*Roxb.	Moraceae	Fruits, seeds, and barks have laxative properties. The sap and bark juice are applied externally to get rid of pimples, boils, cuts, and wounds. The root is astringent and is also used as a purgative. The crushed bark is used as a poultice in treating skin disorders. To treat headache, the bark is used.	Yusuf et al. (1994); Ghani (2003); Mukul et al. (2007); Bhowmik et al. (2014); Haque et al. (2014); Hasan et al. (2014); Kabir & Saha (2014); Rahman et al. (2013); Tuhin et al. (2013)
25.	*Shatamuli*	Asparagus	*Asparagus racemosus*Willd	Asparagaceae	Root extract is used in indigestion, constipation, and intestinal ulcers. Besides, it is used for retention of fluid, nervousness, cancer, diarrhea, asthma, tuberculosis, and diabetes. Also taken in edema, contagious infections and to rises lactation.	Yusuf et al. (1994); Ghani (2003); Khan & Rashid (2006); Mukul et al. (2007); Chowdhurya et al. (2009); Rahmatullah & Mollik (2009); Rashid &Eakram (2009); Rahman et al. (2013); Tuhin et al. (2013); Bhowmik et al. (2014); Haque et al. (2014); Hasan et al. (2014); Kabir & Saha (2014); Rahman & Keya (2015); Rahman & Jamila (2016); Bari et al. (2017); Khatun et al. (2018)
26.	*Neem*	Neem tree or Indian lilac.	*Azadirachta indica* A. Juss	Meliaceae	Leaf, bark, seed is used to treatment of diabetes, measles, pox, intestinal worm, scabies, indigestion, cataract, abscess.	Yusuf et al. (1994); Ghani (2003); Khan & Rashid (2006); Mukul et al. (2007); Chowdhurya et al. (2009); Rahmatullah & Mollik (2009); Rashid &Eakram (2009); Tuhin et al. (2013); Bhowmik et al. (2014); Haque et al. (2014); Hasan et al. (2014); Kabir & Saha (2014); Rahman & Keya (2015); Rahman & Jamila (2016); Sarwar (2017); Khatun et al. (2018)
27.	*Bash*	Bamboo	*Bambusaarundinacea* (Retz.)	Poaceae	Leaf, stem, root is used in cough, asthma, joint pain, stops bleeding.	Yusuf et al. (1994); Ghani (2003); Chowdhurya et al. (2009); Tuhin et al. (2013); Bhowmik et al. (2014); Hasan et al. (2014); Kabir & Saha (2014)

TABLE 7.2 *(Continued)*

Sl. No.	Vernacular Name	English Name	Botanical Name	Family	Uses	References
28.	*Jeuti, Jhingti*	Porcupine Flower	*Barleriaprionitis*L.	Acanthaceae	Leaf is used for the treatment of rheumatism, skin disorders and pain.	Yusuf et al. (1994); Ghani (2003); Chowdhurya et al. (2009); Tuhin et al. (2013); Bhowmik et al. (2014); Haque et al. (2014); Hasan et al. (2014); Kabir & Saha (2014)
29.	*Hijal*	Freshwater mangrove/ mango-pine	*Baringtoniaacutangula* (L.) Gaertn	Lecythidaceae	Leaf, seed, root is used in the treatment of diarrhea, headache, chest pain. Leaf and branches are also used in cough, asthma.	Yusuf et al. (1994); Ghani (2003); Chowdhurya et al. (2009); Tuhin et al. (2013); Bhowmik et al. (2014); Hasan et al. (2014); Bari et al. (2017)
30.	*Puishak*	Indian-spinach	*Basella alba* L.	Basellaceae	Leaf, stem is used for the cure of anemia, cough.	Rashid &Eakram (2009); Rahman et al. (2013); Bhowmik et al. (2014); Hasan et al. (2014); Rahman & Keya (2015); Rahman & Jamila (2016); Ahmmed et al. (2017); Khatun et al. (2018)
31.	*Shialmutra, Shiyalmodi*	Kakronda	*Blumealacera*DC	Asteraceae	Leaf and root are used for the treatment of weakness, edema, piles, cholera, diuretic, microbial infections.	Yusuf et al. (1994); Ghani (2003); Mukul et al. (2007); Rahmatullah & Mollik (2009); Rashid &Eakram (2009); Bhowmik et al. (2014); Haque et al. (2014); Hasan et al. (2014); Kabir & Saha (2014); Rahman (2014); Bari et al. (2017); Khatun et al. (2018)
32.	*Punarnava*	Red hogweed, red spiderling, wine flower.	*Boerhaviadiffusa*L.	Nyctaginaceae	Leaf, root, and stem is used against jaundice, anemia, elephantiasis, gonorrhea, intestinal pain and dysentery, Leaves are used in treating asthma, intestinal injury, and diarrhea.	Ghani (2003); Khan & Rashid (2006); Chowdhurya et al. (2009); Rahmatullah & Mollik (2009); Rahman et al. (2012, 2013); Bhowmik et al. (2014); Haque et al. (2014); Hasan et al. (2014); Kabir & Saha (2014); Rahman & Keya (2015); Bari et al. (2017); Khatun et al. (2018)

TABLE 7.2 *(Continued)*

Sl. No.	Vernacular Name	English Name	Botanical Name	Family	Uses	References
33.	*Shimul*	Silk cotton tree.	*Bombax ceiba* L.	Bombacaceae	Bark and flower extract used for sexual diseases and menstrual complaints in women. Root and bark sap is used to increase sperm amount, treatment of impotency and hydrocele in human being.	Yusuf et al. (1994); Ghani (2003); Mukul et al. (2007); Rahmatullah & Mollik (2009); Rashid &Eakram (2009); Bhowmik et al. (2014); Hasan et al. (2014); Rahman & Jamila (2016); Bari et al. (2017); Kabir & Saha (2014)
34.	*Patharkuchi*	Air plant, cathedral bells.	*Bryophyllumpinnatum/ Kalanchoepinnata*	Crassulaceae	Leaves are used as kidney and bladder stone remover, cough, and fever. Also used to cure urinary complaints. Juice of crushed leaves is orally taken against rheumatism, body sting, inflammation, duodenal ulcer, diabetes, and various contagious infections.	Ahsan et al. (1997); Ghani (2003); Mukul et al. (2007); Chowdhurya et al. (2009); Rahmatullah & Mollik (2009); Rashid &Eakram (2009); Rahman et al. (2013); Tuhin et al. (2013); Bhowmik et al. (2014); Haque et al. (2014); Hasan et al. (2014); Kabir & Saha (2014); Rahman & Keya (2015); Ahmmed et al. (2017); Bari et al. (2017); Khatun et al. (2018)
35.	*Palash*	Flame of forest	*Butea monosperma* (Lam.) Taub	Fabaceae	Leaves, flower, bark, seed used for worm, bleeding, cough, fever, menstruation problems.	Yusuf et al. (1994); Ghani (2003); Khan & Rashid (2006); Rahman et al. (2013); Haque et al. (2014); Hasan et al. (2014); Kabir & Saha (2014); Bari et al. (2017)
36.	*Arhar*	Pigeon pea	*Cajanus cajan*(L.) Millsp	Fabaceae	Leaf, seed is used in jaundice, helminthiasis, lower cholesterol level, diabetes treatment.	Mukul et al. (2007); Chowdhurya et al. (2009); Rahman et al. (2013); Bhowmik et al. (2014); Bari et al. (2017); Ghani (2003); Hasan et al. (2014); Kabir & Saha (2014)
37.	*Akanda*	Aak, Giant Milkweed	*Calotropis gigantea* (L.)	Aslepiadaceae	Leaves and exudates are used for joint inflammation, cut, and wounds. Leaf and flower used for the treatment of fever, arthritis, asthma, and diabetes.	Yusuf et al. (1994); Ghani (2003); Mukul et al. (2007); Chowdhurya et al. (2009); Rashid &Eakram (2009); Rahman et al. (2013); Bhowmik et al. (2014); Hasan et al. (2014); Kabir & Saha (2014); Rahman & Jamila (2016)

TABLE 7.2 *(Continued)*

Sl. No.	Vernacular Name	English Name	Botanical Name	Family	Uses	References
38.	*Kachamorich*	Tabasco pepper/ cayenne pepper	*Capsicum frutescens* L.	Solanaceae	Fruit is used for the treatment of helminthiasis, loss of appetite, urinary disorder, and indigestion.	Ghani (2003); Chowdhurya et al. (2009); Ahsan et al. (1997); Bhowmik et al. (2014); Haque et al. (2014); Ahmmed et al. (2017); Bari et al. (2017); Hasan et al. (2014); Kabir & Saha (2014)
39.	*Papae*	Papaya	*Carica papaya* L.	Caricaceae	Roots are used as medicine for renal and urinary bladder problem. Flowers are used to treat jaundice. Leaves are used for colic, fever, beriberi, abortion, asthma, malaria, dengue, and antiviral activity. Fruits are used in indigestion, jaundice, toothache, and kidney stones. The bark and twig tissues are found to possess anti-tumor and pesticidal properties.	Yusuf et al. (1994); Rahmatullah & Mollik (2009); Rashid &Eakram (2009); Rahman et al. (2013); Tuhin et al. (2013); Bhowmik et al. (2014); Haque et al. (2014); Hasan et al. (2014); Ahmmed et al. (2017); Bari et al. (2017); Kabir & Saha (2014); Rahman & Jamila (2016); Khatun et al. (2018)
40.	*Karamcha*	Karanda	*Carissa caranda* Wight.	Apocynaceae	Roots are histamine releasing and paste of bark is used in diabetic ulcer.	Ghani (2003); Rashid &Eakram (2009); Tuhin et al. (2013); Bhowmik et al. (2014); Haque et al. (2014); Hasan et al. (2014); Kabir & Saha (2014); Rahman & Jamila (2016); Bari et al. (2017)
41.	*Daudpata*	Candle bush	*Cassia alata*(L.) Gaertn.	Fabaceae	Leaf is used for curing skin diseases, microbial infections.	Ghani (2003); Chowdhurya et al. (2009); Rahmatullah & Mollik (2009); Rahman et al. (2012, 2013); Bhowmik et al. (2014); Haque et al. (2014); Hasan et al. (2014); Rahman & Keya (2015); Ahmmed et al. (2017); Kabir & Saha (2014); Khatun et al. (2018)
42.	*Bandar lathi/ Sunalo*	Indian laburnum	*Cassia fistula* L.	Fabaceae	Leaf, bark, root, fruit is used against helminthiasis, constipation, common cold, fever, intestinal disorder, mycological infection, and dermatitis.	Yusuf et al. (1994); Rahmatullah & Mollik (2009); Rashid &Eakram (2009); Rahman et al. (2012, 2013); Tuhin et al. (2013); Bhowmik et al. (2014); Haque et al. (2014); Hasan et al. (2014); Ahmmed et al. (2017); Bari et al. (2017); Kabir & Saha (2014); Khatun et al. (2018)

TABLE 7.2 *(Continued)*

Sl. No.	Vernacular Name	English Name	Botanical Name	Family	Uses	References
43.	*Nayantara*	Madagascar periwinkle	*Catharanthus roseus* (L.)/ *Vinca rosea*	Apocynaceae	Root, leaves are used for leukemia, diabetes, hypertension. The plant is used in insomnia and cancer. Extract of boiled plant is used to arrest bleeding.	Yusuf et al. (1994); Ghani (2003); Khan & Rashid (2006); Mukul et al. (2007); Rahman et al. (2013); Kabir & Saha (2014); Bari et al. (2017); Khatun et al. (2018)
44.	*Thankuni*	Centella or Indian pennywort.	*Centella asiatica* L.	Umbelliferae/ Apiaceae	Whole plant or leaf is used against weakness, dysentery, cataract, skin problem, gonorrhea, and leucorrhea. Improves appetite and memory. Also used for the treatment of urinary discharges, asthma, inflammations, fevers, convulsive complaints, senselessness, and syphilis. It is used to reduce swelling, anxiety, and stress, improve blood circulation.	Yusuf et al. (1994); Ghani (2003); Khan & Rashid (2006); Mukul et al. (2007); Chowdhurya et al. (2009); Rahmatullah & Mollik (2009); Rahman et al. (2013); Tuhin et al. (2013); Bhowmik et al. (2014); Hasan et al. (2014); Kabir & Saha (2014); Rahman & Keya (2015); Rahman & Jamila (2016); Bari et al. (2017); Khatun et al. (2018)
45.	*Hasnahena*	Night-blooming jasmine.	*Cestrum nocturnum*L.	Solanaceae	Leaf is used to stop bleeding. Effective against asthma, headache, dengue, epilepsy, throat irritation and nausea.	Ghani (2003); Mukul et al. (2007); Chowdhurya et al. (2009); Rahman et al. (2013); Bhowmik et al. (2014); Hasan et al. (2014); Kabir & Saha (2014); Rahman & Jamila (2016)
46.	*Tejpata*	Bay leaf	*Cinnamomum tamala*(Buch.-Ham.)	Lauraceae	Leaves and bark are used in gonorrhea and as carminative.	Ghani (2003); Khan & Rashid (2006); Rahmatullah &Mollik (2009); Rashid &Eakram (2009); Rahman et al. (2013); Haque et al. (2014); Hasan et al. (2014); Kabir & Saha (2014); Rahman & Jamila (2016); Bari et al. (2017); Khatun et al. (2018)
47.	*Harjoralota*	Veld grape, winged treebine	*Cissus quadrangularis*	Vitaceae	Paste of stem is used on fractured bones and rheumatic fever. Also used to cure asthma, diabetes, hemorrhoids, high cholesterol, joint pain, menstrual convulsions, metabolic disorder, and osteoporosis.	Yusuf et al. (1994); Rahmatullah & Mollik (2009); Rahman et al. (2013); Tuhin et al. (2013); Bhowmik et al. (2014); Haque et al. (2014); Hasan et al. (2014); Kabir & Saha (2014); Bari et al. (2017); Khatun et al. (2018)

TABLE 7.2 *(Continued)*

Sl. No.	Vernacular Name	English Name	Botanical Name	Family	Uses	References
48.	Tormuj	Watermelon	*Citrullus lanatus*(Thunb.)	Cucurbitaceae	Ripe fruits are used as cooling, diuretic, stomachache, blood purifiers, aphrodisiac, biliousness, astringent, aching eyes, dermatitis, and itching. Fruit juice is used as an anti-septic in typhoid fever and laxative. The seeds can be used as brain tonic.	Ghani (2003); Chowdhurya et al. (2009); Rahman (2013); Rahman et al. (2013); Kabir & Saha (2014)
49.	*Komla*	Sour orange	*Citrus aurantium* L.	Rutaceae	Dried rind of orange and betel leaf is chewed, and extract is swallowed to get rid of stomach pain.	Yusuf et al. (1994); Ghani (2003); Khan & Rashid (2006); Chowdhurya et al. (2009); Rashid &Eakram (2009); Rahman et al. (2013); Bhowmik et al. (2014); Kabir & Saha (2014); Rahman & Jamila (2016); Ahmmed et al. (2017); Khatun et al. (2018)
50.	*Batabilebu/ Jaambura*	Pummelo	*Citrus grandis* (L.) Osbeck	Rutaceae	Fruit is used for the treatment of diabetes, halitosis, dyspepsia, insect bite, oral wounds, constipation, cancer, asthma, epilepsy, headache, and fever and increases eye vision.	Ghani (2003); Khan & Rashid (2006); Mukul et al. (2007); Chowdhurya et al. (2009); Rahmatullah & Mollik (2009); Rashid &Eakram (2009); Rahman et al. (2013); Bhowmik et al. (2014); Haque et al. (2014); Hasan et al. (2014); Kabir & Saha (2014); Rahman & Keya (2015); Khatun et al. (2018)
51.	*Lebu*	Lemon	*Citrus limon* Roxb.	Rutaceae	Fruit juice possesses antiseptic, antibiotic, and anthelmintic properties and used for the treatment of tonsillitis, prevention of tooth cavities, sore throats, bad breath, fevers, rheumatism, high blood pressure and chest pain.	Yusuf et al. (1994); Ghani (2003); Khan & Rashid (2006); Mukul et al. (2007); Chowdhurya et al. (2009); Rahmatullah & Mollik (2009); Rashid &Eakram (2009); Rahman et al. (2013); Tuhin et al. (2013); Bhowmik et al. (2014); Hasan et al. (2014); Kabir & Saha (2014); Rahman & Keya (2015); Rahman & Jamila (2016); Khatun et al. (2018)

TABLE 7.2 *(Continued)*

Sl. No.	Vernacular Name	English Name	Botanical Name	Family	Uses	References
52.	*Hurhuria*	Yellow spider flower.	*Cleome viscosa* Linn	Capparidaceae	Leaves are useful in hypertension, fever, arthritis, dysentery, malaria, and bronchitis. It also possesses wound healing properties.	Ghani (2003); Rashid &Eakram (2009); Tuhin et al. (2013); Bhowmik et al. (2014); Haque et al. (2014); Hasan et al. (2014); Bari et al. (2017); Kabir & Saha (2014); Rahman & Jamila (2016)
53.	*Bamonhaki*	Glory bower, bag flower.	*Clerodendrum indicum* (L.) Kuntze	Verbenaceae	Root is used for the treatment of asthma and cough. Root increases appetite and lowers temperature in fever. Leaves and roots are externally used for the treatment of tumors and skin diseases.	Tuhin et al. (2013); Bhowmik et al. (2014); Haque et al. (2014); Hasan et al. (2014); Kabir & Saha (2014)
54.	*Vait, Bandigach*	Glory bower	*Clerodendrum viscosum*Vent.	Verbenaceae	Leaves are good medicine for diarrhea, gonorrhea, liver disorders, asthma, and headache. Leaf and root is used as anti-dandruff, stomach pain remover, Ascaricide, and laxative. Also used for the treatment of skin diseases, scabies, fever, sores, tumors, and snake bite.	Yusuf et al. (1994); Ghani (2003); Mukul et al. (2007); Rahmatullah & Mollik (2009); Rashid &Eakram (2009); Bhowmik et al. (2014); Hasan et al. (2014); Kabir & Saha (2014); Bari et al. (2017)
55.	*Aparajita*	Butterfly pea	*Clitoreaternatea*(L.) Gaertn	Fabaceae	Leaf extract is used to improve eyesight, promote hair growth, and get rid of different menstruation problems and white discharge. Root is used as laxative and to inhibit glucose intake from diet. Seed is used against sore throat, tumors, ascites, dropsy, and various skin diseases.	Mukul et al. (2007); Rashid &Eakram (2009); Akter et al. (2013);Rahman et al. (2013); Bhowmik et al. (2014); Hasan et al. (2014); Rahman & Jamila (2016); Khatun et al. (2018)
56.	*Telakucha*	Scarlet gourd/Ivy gourd.	*Coccinia indica* L.	Cucurbitaceae	It has anti-helminthic, antioxidant, anti-inflammatory, and analgesic, anti-microbial, anti-hyperglycemic, and hepato-protective activities. Therefore, the plant and its parts are used for a wide range of diseases and disorders. Green leaves and fruits are used in cold ailments, diabetes, earache, jaundice, tongue sores, asthma, skin disease, burns, and fever.	Ghani (2003); Mukul et al. (2007); Rahman (2013); Rahman et al. (2013); Tuhin et al. (2013); Bhowmik et al. (2014); Bari et al. (2017); Kabir & Saha (2014); Rahman & Jamila (2016); Khatun et al. (2018)

TABLE 7.2 *(Continued)*

Sl. No.	Vernacular Name	English Name	Botanical Name	Family	Uses	References
57.	*Narikel*	Coconut	*Cocos nucifera* L.	Arecaceae	Coconut water is used for oral rehydration purposes, blood transfusion, diarrhea, and gastric ulcer. The roots possess anti-pyretic and diuretic properties. Coconut milk is use as laxative and neutralizes the effects of various poisons. The coconut oil is used to cure skin and teeth disease.	Ghani (2003); Mukul et al. (2007); Rahmatullah & Mollik (2009); Rashid &Eakram (2009); Bhowmik et al. (2014); Hasan et al. (2014); Kabir & Saha (2014); Rahman & Jamila (2016); Bari et al. (2017)
58.	*Kola duma*	Day Flower	*Commelinabenghalensis* L.	Commelinaceae	Leaf and root are used for the treatment of external poisoning and burns. The plant is also used to reduce blood pressure and as a poultice in different inflammations.	Mukul et al. (2007); Rahmatullah & Mollik (2009); Rahman et al. (2013); Bhowmik et al. (2014); Kabir & Saha (2014); Rahman & Jamila (2016)
59.	*Shankhahul*	Bindweed	*Convolvulus pluricaulis*	Gentianaceae	Leaf and flower are used in tumor, anxiety, insomnia, neurosis, and epilepsy treatment. Also used in nervous debility, loss of memory and as a brain tonic.	Yusuf et al. (1994); Ghani (2003); Khan & Rashid (2006); Chowdhurya et al. (2009); Rahmatullah & Mollik (2009); Rashid &Eakram (2009); Bhowmik et al. (2014); Mukul et al. (2007); Rahman et al. (2013); Rahman & Jamila (2016)
60.	*Paat*	Jute	*Corchorus capsularis* L.	Tiliaceae	Leaf, stem is used for the treatment of constipation, stomachic, loss of appetite, flatulence.	Ghani (2003); Chowdhurya et al. (2009); Rahmatullah & Mollik (2009); Rashid &Eakram (2009); Bhowmik et al. (2014); Hasan et al. (2014); Kabir & Saha (2014); Rahman & Keya (2015); Rahman & Jamila (2016); Khatun et al. (2018)
61.	*Dhonia*	Coriander	*Coriandrum sativum* L.	Apiaceae	Seed is used for treating different digestion problems like loss of appetite, upset stomach and intestinal gas. Also used against toothaches, worms, and joint ache along with contagions caused by fungus and bacteria.	Yusuf et al. (1994); Rahmatullah & Mollik (2009); Rahman et al. (2013); Tuhin et al. (2013); Bhowmik et al. (2014); Haque et al. (2014); Hasan et al. (2014); Kabir & Saha (2014); Rahman & Jamila (2016); Ahmmed et al. (2017); Bari et al. (2017); Khatun et al. (2018)

TABLE 7.2 *(Continued)*

Sl. No.	Vernacular Name	English Name	Botanical Name	Family	Uses	References
62.	*Mishtikumra*	Sweet gourd	*Cucurbita maxima* Duchesne	Cucurbitaceae	The fruit is used as diuretic, stimulant, infections, and carbuncles. Seeds are used to treat intestinal disorders and kidney complications and to expel worms. Fruit stalk paste is used to heal carbuncles and in earache.	Ghani (2003); Mukul et al. (2007); Rahmatullah & Mollik (2009); Rahman (2013); Bhowmik et al. (2014); Hasan et al. (2014); Kabir & Saha (2014)
63.	*Kachaholud*	Turmeric	*Curcuma longa* L.	Zingiberaceae	Rhizomes possess anti-septic, anti-inflammatory, antioxidant, anti-carcinogenic, anti-mutagenic, anti-coagulant, anti-diabetic, anti-fungal, anti-bacterial, anti-viral, anti-fertility, and anti-venom activities. The plant is used to help digestion, improve intestinal disorders and to treat various skin diseases.	Yusuf et al. (1994); Ghani (2003); Khan & Rashid (2006); Mukul et al. (2007); Rahmatullah & Mollik (2009); Rashid &Eakram (2009); Bhowmik et al. (2014); Hasan et al. (2014); Kabir & Saha (2014); Rahman & Jamila (2016); Bari et al. (2017)
64.	*Shoti*	White turmeric/ zedoary	*Curcuma zedoaria*(Berg) Rosc. E	Zingiberaceae	Rhizome extraction is used for the remedy of stomach pain and other digestive disorders. External application of rhizome as poultice cures wounds and other skin ailments.	Ghani (2003); Khan & Rashid (2006); Mukul et al. (2007); Rahmatullah & Mollik (2009); Bhowmik et al. (2014); Hasan et al. (2014); Kabir & Saha (2014); Rahman & Jamila (2016)
65.	*Swarnalata*	Giant dodder	*Cuscutareflexa*Roxb	Convolvulaceae	Leaf and stem are used for the treatment of hypertension, skin disease and liver cirrhosis.	Rashid &Eakram (2009); Bhowmik et al. (2014); Haque et al. (2014); Rahman & Jamila (2016); Bari et al. (2017)
66.	*Durba*	Bermuda grass	*Cynodondactylon*L.	Poaceae	Used in treatment of convulsions, measles, tumors, cut injuries, warts, fever, and rheumatism.	Yusuf et al. (1994); Ghani (2003); Mukul et al. (2007); Rahmatullah & Mollik (2009); Kabir & Saha (2014); Rahman (2014); Rahman & Jamila (2016)
67.	*Mutha*	Nut grass	*Cyperus rotundus*L.	Cyperaceae	Root is used for treatment of weakness, fever, diarrhea, wound, cholera, vomiting.	Yusuf et al. (1994); Bhowmik et al. (2014); Haque et al. (2014); Kabir & Saha (2014); Rahman & Jamila (2016)

TABLE 7.2 *(Continued)*

Sl. No.	Vernacular Name	English Name	Botanical Name	Family	Uses	References
68.	*Shishu*	Indian rosewood	*Dalbergia sissoo* Roxb	Fabaceae	Leaf, root, and bark is used in inflammation and astringent treatment. Besides, it is used as stimulant and for the remedy of various skin diseases. Leaf extract is used in eye ailments and bark paste is used as anti-pyretic, anti-helminthic, and analgesic.	Rahmatullah & Mollik (2009); Rahman et al. (2013); Tuhin et al. (2013); Bhowmik et al. (2014); Haque et al. (2014); Hasan et al. (2014); Rahman & Keya (2015); Khatun et al. (2018)
69.	*Kaldhutura*	Datura metel/ devil's trumpet	*Datura metel*fastuosa	Solanaceae	Leaf, root, and seed are used for insanity, rheumatic swellings, abscess, and piles.	Ghani (2003); Khan & Rashid (2006); Mukul et al. (2007); Chowdhurya et al. (2009); Bhowmik et al. (2014); Haque et al. (2014); Hasan et al. (2014); Khatun et al. (2018)
70.	*Dhutura*	Jimsonweed/ Devil's snare	*Datura stramonium* L.	Solanaceae	Leaf, root, and seeds are used in asthma, scabies, diarrhea, mental illness and induce unconsciousness.	Ghani (2003); Mukul et al. (2007); Rashid &Eakram (2009); Rahman et al. (2013); Bhowmik et al. (2014); Kabir & Saha (2014); Rahman & Jamila (2016)
71.	*Gajor*	Carrot	*Daucus carota* L.	Apiaceae	Carrot plants is cytotoxic, antioxidant, anti-diabetic, anti-microbial, smooth muscle relaxant, hypotensive effect and decrease intraocular pressure, wound healing, and hear induction and many other effects. Root is appetite stimulant.	Yusuf et al. (1994); Ghani (2003); Rahmatullah & Mollik (2009); Rahman et al. (2013); Bhowmik et al. (2014); Haque et al. (2014); Hasan et al. (2014); Rahman & Keya (2015); Bari et al. (2017); Khatun et al. (2018)
72.	*Chalta*	Elephant apple	*Dillenia indica* Linn.	Dilleniaceae	The fruits are tonic and laxative; used in diarrhea, relief of indigestion, arthritis, asthma, viral infection, jaundice, and physical weakness.	Ghani (2003); Khan & Rashid (2006); Rashid &Eakram (2009); Hasan et al. (2014); Rahman & Jamila (2016); Bari et al. (2017)
73.	*Kalokeshi*	False daisy, Bhringaraj, Trailing eclipta	*Eclipta alba*	Asteraceae	Whole plant is used in various inflammations, bronchitis hernias and eye diseases. Plant extract is used as liver tonic, for diabetes as it lowers blood glucose level and for promotion of hair growth.	Ghani (2003); Mukul et al. (2007); Bhowmik et al. (2014); Kabir & Saha (2014); Rahman & Keya (2015); Rahman & Jamila (2016); Bari et al. (2017)

TABLE 7.2 *(Continued)*

Sl. No.	Vernacular Name	English Name	Botanical Name	Family	Uses	References
74.	*Elachi*	Cardamom	*Elettaria cardamomum* (L.)	Zingiberaceae	Seeds are used as abortifacient, carminative, cardiac tonic, digestive, expectorant, stimulant, and beneficial in bronchitis, hemorrhoids, vesical calculi, anorexia, indigestion, and other gastro-intestinal disorders and in burning sensation.	Khan & Rashid (2006); Mukul et al. (2007); Rashid &Eakram (2009); Rahman et al. (2013); Haque et al. (2014); Kabir & Saha (2014);Rahman & Jamila (2016)
75.	*Mandar*, *madar*	Indian coral tree	*Erythrina indica* L.	Fabaceae	Bark and leaf are used in stomach ailments, anti-abortion medication, malaria fever and liver complications. Barks of *Erythrina indica* and *Terminalia arjuna* are powdered after drying and taken at night as sex stimulant with equal amount of honey.	Ghani (2003); Rahmatullah & Mollik (2009); Rashid &Eakram (2009); Tuhin et al. (2013); Hasan et al. (2014); Kabir & Saha (2014); Rahman & Jamila (2016); Bari et al. (2017)
76.	Khir-*kakoli*	Prostrate spurge	*Euphorbia prostrata*Aiton.	Euphorbiaceae	Whole plant is used to get rid of piles and other hemorrhoids difficulties.	Ghani (2003); Mukul et al. (2007); Rahmatullah & Mollik (2009); Bhowmik et al. (2014); Hasan et al. (2014); Bari et al. (2017); Kabir & Saha (2014)
77.	*Kathbel*	Elephant's foot apple	*Feronia limonia* (L.)	Rutaceae	Fruit and leaves are used as laxative and cures various intestinal disorders. Young leaf juice along with milk and sugar is effective against motion sickness and bowel problems. The unripe fruit is prescribed against diarrhea and dysentery.	Ghani (2003); Khan & Rashid (2006); Mukul et al. (2007); Bhowmik et al. (2014); Haque et al. (2014); Rahman & Jamila (2016); Bari et al. (2017)
78.	*Bot*	Banyan, banyan fig	*Ficus benghalensis*L.	Moraceae	Leaves after crushing are applied for the treatment of boils, sores, and carbuncles. Crumpled roots are applied to hair for lengthening hair in women.	Ghani (2003); Chowdhurya et al. (2009); Rahmatullah & Mollik (2009); Rashid &Eakram (2009); Haque et al. (2014); Hasan et al. (2014); Kabir & Saha (2014); Rahman & Jamila (2016)
79.	*Jogh-dumur*	Cluster fig tree, Indian fig tree	*Ficus racemosa*L	Moraceae	Root, fruit, and bark is prescribed for the treatment of leucorrhea, piles, and diabetes.	Ghani (2003); Bhowmik et al. (2014); Haque et al. (2014); Hasan et al. (2014); Kabir & Saha (2014); Rahman & Jamila (2016)

TABLE 7.2 *(Continued)*

Sl. No.	Vernacular Name	English Name	Botanical Name	Family	Uses	References
80.	*Ashoth*	Sacred fig	*Ficus religiosa* L.	Moraceae	Bark, leaf, and fruit is used for the treatment of asthma, gonorrhea, diabetes, constipation, inflammatory complaints, gastric complications, epilepsy, and sexual ailments.	Yusuf et al. (1994); Rahmatullah & Mollik (2009); Rahman et al. (2013); Tuhin et al. (2013); Bhowmik et al. (2014); Haque et al. (2014); Hasan et al. (2014); Ahmmed et al. (2017); Bari et al. (2017); Kabir & Saha (2014); Khatun et al. (2018)
81.	*Titadima*	Sweet juice	*Glinusoppositifolius*(L.)	Aizoaceae	It is prescribed for treating joint soreness, inflammation, diarrhea, colonic parasites, fever, and malaria. Also used in skin ailments, earache, dyspepsia, and loss of appetite.	Ghani (2003); Mukul et al. (2007); Chowdhurya et al. (2009); Bhowmik et al. (2014); Haque et al. (2014); Hasan et al. (2014); Kabir & Saha (2014); Khatun et al. (2018)
82.	*Motkila, Hotikkhira*	Orange berry and gin berry	*Glycosmis pentaphylla* (Retz.) DC.	Rutaceae	Leaf, stem, and fruit is prescribed for the treatment of liver disorder, anemia, fever, cough, skin disease, boil, arthritis, dysentery, and dental ailments.	Ghani (2003); Rahman et al. (2013); Bhowmik et al. (2014); Kabir & Saha (2014); Bari et al. (2017); Khatun et al. (2018)
83.	*Jastimadhu*	Liquorice	*Glycyrrhiza glabra*	Fabaceae	Root is used to reduce infection of the upper respiratory tract, fever. It reduces throat irritation and helps cases of a chronic cough.	Ghani (2003); Dixie et al. (2003); Mukul et al. (2007); Chowdhurya et al. (2009); Rahmatullah & Mollik (2009); Rahman et al. (2013); Bhowmik et al. (2014); Bari et al. (2017); Haque et al. (2014); Hasan et al. (2014); Kabir & Saha (2014)
84.	*Gamir*	Yemane/Gmelina	*Gmelina arborea* Roxb.	Verbenaceae	Bark and root are used in stomach pain, loss of appetite, piles, burning sensations and leucorrhea. Application of leaf paste remove headache and also beneficial for wound healing. Leaf sap is used for the treatment of gonorrhea and cough. Roots are thought to have tonic, stomachic, and purgative properties. Flowers are used to treat various blood diseases and leprosy.	Rahmatullah & Mollik (2009); Rahman et al. (2013); Bhowmik et al. (2014); Haque et al. (2014); Hasan et al. (2014); Kabir & Saha (2014)

TABLE 7.2 *(Continued)*

Sl. No.	Vernacular Name	English Name	Botanical Name	Family	Uses	References
85.	*Khetroa*	Parpat	*Hedyotiscorymbosa*(L.) Lam.	Rubiaceae	Whole plant is used for the treatment of appendicitis, jaundice, pneumonia, liver complaints, cholecystitis, fever, hepatitis, and urinary infection.	Yusuf et al. (1994); Khan & Rashid (2006); Chowdhurya et al. (2009); Rahman et al. (2013); Bhowmik et al. (2014); Haque et al. (2014); Kabir & Saha (2014); Rahman & Jamila (2016)
86.	*Hatisur*	Indian heliotrope, devil weed, scorpion weed.	*Heliotropium indicum*	Boraginaceae	Whole plant is used for treating warts, ulcers, inflammations, sores, tumors, wounds, gum boils, various skin infections, stings of insects and rheumatism.	Yusuf et al. (1994); Ghani (2003); Khan & Rashid (2006); Mukul et al. (2007); Rashid &Eakram (2009); Rahman et al. (2013); Rahman & Keya (2015); Rahman & Jamila (2016); Bari et al. (2017)
87.	*Sthol-padma*	Changeable rose/Cotton rose mallow	*Hibiscus mutabilis*L.	Malvaceae	Leaf extract is used to reduce inflammation and to control excessive bleeding in menstruation period. Mashed flowers are applied to burns, bulges, and other skin complications.	Yusuf et al. (1994); Ghani (2003); Mukul et al. (2007); Rashid &Eakram (2009); Rahman et al. (2013); Bhowmik et al. (2014); Kabir & Saha (2014); Rahman & Jamila (2016); Das & Islam (2019)
88.	*Joba*	China rose	*Hibiscus rosa-chinensis*	Malvaceae	It is used in the treatment of kidney disease and promotion of hair growth. Roots are medicated in cough, excessive, and painful menstruation, and bronchitis. Flowers and leaves are used as contraceptive, for abortion, checks bleeding and calms irritated skins.	Ghani (2003); Mukul et al. (2007); Rashid &Eakram (2009); Chowdhurya et al. (2009); Rahman et al. (2012, 2013); Kabir & Saha (2014); Rahman & Jamila (2016); Das & Islam (2019)
89.	*Roktojoba*	China rose	*Hibiscus rosa-sinensis* L.	Malvaceae	Leaf, flower is used in the treatment of dysentery, leucorrhea, and other female menstrual problem.	Yusuf et al. (1994); Ghani (2003); Khan & Rashid (2006); Mukul et al. (2007); Rashid &Eakram (2009); Rahman et al. (2013); Bhowmik et al. (2014); Kabir & Saha (2014); Das & Islam (2019)
90.	*Tokma*	Pignut/Chia seed	*Hyptissuaveolens* (Linn) Poit.	Lamiaceae	Seed, leaf is used for gonorrhea, microbial infections treatment. Also used as stimulant, carminative, and (galactagogue) a food or drug that promotes or increases the flow of a mother's milk.	Rashid &Eakram (2009); Rahman et al. (2013); Bhowmik et al. (2014); Haque et al. (2014); Hasan et al. (2014); Kabir & Saha (2014); Rahman & Jamila (2016)

TABLE 7.2 *(Continued)*

Sl. No.	Vernacular Name	English Name	Botanical Name	Family	Uses	References
91.	*Kalmilota*	Water spinach	*Ipomoea aquatic* Forssk	Convolvulaceae	It is used as a laxative in piles, insomnia, and headache. Also used in liver disorders, constipations, diabetes, mental disorder, nose bleeding, hypertension, intestinal disorders, and diabetes.	Yusuf et al. (1994); Mukul et al. (2007); Rahman et al. (2013); Bhowmik et al. (2014); Haque et al. (2014); Hasan et al. (2014); Rahman & Jamila (2016)
92.	*Arenda*, Ban *Verenda*	Barbados nut, poison nut, bubble bush or purging nut.	*Jatropha curcus* Linn.	Euphorbiaceae	Root, stem, leaf, bark, fruit, seed, and latex are used for the treatment of a wide range of diseases and disorders. Leaves are used in rheumatism, fever, jaundice, malaria, in increasing lactation, white discharge and wound healing. Latex is used for wound healing, to treat gum bleeding and tooth ache. Fruit, seed, and seed oil is used as laxative and contraceptive, for arthritis, gout, eczema, and syphilis. Roots are prescribed against eczema, ringworm, dysentery, gonorrhea, and other sexually transmitted diseases.	Yusuf et al. (1994); Khan & Rashid (2006); Chowdhurya et al. (2009); Rahmatullah & Mollik (2009); Rashid &Eakram (2009); Rahman et al. (2013); Bari et al. (2017); Hasan et al. (2014); Rahman & Jamila (2016)
93.	*Bashok*	Malabar nut	*Justicia adhatoda L.*	Acanthaceae	Widely used in cough, bronchitis, and other respiratory disorders. Leaf extract is used in the treatment of menstrual problems, sore throat, asthma, dysentery, jaundice, indigestion, and hepatitis.	Yusuf et al. (1994); Dixie et al. (2003); Ghani (2003); Mukul et al. (2007); Chowdhurya et al. (2009); Rashid &Eakram (2009); Rahman et al. (2013); Bhowmik et al. (2014); Haque et al. (2014); Kabir & Saha (2014); Rahman & Jamila (2016); Bari et al. (2017); Khatun et al. (2018)
94.	*Bishdorun*, Jagatmardan	Willow-Leaf	*Justicia gendarussa* L	Acanthaceae	The leaves are used in the folk medicine for the treatment of various ailments like rheumatism, jaundice, head and earache, dysentery, helminthiasis, wounds, and infections, facial paralysis, and eczema. The roots are used as laxative.	Yusuf et al. (1994); Ghani (2003); Mukul et al. (2007); Chowdhurya et al. (2009); Rashid &Eakram (2009); Rahman et al. (2013); Haque et al. (2014); Rahman & Jamila (2016)

TABLE 7.2 *(Continued)*

Sl. No.	Vernacular Name	English Name	Botanical Name	Family	Uses	References
95.	*Lau*	Bottle gourd	*Lagenaria siceraria* (Molina)	Cucurbitaceae	Pulp of fruit possesses freshening, emetic, purging, diuretic, and anti-bilious properties. Seed oil is cooling and used to remove headache. Decoction of leaves with sugar is used in jaundice. Tender stem is a good remedy for earache. Fruit curry is used in cholera.	Rahmatullah & Mollik (2009); Rashid &Eakram (2009); Rahman (2013); Tuhin et al. (2013); Haque et al. (2014); Rahman & Keya (2015); Rahman & Jamila (2016)
96.	*Jigar*	Indian ash tree	*Lannea grandis*	Anacardiaceae	Meristem is for gum venereal diseases treatment.	Yusuf et al. (1994); Ghani (2003); Khan & Rashid (2006); Rahman et al. (2013); Bhowmik et al. (2014); Haque et al. (2014); Bari et al. (2017); Kabir & Saha (2014); Rahman & Jamila (2016); Rashid &Eakram (2009);
97.	*Bonadi*	Lasia	*Lasia spinosa* (L.) Thwaites	Araceae	Leaf, Root is used for Joint pain, edema, dysentery, increase lactation.	Yusuf et al. (1994); Ghani (2003); Rahmatullah & Mollik (2009); Mukul et al. (2007); Bhowmik et al. (2014); Hasan et al. (2014); Kabir & Saha (2014); Bari et al. (2017);
98.	*Mehandi*	Henna/Henna tree	*Lawsoniainermis*L.	Lythraceae	Leaf keeps head cool and used as anti-dandruff. Seeds used as antipyretic, brain tonic, constipating, intermittent fevers, diarrhea, dysentery, and gastropathy. Root used for burning sensation, leprosy, skin diseases and premature graying of hair treatment.	Yusuf et al. (1994); Ghani (2003); Chowdhurya et al. (2009); Rahmatullah & Mollik (2009); Rashid &Eakram (2009); Rahman et al. (2013); Haque et al. (2014); Rahman & Jamila (2016); Khatun et al. (2018)
99.	*Huria*	Bandicoot berry	*Leea indica* (Burm. f.) Merr.	Vitaceae	Leaf and root are used against eczema and skin diseases, intestinal disorders, and uterus cancer.	Yusuf et al. (1994); Ghani (2003); Khan & Rashid (2006); Rashid &Eakram (2009); Rahman et al. (2013); Bhowmik et al. (2014); Haque et al. (2014); Kabir & Saha (2014); Khatun et al. (2018)

TABLE 7.2 *(Continued)*

Sl. No.	Vernacular Name	English Name	Botanical Name	Family	Uses	References
100.	*Roktodrone*	Honey weed	*Leonurus sibiricus*L	Lamiaceae	Leaf and root are used against diabetes, cancer, and helminthiasis. Leaves are also used for curing menstrual disorders and curing uterus contraction.	Yusuf et al. (1994); Ghani (2003); Khan & Rashid (2006); Rashid &Eakram (2009); Bhowmik et al. (2014); Rahman & Keya (2015); Rahman & Jamila (2016); Bari et al. (2017); Khatun et al. (2018)
101.	*Setodron, Dolkolosh*	Leucas, Thumbe or Thumbai	*Leucas aspera* Willd	Lamiaceae	Leaf, stem, and flower are used in the treatment of eye infection, external hemorrhage, and dysentery. Leaves are used against rheumatism and other skin diseases.	Yusuf et al. (1994); Ghani (2003); Rashid &Eakram (2009); Rahman et al. (2013); Bhowmik et al. (2014); Rahman & Keya (2015)
102.	*Karpur, Ambuja*	Ambulia	*Limnophila indica* (L.)	Scrophulariaceae	Plant is thought to be an antiseptic. Used for the treatment of elephantiasis, pestilent fever, and dysentery.	Yusuf et al. (1994); Ghani (2003); Rahman et al. (2012); Bhowmik et al. (2014); Hasan et al. (2014)
103.	*Kothbel*	Wood-apple and elephant-apple	*Limoniaacidissima*	Rutaceae	Fruit is unique as a tonic for liver, heart, and lungs. It cures cough, asthma, constipation, hiccup, dysentery, tumors, eye disorders and white discharge in women. Spines and barks are used for the treatment of excessive bleeding during menstruation, liver ailments, insect bites and nausea.	Yusuf et al. (1994); Ghani (2003); Rashid &Eakram (2009); Rahman et al. (2013); Bhowmik et al. (2014); Rahman & Keya (2015); Bari et al. (2017); Khatun et al. (2018)
104.	*Lichu*	Lychee	*Litchi chinensis* Sonn.	Sapindaceae	Seed and other plant part is used for indigestion, stomach ulcers, hernia, and loss of appetite, diabetes, obesity, and testicular swelling.	Ghani (2003); Mukul et al. (2007); Chowdhurya et al. (2009); Rahman et al. (2013); Bhowmik et al. (2014); Rahman & Keya (2015)
105.	*Kharajora*	Not available (Hindi – *Meda*)	*Litseamonopetela*	Lauraceae	The leaves are used for the treatment of arthritis, gonorrhea, various skin diseases, carbuncles, chronic dysentery, and diarrhea.	Yusuf et al. (1994); Rahmatullah & Mollik (2009); Rashid &Eakram (2009); Bari et al. (2017); Sarwar (2017)

TABLE 7.2 *(Continued)*

Sl. No.	Vernacular Name	English Name	Botanical Name	Family	Uses	References
106.	*At-tika*	Seedbox	*Ludwigiahyssopifolia*(G Don) Exell.	Onagraceae	Stem is used for the treatment of liver and gallbladder complaints.	Ghani (2003); Mukul et al. (2007); Bhowmik et al. (2014); Haque et al. (2014); Kabir & Saha (2014); Rahman & Jamila (2016); Khatun et al. (2018)
107.	*Kalo-keshari, agni-dolon*	Creeping water primrose	*Ludwigiaprostrata*Roxb.	Onagraceae	Whole plant is used in headache, open wounds, abscess, stomach disorder, leucorrhea, fevers, and bone fractures, measles, and colds ailments.	Yusuf et al. (1994); Ghani (2003); Mukul et al. (2007); Rahmatullah & Mollik (2009); Bhowmik et al. (2014); Hasan et al. (2014); Kabir & Saha (2014); Bari et al. (2017)
108.	Tomato	Tomato	*Lycopersicon esculentum* Mill.	Solanaceae	Tomatoes are used to treat cancer, tumor, osteoporosis, cardiac diseases, rheumatism, and headache. Root decoction is swallowed for getting relief from tooth ache. Used to reduce risk of different types of cancer.	Yusuf et al. (1994); Ghani (2003); Khan & Rashid (2006); Rahmatullah & Mollik (2009); Bhowmik et al. (2014); Kabir & Saha (2014)
109.	*Dhekishak*	Fern	*Lygodiumflexuosum*(L.) Sw.	Lygodiaceae	Whole plant, root, and leaf is used for rheumatic pain, skin disorders, cough, and injuries.	Yusuf et al. (1994); Ghani (2003); Rashid &Eakram (2009); Rahman et al. (2013); Bhowmik et al. (2014); Kabir & Saha (2014); Rahman & Jamila (2016); Khatun et al. (2018)
110.	*Morich-phool*	Turkcap/Turk's turban	*Malvaviscusarboreus* Dill. Ex Cav	Malvaceae	Leaf and flower are used for the treatment of digestive tract inflammation, skin diseases and menstrual disorders.	Yusuf et al. (1994); Rahmatullah & Mollik (2009); Rahman et al. (2013); Tuhin et al. (2013); Bhowmik et al. (2014); Haque et al. (2014); Hasan et al. (2014); Kabir & Saha (2014); Ahmmed et al. (2017); Bari et al. (2017); Das & Islam (2019); Khatun et al. (2018)
111.	*Aam*	Mango	*Mangifera indica* L.	Anacardiaceae	Fruit, root, bark, and leaves are used to cure diarrhea, syphilis, diabetes, boils, kidney stone, tuberculosis, duodenal disorder, blood purification, nasal bleeding, amoebiasis, piles, and heart ailments.	Yusuf et al. (1994); Ghani (2003); Rahman et al. (2013); Bhowmik et al. (2014); Kabir & Saha (2014); Rahman &Jamila (2016); Khatun et al. (2018)

TABLE 7.2 *(Continued)*

Sl. No.	Vernacular Name	English Name	Botanical Name	Family	Uses	References
112.	*Nageswer* Champa	Indian rose chestnut	*Mesua ferrea*	Guttiferae	The plant is used in various infection and septic disorders. Also used for its anti-bacterial, laxative, blood purifier, worm control and tonic properties.	Yusuf et al. (1994); Ghani (2003); Chowdhurya et al. (2009); Rahmatullah & Mollik (2009); Rahman et al. (2013); Tuhin et al. (2013); Bhowmik et al. (2014); Haque et al. (2014); Hasan et al. (2014); Rahman & Keya (2015); Rahman & Jamila (2016); Bari et al. (2017)
113.	*Libujilota*	Climbing hemp weed	*Mikania cordata* (Burm.f.) B.L. Robinson	Asteraceae	Whole plant and leaf are used in dysentery, diabetes, infections, and abdominal ulcer treatment. Crushed leaf is used as poultice in wounds, itches, swelling, and insect bites.	Rashid &Eakram (2009); Rahman et al. (2013); Bhowmik et al. (2014); Haque et al. (2014); Hasan et al. (2014); Kabir & Saha (2014); Rahman & Keya (2015); Bari et al. (2017); Khatun et al. (2018); Das &Islam (2019)
114.	*Lojjaboti*	Sensitive plant/ Touch-me-not	*Mimosa pudica*L.	Fabaceae	Whole plant or root is used for the treatment of piles, infections, dysentery, sinus, prevent gums decaying, urogenital disorders, and scabies.	Yusuf et al. (1994); Ghani (2003); Khan & Rashid (2006); Rahmatullah & Mollik (2009); Rashid &Eakram (2009); Rahman et al. (2013); Tuhin et al. (2013); Haque et al. (2014); Rahman & Jamila (2016); Khatun et al. (2018)
115.	*Korolla*	Bitter gourd	*Momordica charantia*L.	Cucurbitaceae	Regular consumption of bitter gourd extract increases body stamina and prevents prolonged fatigue. Leaf and fruit juice is used in diabetes, helminthiasis, fever, HIV, and AIDS, infectious jaundice, intestinal disorders, and constipation.	Yusuf et al. (1994); Rahman et al. (2013); Haque et al. (2014); Kabir & Saha (2014); Rahman & Jamila (2016); Bari et al. (2017); Khatun et al. (2018)
116.	*Kakroal*	Chinese cucumber/Spiny bitter cucumber.	*Momordica cochinchinensis* (Lour.) Spreng.	Cucurbitaceae	Leaf and fruit paste is applied externally for back pain, ulceration, and bone fracture. Crushed roots are medicated in rheumatism. Seeds are used for ulcers, wounds, and obstructions of liver and spleen.	Rahmatullah & Mollik (2009); Rahman (2013); Bhowmik et al. (2014); Haque et al. (2014); Hasan et al. (2014); Kabir &Saha (2014); Rahman & Jamila (2016); Ahmmed et al. (2017)

TABLE 7.2 *(Continued)*

Sl. No.	Vernacular Name	English Name	Botanical Name	Family	Uses	References
117.	*Sajna*	Drumstick tree	*Moringa oleifera*	Moringaceae	Leaves after frying with garlic, cinnamon, and cloves are taken as sex tonic, headache, coughs, and discharge of mucus. Fruits are used to protect and nourish skin and hair, cardiovascular system, helps in wounds healing. Also cure diabetes, edema, and asthma, reduces blood pressure, and makes healthier bones. Used against bacterial diseases, various stomach ailments and cancer.	Yusuf et al. (1994); Ghani (2003); Khan & Rashid (2006); Mukul et al. (2007); Rashid &Eakram (2009); Rahman et al. (2013); Hasan et al. (2014); Kabir & Saha (2014); Rahman & Jamila (2016); Ahmmed et al. (2017); Khatun et al. (2018)
118.	Atia Kala	Banana, plantain	*Musa sapientum* Linn.	Musaceae	Roots, leaves, and fruits are medicated against cholera, dysentery, diarrhea, and constipation. Several disorders and ailments such as hypertension, headache, cancer, high cholesterol, and high blood sugar are controlled through regular uptake of fruit.	Yusuf et al. (1994); Ghani (2003); Mukul et al. (2007); Rahmatullah & Mollik (2009); Bhowmik et al. (2014); Hasan et al. (2014); Kabir & Saha (2014); Rahman & Jamila (2016); Bari et al. (2017)
119.	*Korobi*	Oleander	*Nerium oleander*	Apocynaceae	Root and root bark extract is effective against different types of cancers and ulcers on the male genitals. Also used in abdominal and joint pain, painful menstrual periods, asthma, ringworm, dyspepsia, and to cause abortions.	Ghani (2003); Mukul et al. (2007); Kabir & Saha (2014); Rahman & Keya (2015); Rahman & Jamila (2016); Sarwar (2017)
120.	*Bontamak*	Tex-Mex tobacco	*Nicotiana plumbaginifolia*var. bigelovii	Solanaceae	It is used in several human complaints like in treating piles, swelling, and rheumatic fever. It acts as analgesic that possesses larvicidal and anti-bacterial characteristics.	Ghani (2003); Mukul et al. (2007); Nawaz et al. (2009); Rahmatullah & Mollik (2009); Rahman et al. (2013); Bhowmik et al. (2014);Haque et al. (2014); Kabir & Saha (2014);Khatun et al. (2018)

TABLE 7.2 *(Continued)*

Sl. No.	Vernacular Name	English Name	Botanical Name	Family	Uses	References
121.	*Kalojira*, *Kalijira*	Black cumin	*Nigella sativa* Linn.	Ranunculaceae	The seeds have stimulant, anti-carcinogenic, anti-microbial, anti-diabetic, anti-helminthic, analgesic, anti-inflammatory, and diuretic characters. The seeds and oils are used mainly for a wide spectrum of ailments like asthma, arthritis, cough, various inflammation, liver, and gastro-intestinal disorders. Besides, it is also used in reducing blood sugar and cancers.	Yusuf et al. (1994); Ahsan et al. (1997); Ghani (2003); Khan & Rashid (2006); Chowdhurya et al. (2009); Rashid &Eakram (2009); Rahman et al. (2013); Bhowmik et al. (2014); Haque et al. (2014); Hasan et al. (2014); Rahman & Jamila (2016); Ahmmed et al. (2017); Bari et al. (2017); Khatun et al. (2018)
122.	*Shefali*/*Sheuli*	Parijat, Hengrabubar, night-flowering jasmine	*Nyctanthesarboritristis*L.	Oleaceae	It cures allergy, arthritis, spleen enlargement, fever, blood dysentery, cough, gastritis, and joint inflammation. Also acts as immuno-stimulant, anti-viral, anti-fungal, and as a laxative. Seeds are used to cure scalp infection and various skin ailments.	Ghani (2003); Mukul et al. (2007); Chowdhurya et al. (2009); Rahman et al. (2013); Bhowmik et al. (2014); Haque et al. (2014); Kabir &Saha (2014); Ahmmed et al. (2017)
123.	*Lal shapla*	Water lily	*Nymphaea nouchali*Burm.f.	Nymphaeaceae	Flowers are used in menstruation disorders. Flower decoction is used in heart ailments and as a blood purifier. Raw or powdered rhizome is used for the treatment of fevers, diabetes, cystitis, dysentery, nephritis, and insomnia. Powdered root stock is used for stomachache, diarrhea, piles, and urinary infections. Decanted water of overnight soaked flower is prescribed for different cardiac complications. Poultice of soaked seeds is used for various skin infection and wounds.	Yusuf et al. (1994); Ghani (2003); Rahman et al. (2013); Hasan et al. (2014); Rahman & Jamila (2016); Bari et al. (2017); Khatun et al. (2018)

TABLE 7.2 *(Continued)*

Sl. No.	Vernacular Name	English Name	Botanical Name	Family	Uses	References
124.	*Tulsi*	Basil/Holy basil	*Ocimum sanctum* L.	Lamiaceae	Stem, leaf, flower, seed, and root of this plant are used as medicine. Leaf extract is used to prevent fever, cough, cold, asthma, dysentery, and indigestion. It is also prescribed in diabetes, heart ailments, to cure various infectious diseases and to lower cholesterol level. Tulsi is also prescribed as anti-carcinogenic and anti-inflammatory medications.	Khan & Rashid (2006); Rashid &Eakram (2009); Haque et al. (2014); Rahman & Jamila (2016); Khatun et al. (2018)
125.	*Thona*	Midnight horror, Oroxylum	*Oroxylum indicum* Vent.	Bignoniaceae	Bark, seed, stem for jaundice, diarrhea, scabies diseases.	Ghani (2003); Mukul et al. (2007); Rahmatullah & Mollik (2009); Rahman et al. (2013); Bhowmik et al. (2014); Hasan et al. (2014); Kabir & Saha (2014)
126.	*Dhanpata*	Rice	*Oryza sativa* L.	Poaceae	Leaf used for piles, loss of appetite, scurvy, flatulence.	Yusuf et al. (1994); Ghani (2003); Mukul et al. (2007); Rashid &Eakram (2009); Rahman et al. (2013); Bhowmik et al. (2014); Hasan et al. (2014)
127.	*Gondhobadali*	Skunk vine, Chinese fever vine	*Paederiafetida*L.	Rubiaceae	Leaf, root, fruit is used in gastritis, weakness, cancer, toothache, paralysis, abscesses, gout, dysentery, rheumatism. Leaves are used widely as poultice for relieving various infections.	Ghani (2003); Mukul et al. (2007); Rahmatullah & Mollik (2009); Rahman et al. (2013); Bhowmik et al. (2014); Hasan et al. (2014); Kabir & Saha (2014); Rahman & Jamila (2016); Bari et al. (2017)
128.	*Keyakanta*	Walnut	*Pandanus foetidusRoxb.*	Pandanaceae	Root is for diarrhea, central nervous system stimulant.	Ghani (2003); Mukul et al. (2007); Chowdhurya et al. (2009); Rahman et al. (2013); Bhowmik et al. (2014); Haque et al. (2014); Hasan et al. (2014); Kabir & Saha (2014); Rahman & Jamila (2016)

TABLE 7.2 *(Continued)*

Sl. No.	Vernacular Name	English Name	Botanical Name	Family	Uses	References
129.	*Amloki*	Aonla, Emblica	*Phyllanthus emblica*	Euphorbiaceae	Bark, Fruits are used in diabetes, dysentery, venereal diseases. Also used for skin diseases, hair falls, indigestion. It is an ingredient of 'Triphala.'	Yusuf et al. (1994); Khan & Rashid (2006); Mukul et al. (2007); Rashid &Eakram (2009); Bhowmik et al. (2014); Rahman & Jamila (2016); Sarwar (2017); Khatun et al. (2018)
130.	*Vuiamloki*	Gulf Leaf-Flower	*Phyllanthus fraternus*G.L. Webster	Euphorbiaceae	Whole plant, root, and leaf is used for treatment of fever, joint ache, gonorrhea, malaria, diarrhea, and liver or gall bladder complaint. External application of leaf chucks out skin infections.	Yusuf et al. (1994); Rahmatullah & Mollik (2009); Rahman et al. (2013); Tuhin et al. (2013); Bhowmik et al. (2014); Haque et al. (2014); Hasan et al. (2014); Kabir & Saha (2014); Ahmmed et al. (2017); Bari et al. (2017); Khatun et al. (2018)
131.	*Paanpata*	Betel leaf	*Piper betle* L.	Piperaceae	Use of betel leaf makes gum stronger, treat dyspepsia, remove constipation, and get rid of congestion, coughs, and asthma. Fresh leaves are chewed for improving digestion and to eliminate filthy odor from mouth. Macerated leaf or juice when applied on wounds, boils or bruises acts as antibiotic and stops bleeding.	Yusuf et al. (1994); Khan & Rashid (2006); Chowdhurya et al. (2009); Rahmatullah & Mollik (2009); Rashid &Eakram (2009); Tuhin et al. (2013); Rahman & Jamila (2016)
132.	*Bonpipul*	Choi Jhal	*Piper chaba*W. Hunter.	Piperaceae	Whole plant, fruit, and root is used in urological problems, weakness, for wound healing, bronchitis, skin, liver, and stomach ailments and rheumatic pain.	Yusuf et al. (1994); Ghani (2003); Mukul et al. (2007); Rahmatullah & Mollik (2009); Bhowmik et al. (2014); Hasan et al. (2014); Kabir & Saha (2014); Bari et al. (2017)
133.	*Kabab-chini*	Cubeb or tailed pepper	*Piper cubeba* L.	Piperaceae	Whole plant is used in irregular and painful bleeding during menstruation, amoebic dysentery, abdominal gas, gonorrhea, and cancer.	Yusuf et al. (1994); Ghani (2003); Khan & Rashid (2006); Mukul et al. (2007); Rahmatullah & Mollik (2009); Rahman et al. (2013); Bhowmik et al. (2014); Sarwar (2017)
134.	*Pipul*, *Vutsan*	Indian long pepper or Pipli	*Piper longum* L	Piperaceae	Plant, fruit, and root is used for cough, gonorrhea, paralysis of the tongue, bronchitis, dyspepsia, constipation, hepatitis, respiratory tract infections and tuberculosis treatment.	Ghani (2003); Khan & Rashid (2006); Rahman et al. (2013); Bhowmik et al. (2014); Haque et al. (2014); Hasan et al. (2014); Kabir & Saha (2014)

TABLE 7.2 *(Continued)*

Sl. No.	Vernacular Name	English Name	Botanical Name	Family	Uses	References
135.	*Golmorich, Kalimarich*	Black pepper	*Piper nigrum* L.	Piperaceae	Pepper is used to cure constipation, diarrhea, ear pain, gangrene, heart ailments, hernia, indigestion, insect bites, insomnia, joint ache, liver ailments, lung disorders, sunburn, tooth decay and aches.	Yusuf et al. (1994); Ahsan et al. (1997); Ghani (2003); Khan & Rashid (2006); Chowdhurya et al. (2009); Rashid &Eakram (2009); Rahman et al. (2013); Bhowmik et al. (2014); Haque et al. (2014); Hasan et al. (2014); Rahman & Jamila (2016); Bari et al. (2017); Khatun et al. (2018)
136.	*Chitamool, Chutrapata*	Ceylon leadwort	*Plumbago zeylanica*L.	Plumbaginaceae	Root is used for ulcer, gastro-intestinal disorder and as antiseptic. Leaves are used in antifertility.	Ghani (2003); Mukul et al. (2007); Rahmatullah & Mollik (2009); Rahman et al. (2013); Bhowmik et al. (2014); Haque et al. (2014); Hasan et al. (2014); Rahman & Jamila (2016); Bari et al. (2017)
137.	*Biskatali*	Water pepper	*Polygonum hydropiper* L.	Polygonaceae	Green leaves are used against stomach pain and insect bites. Crushed plant as a poultice is used in inflamed areas. Plant is used to regulate menstrual disorders, dyspepsia, scabies, and termination of pregnancy. Leaf extract is used for treating head and toothache, liver expansion, duodenal ulcer, and loss of appetite. Roots are used as tonic and their extract is medicated to cure infections, various skin disorder and painful sores, carbuncles, or boils.	Ghani (2003); Khan & Rashid (2006); Mukul et al. (2007); Rashid &Eakram (2009); Haque et al. (2014); Hasan et al. (2014); Kabir & Saha (2014); Rahman & Jamila (2016); Bari et al. (2017); Khatun et al. (2018)
138.	*Goniari*	Kulawin	*Premna integrifolia* L.	Verbenaceae	Leaf, root fever is used for the treatment of cold, inflammation, rheumatism.	Ghani (2003); Rahmatullah & Mollik (2009); Rahman et al. (2013); Bhowmik et al. (2014); Haque et al. (2014); Hasan et al. (2014)
139.	*Piyara*	Guava	*Psidium guajava* L.	Myrtaceae	Fruits as well as leaves has many health benefits viz, anti-diarrheal, anti-hypertensive, anti-lipidemic, anti-carcinogenic.	Yusuf et al. (1994); Bhowmik et al. (2014); Hasan et al. (2014); Kabir & Saha (2014); Rahman & Jamila (2016); Sarwar (2017)

TABLE 7.2 *(Continued)*

Sl. No.	Vernacular Name	English Name	Botanical Name	Family	Uses	References
140.	*Raktochandan*	Red sandalwood	*Pterocarpus santalinus*L.f.	Fabaceae	Fruits and seeds are used in skin diseases, cancer, wounds, digestive tract problems, fluid retention. Bark is used in tuberculosis, debility, burning sensations during urination.	Yusuf et al. (1994); Ghani (2003); Mukul et al. (2007); Rahmatullah & Mollik (2009); Bhowmik et al. (2014); Hasan et al. (2014); Kabir & Saha (2014); Bari et al. (2017); Khatun et al. (2018)
141.	*Dalim*	Pomegranate	*Punica granatum* L.	Lythraceae	Leaf and bark are used for inflammation, heart disorder, loss of appetite, blood dysentery, breast cancer and skin disease treatment.	Ghani (2003); Rahman et al. (2012); Kabir & Saha (2014); Rahman & Jamila (2016); Sarwar (2017); Khatun et al. (2018)
142.	Palik, *Podika, Kandir*	Celery-leaved buttercup	*Ranunculus sceleratus L.*	Ranunculaceae	Stimulant, diuretic, vermifuge, kidney troubles, sciatica, rheumatism, rubefacient, antispasmodic, tonic, asthma, pneumonia.	Ghani (2003); Mukul et al. (2007); Rahman et al. (2012, 2013); Haque et al. (2014); Kabir & Saha (2014); Khatun et al. (2018); Rahman & Keya (2015); Rahman &Jamila (2016)
143.	*Mula*	Radish	*Raphanus sativus* L	Cruciferae	Regular consumption increases digestion and reduces acidity. Crushed roots are used as a poultice on burn and bruises area.	Yusuf et al. (1994); Ghani (2003); Rahmatullah & Mollik (2009); Rashid &Eakram (2009); Rahman et al. (2013); Bhowmik et al. (2014); Rahman & Jamila (2016)
144.	*Shorpogonda*	Serpentine	*Rauvolfia serpentina* L.	Apocynaceae	It is effective against systolic hypertension and psychiatric disarrays like nervousness, schizophrenia, epilepsy, insomnia, and insanity. Also used as sedative, in irregular and painful menstruation, joint pain, edema, snake bite, constipation, and liver problems.	Yusuf et al. (1994); Khan & Rashid (2006); Rahmatullah & Mollik (2009); Sajib &Uddin (2013); Haque et al. (2014); Hasan et al. (2014); Rahman & Jamila (2016)
145.	*Rashna*	Cat-tail Orchid	*Rhynchostylis retusa* Blume.	Orchidaceae	Leaf and root are used for rheumatism, wound, cut, fungus infections and blood dysentery.	Chowdhurya et al. (2009); Bhowmik et al. (2014); Haque et al. (2014); Hasan et al. (2014); Kabir & Saha (2014)

TABLE 7.2 *(Continued)*

Sl. No.	Vernacular Name	English Name	Botanical Name	Family	Uses	References
146.	*Arondomul, Venna, Reri, Bherenda*	Castor bean	*Ricinus communis* L.	Euphorbiaceae	Leaf and seed are used to cure from gastric ulcer, liver or gallbladder ailment, constipation, head, and joint ache.	Yusuf et al. (1994); Ghani (2003); Mukul et al. (2007); Chowdhurya et al. (2009); Rahman et al. (2013); Bhowmik et al. (2014); Haque et al. (2014); Hasan et al. (2014); Kabir & Saha (2014); Rahman & Jamila (2016)
147.	*Golapphool*	Damask rose	*Rosa damascene* Mill.	Rosaceae	Leaf and flower are used for abdominal and chest pains, menstrual problems, strengthening heart, skin spots, digestive disorders, and constipation.	Yusuf et al. (1994); Ghani (2003); Khan & Rashid (2006); Mukul et al. (2007); Rahmatullah &Mollik (2009); Rahman et al. (2013); Bhowmik et al. (2014); Hasan et al. (2014)
148.	*Moyda*	Kans grass	*Saccharum spontaneum*L.	Poaceae	Plant extract is used in digestive disorder, sexual weakness, burning sensation, piles, to stop vomiting, gynecological, and respiratory troubles.	Yusuf et al. (1994); Mukul et al. (2007); Rahmatullah & Mollik (2009); Rahman et al. (2013); Bhowmik et al. (2014); Kabir & Saha (2014); Rahman & Jamila (2016)
149.	*Shethchondon*	Indian sandalwood	*Santalum album* L.	Santalaceae	Bark is used for tuberculosis, weakness, and burning sensations during urine discharge.	Ghani (2003); Khan & Rashid (2006); Chowdhurya et al. (2009); Rahman et al. (2013); Bhowmik et al. (2014); Haque et al. (2014); Hasan et al. (2014); Bari et al. (2017); Kabir & Saha (2014)
150.	*Ashok*	Ashoka tree	*Saracaasoca*	Caesalpinaceae	Barks are used in anti-tumor, menorrhagia, bleeding hemorrhoids, dysentery, and ulcers. Leaves are used for purification of blood. Flowers are used in dysentery, unsettled stomach, syphilis, and as uterine tonic.	Ghani (2003); Khan & Rashid (2006); Mukul et al. (2007); Rahmatullah & Mollik (2009); Akter et al. (2013);Bhowmik et al. (2014); Hasan et al. (2014); Khatun et al. (2018); Rahman et al. (2013); Tuhin et al. (2013)
151.	*Modhumaloti, chinipata*	Licorice weed, goat weed,	*Scoparia dulcis* L.	Scrophulariaceae	Diarrhea, dysentery in adults or children. Leaves are slightly crushed, and the resultant juice is orally taken.	Yusuf et al. (1994); Ghani (2003); Mukul et al. (2007); Rahman et al. (2013); Bhowmik et al. (2014); Hasan et al. (2014); Khatun et al. (2018)

TABLE 7.2 *(Continued)*

Sl. No.	Vernacular Name	English Name	Botanical Name	Family	Uses	References
152.	*Kalkasunda*	Coffee Senna	*Senna occidentalis* L.	Fabaceae	The plant is used as sedative and alterative. Also used for rheumatism, fever, and other venereal diseases. Lotion prepared from the plant is used to treat ringworm and other skin ailments.	Yusuf et al. (1994); Ghani (2003); Mukul et al. (2007); Rahman et al. (2013); Hasan et al. (2014); Kabir & Saha (2014); Bari et al. (2017); Khatun et al. (2018)
153.	*Chotokalkasunda*	Sophera	*Senna sophera* L.	Fabaceae	Leaves are used in coughs, asthma, bronchitis, and hiccup.	Ghani (2003); Rahmatullah & Mollik (2009); Hasan et al. (2014); Rahman & Jamila (2016); Ahmmed et al. (2017); Bari et al. (2017)
154.	*Bailodi*	Rhombus-leaved Sida, Paddy's lucerne, jelly leaf	*Sidarhombifolia*L.	Malvaceae	Crushed leaves are used to get relief from inflammation. Stem and root are used treat rheumatism. Fruit is used against headache and the mucilage is prescribed as a balm.	Chowdhurya et al. (2009); Rahmatullah & Mollik (2009); Bhowmik et al. (2014); Haque et al. (2014); Hasan et al. (2014); Kabir & Saha (2014); Das & Islam (2019)
155.	*Komariamul/ Kumari lota*	Kumarika	*Smilax ovalifolia*Roxb.	Smilacaceae	Root is used for the treatment of wounds and to keep liver healthy. Young leaves after fried in ghee are taken orally for 7 days as sex stimulant.	Ghani (2003); Mukul et al. (2007); Rahman et al. (2012, 2013); Tuhin et al. (2013); Bhowmik et al. (2014); Haque et al. (2014); Kabir & Saha (2014); Rahman & Keya (2015); Khatun et al. (2018)
156.	*Brihoti*	Black nightshade	*Solanum indicum* L.	Solanaceae	Leaf and stem are used for the treatment of cold, cough, rheumatism, sore throat, skin diseases and asthma. Also used as expectorant and stimulant.	Yusuf et al. (1994); Rahmatullah & Mollik (2009); Rahman et al. (2013); Tuhin et al. (2013); Bhowmik et al. (2014); Haque et al. (2014); Hasan et al. (2014); Kabir & Saha (2014); Bari et al. (2017); Khatun et al. (2018)
157.	*Begun*	Eggplant	*Solanum melongena* L.	Solanaceae	Leaf, fruit is used in the treatment of obesity, vomiting, cough, loss of appetite.	Ghani (2003); Bhowmik et al. (2014); Haque et al. (2014); Hasan et al. (2014); Rahman et al. (2013); Kabir & Saha (2014); Rahman & Keya (2015)
158.	*Kakmachi*	Black night shade	*Solanum nigrum* L.	Solanaceae	Leaf, fruit is used for weakness, edema, fever, and diarrhea.	Yusuf et al. (1994); Mukul et al. (2007); Rahman et al. (2013); Bhowmik et al. (2014); Hasan et al. (2014); Kabir & Saha (2014)

TABLE 7.2 *(Continued)*

Sl. No.	Vernacular Name	English Name	Botanical Name	Family	Uses	References
159.	Kondikari	Surattense nightshade/ yellow-fruit nightshade	*Solanum SurattenseBurm.f.*	Solanaceae	Whole plant, leaf, seed, and root is used for the treatment of venereal disorders, bronchitis, chest, and joint pain.	Ghani (2003); Chowdhurya et al. (2009); Rahmatullah & Mollik (2009); Rashid &Eakram (2009); Rahman et al. (2013); Bhowmik et al. (2014); Haque et al. (2014); Rahman & Jamila (2016)
160.	*Bichuti*	Poison berry	*Solanum violaceum*Ortega.	Solanaceae	Leaf, root is used for vomiting, asthma, toothache, fever, helminthiasis.	Ghani (2003); Khan & Rashid (2006); Mukul et al. (2007); Rahman et al. (2012, 2013); Haque et al. (2014); Kabir & Saha (2014); Rahman & Keya (2015); Khatun et al. (2018)
161.	*Bishkochu*	Wild lettuce/Great Lettuce	*Stendneravirosa*Prain.	Araceae	Root is used for the treatment of rheumatism. Besides, it is prescribed to get rid of cholera, dysentery, diarrhea, digestive complications, menstrual irregularities, infertility problem, hernia, and venereal ailments.	Ghani (2003); Khan & Rashid (2006); Mukul et al. (2007); Rahmatullah & Mollik (2009); Rahman et al. (2013); Tuhin et al. (2013); Bhowmik et al. (2014); Hasan et al. (2014); Khatun et al. (2018)
162.	*Muchilat*	Snake vine	*Stephania japonica* (Thumb.) Miers	Menispermaceae	Whole plants of the climber are used as plaster on injury and in jaundice.	Mukul et al. (2007); Rahmatullah & Mollik (2009); Rahman et al. (2013); Rahman & Jamila (2016); Bari et al. (2017)
163.	*Modiani, mucchani*	Snake vine	*Stephania japonica* (Thunb. Ex Murray) Miers.	Menispermaceae	Whole plant, leaf, and root are used for the treatment of fever, diarrhea, and urinary ailments.	Yusuf et al. (1994); Ghani (2003); Mukul et al. (2007); Bhowmik et al. (2014); Kabir & Saha (2014); Rahman & Jamila (2016)
164.	*Sheora, Hapra*	Siamese rough bush, Khoi, serut, and toothbrush tree	*Streblus asper* Lour.	Moraceae	Leaves are used for skin diseases, piles, and urinary disorders. Whole plant is used in the treatment of fever, filaria, chickenpox, lymphedema, diarrhea, and rheumatic fever.	Yusuf et al. (1994); Ghani (2003); Mukul et al. (2007); Rahman et al. (2013); Bhowmik et al. (2014); Kabir & Saha (2014); Rahman & Keya (2015); Bari et al. (2017)

TABLE 7.2 *(Continued)*

Sl. No.	Vernacular Name	English Name	Botanical Name	Family	Uses	References
165.	*Chirata*	Bitter stick, Chirayta	*Swertia chirata*Buch.-Ham. ex Wall	Gentianaceae	Stems and leaves after soaking overnight into water, the decanting is used for removing worms and also prescribed against allergy, dysentery, blood purifier, fever, loss of appetite, gastro-intestinal disorders, diabetes, and skin diseases.	Khan & Rashid (2006); Mukul et al. (2007); Rahmatullah & Mollik (2009); Rashid et al. (2014); Rahman & Jamila (2016); Bari et al. (2017)
166.	*Jam*	Black berry	*Syzygiumcumini* (L.)	Myrtaceae	Black berry possesses anti-carcinogenic, anti-microbial, anti-inflammatory, anti-diabetic, anti-pyretic, antioxidant, anti-viral, anti-diarrheal, and antihistamine activities. Barks, leaves, and seed-powder are given orally to cancer patient and to diabetic patients for reducing blood sugar level.	Ghani (2003); Mukul et al. (2007); Rashid &Eakram (2009); Rahman et al. (2013); Sajib & Uddin (2013); Haque et al. (2014); Rahman & Jamila (2016); Bari et al. (2017)
167.	*Gadaful*	African marigold	*Tagetes erecta*L.	Asteraceae	Leaves and flowers are used to cure fever, skin diseases, liver ailments and eye diseases. Flower extract is used as a remedy for bleeding piles, colds, rheumatism, and asthma. Poultice of leaf stops hemorrhage and soreness.	Ghani (2003); Khan & Rashid (2006); Mukul et al. (2007); Rahmatullah & Mollik (2009); Rahman et al. (2013); Tuhin et al. (2013); Bhowmik et al. (2014); Haque et al. (2014); Hasan et al. (2014); Rahman & Jamila (2016); Khatun et al. (2018)
168.	*Tetul*	Tamarind	*Tamarindus indica* L.	Fabaceae	Leaf extract is taken as a remedy for burning sensations at the time of urination. Young leaves after boiling in water and the decanted water is taken for the treatment of dysentery.	Yusuf et al. (1994); Rahmatullah & Mollik (2009); Bhowmik et al. (2014); Haque et al. (2014); Hasan et al. (2014); Kabir & Saha (2014); Rahman & Jamila (2016); Bari et al. (2017); Sarwar (2017); Khatun et al. (2018)
169.	*Arjun*	Arjuna/arjun tree	*Terminalia arjuna*	Combretaceae	Bark is used as cardio-tonic and treatment of asthma, dysentery, heart disorder, sexual diseases, skin disease.	Khan & Rashid (2006); Mukul et al. (2007); Rahmatullah & Mollik (2009); Rashid &Eakram (2009); Sajib & Uddin (2015); Rahman & Jamila (2016); Sarwar (2017)

TABLE 7.2 *(Continued)*

Sl. No.	Vernacular Name	English Name	Botanical Name	Family	Uses	References
170.	*Bohera*	Bedda nut tree, Beach almond	*Terminalia belerica*(Gaertn.) Roxb.	Combretaceae	Fruit is used against different microbial infections, asthma, allergy, liver complaints, bronchitis, cough, and sore throat. Besides, fruit powder or soaked water is effective for constipation, various stomach disorders, eye infection and loss of appetite. It is an ingredient of 'Triphala.'	Ghani (2003); Mukul et al. (2007); Yusuf et al. (1994); Khan & Rashid (2006); Chowdhurya et al. (2009); Rahmatullah & Mollik (2009); Rashid &Eakram (2009); Rahman et al. (2013); Bhowmik et al. (2014); Haque et al. (2014); Hasan et al. (2014); Rahman & Jamila (2016); Ahmmed et al. (2017); Bari et al. (2017); Khatun et al. (2018)
171.	*Horitaki*	Myrobalan	*Terminalia chebula*Retz	Combretaceae	Fruit is used in diabetes, constipation, fever, heart disease, cough, dysentery, urinary problems, and loss of appetite. Also used widely in many contagious diseases such as prolonged ulcers, white discharge, pyorrhea, and mycological infections of skin. Fruit, bark, and leaves are used as medicine. It is an ingredient of 'Triphala.'	Yusuf et al. (1994); Ghani (2003); Khan & Rashid (2006); Mukul et al. (2007); Chowdhurya et al. (2009); Rahmatullah & Mollik (2009); Rashid &Eakram (2009); Rahman et al. (2013); Bhowmik et al. (2014); Haque et al. (2014); Hasan et al. (2014); Kabir & Saha (2014); Rahman & Jamila (2016); Bari et al. (2017); Sarwar (2017)
172.	*Amgorancha, Poddogurunchi*	Heart-leaved moonseed	*Tinospora cordifolia* (Willd.)	Menispermaceae	Branch of the climber and leaves are used as anti-pyretic, blood purifier and blood sugar level reducer. The stem is cut into small pieces and soaked overnight in water. If the water is taken into empty stomach – it increases immunity of the body, enhances function of liver and kidney, get rid of anemia, reduces inflammation, and increases digestion power. Paste of leaves when used as poultice on wound it fills quickly.	Yusuf et al. (1994); Ghani (2003); Rashid &Eakram(2009); Rahman et al. (2013); Haque et al. (2014); Hasan et al. (2014); Kabir & Saha (2014); Rahman & Jamila (2016); Bari et al. (2017); Khatun et al. (2018)

TABLE 7.2 *(Continued)*

Sl. No.	Vernacular Name	English Name	Botanical Name	Family	Uses	References
173.	*Guloncho*	Heart-leaved moonseed, gaduchi	*Tinospora sinensis* (Lour.) Merrill.	Menispermaceae	Different parts of the plant are used against fever, human malignant cancer, and jaundice. Tablets made from mashed roots and leaves to be taken 3 times daily. Removes disease of urinary tracts infection, fatigue, physical as well as mental weakness.	Yusuf et al. (1994); Ghani (2003); Khan & Rashid (2006); Mukul et al. (2007); Chowdhurya et al. (2009); Rahman et al. (2013); Haque et al. (2014); Hasan et al. (2014); Kabir & Saha (2014); Ahmmed et al. (2017); Sarwar (2017);
174.	*Antamul*	Indian Ipecac/ emetic swallowwort	*Tylophora indica* (Burm.f.) Merr.	Asclepiadaceae	It is commonly used to treat asthma, respiratory infections and also use as a remedy for diabetes. Roots are used in leucorrhea. The plant or plant parts are used for the treatment of bronchial asthma, whooping cough, cancer, dysentery, hay fever, inflammation, rheumatism, arthritis, and dermatitis. Leaves and roots of the plant are known to have emetic, diaphoretic, cathartic, expectorant, laxative, and cleansing properties.	Yusuf et al. (1994); Ghani (2003); Rahmatullah & Mollik (2009); Rashid &Eakram (2009); Rahman et al. (2013); Bhowmik et al. (2014); Haque et al. (2014); Hasan et al. (2014); Kabir & Saha (2014); Rashid et al. (2014); Bari et al. (2017); Khatun et al. (2018)
175.	*Kalakochu*	Giant Typhonium	*Typhoniumgiganteum* Engl	Araceae	Whole plant and stem are used to stop flow of blood and kidney stones treatment.	Yusuf et al. (1994); Rahmatullah & Mollik (2009); Rahman et al. (2013); Tuhin et al. (2013); Bhowmik et al. (2014); Haque et al. (2014); Hasan et al. (2014); Kabir & Saha (2014); Rashid et al. (2014); Ahmmed et al. (2017); Bari et al. (2017); Khatun et al. (2018)
176.	*Bon okra, Bilaikani*	Caesar weed, Congo jute	*Urena lobata* L.	Malvaceae	Root is used as a remedy of back pain and rheumatism. Stem and root decoction is used in indigestion. In dry coughs, flowers are used as expectorant. Infusion of flowers as a gargle is also used for sore-throat. Leaf, root, and flower is used for the treatment of malaria, gonorrhea, infections, tooth, and earache.	Ghani (2003); Mukul et al. (2007); Chowdhurya et al. (2009); Rahmatullah &Mollik (2009); Rashid &Eakram (2009); Rahman et al. (2013); Bhowmik et al. (2014); Haque et al. (2014); Hasan et al. (2014); Kabir & Saha (2014); Rahman & Jamila (2016)

TABLE 7.2 *(Continued)*

Sl. No.	Vernacular Name	English Name	Botanical Name	Family	Uses	References
177.	*Nisinda*	Chinese chaste tree, five-leaved chaste tree, horseshoe vitex	*Vitex negundo*	Verbenaceae	Leaf poultice is used to remove headache. Leaf, seed, and root is prescribed for treatment in too much vaginal discharge, edema, various skin disorders, helminthiasis, and arthritis, puerperal, and rheumatic fever.	Yusuf et al. (1994); Ghani (2003); Khan & Rashid (2006); Mukul et al. (2007); Chowdhurya et al. (2009); Rashid &Eakram (2009); Rahman et al. (2013); Haque et al. (2014); Hasan et al. (2014); Kabir & Saha (2014); Rahman & Keya (2015); Rahman & Jamila (2016); Bari et al. (2017)
178.	*Haptavanga*	Veld grape	*Vitis quadrangularis* L.	Vitaceae	Whole plant is prescribed for the treatment of rheumatic fever and various joint inflammations.	Yusuf et al. (1994); Rahmatullah & Mollik (2009); Rahman et al. (2013); Tuhin et al. (2013); Bhowmik et al. (2014); Haque et al. (2014); Hasan et al. (2014); Kabir & Saha (2014); Ahmmed et al. (2017); Bari et al. (2017); Khatun et al. (2018)
179.	*Ashwagandha*	Indian ginseng, Poison gooseberry	*Withaniasomnifera*	Solanaceae	Used to cool the brain, reduce inflammation, cholesterol level, anxiety, diabetes, hypothyroidism, sexual dysfunction, arthritis, tuberculosis, lever problem and tumor.	Yusuf et al. (1994); Dixie et al. (2003); Ghani (2003); Siddique et al. (2005); Rashid &Eakram (2009); Rahman et al. (2013); Kabir & Saha (2014); Rahman & Jamila (2016); Bari et al. (2017); Khatun et al. (2018); Hoque (2019)
180.	*Ghagra*	Cocklebur, Rough cocklebur, Burweed	*Xanthium indicum*	Asteraceae	Whole plant is used for urinary and other renal complaints, improving appetite and voice, white discharge, and menorrhagia. Also used as laxative, anti-helminthic, diuretic, emollient, tonic, and sedative.	Ghani (2003); Chowdhurya et al. (2009); Rahman et al. (2013); Bhowmik et al. (2014); Kabir & Saha (2014); Rahman & Keya (2015); Rahman & Jamila (2016); Ahmmed et al. (2017)
181.	*Ada*	Ginger	*Zingiber officinale*	Zingiberaceae	The rhizome has been found to be effective for controlling a wide range of bacterial, fungal, virus-related, and parasitic diseases. Rhizome is used to increase digestion, post-operative vomiting, and nausea. It is also indicated for sore throat, cough, and various intestinal, cardio-vascular, and sexual complaints. Very much effective as expectorant and anti-inflammatory agent.	Yusuf et al. (1994); Dixie et al. (2003); Ghani (2003); Khan & Rashid (2006); Mukul et al. (2007); Chowdhurya et al. (2009); Rahmatullah & Mollik (2009); Rahman et al. (2012, 2013); Bhowmik et al. (2014); Haque et al. (2014); Kabir & Saha (2014); Rahman & Keya (2015); Rahman & Jamila (2016); Ahmmed et al. (2017); Bari et al. (2017); Khatun et al. (2018)

TABLE 7.2 *(Continued)*

Sl. No.	Vernacular Name	English Name	Botanical Name	Family	Uses	References
182.	Iswarmul	Indian birthwort	*Aristolochia indica*	Aristolochiaceae	Used as stimulant and tonic. Also used in diarrhea, skin infection, snake bites, inflammation, cough, unusual bleeding during menstruation, indigestion, headache, tonsillitis, and leukoderma.	Ghani (2003); Siddique et al. (2005); Mukul et al. (2007); Nawaz et al. (2009); Rahman et al. (2012, 2013); Tuhin et al. (2013); Bhowmik et al. (2014); Haque et al. (2014); Kabir & Saha (2014); Rahman & Keya (2015); Khatun et al. (2018)
183.	Anantamul	Indian Sarsaparilla	*Hemidesmus indicus*	Apocynaceae	It is used in leprosy, skin ailments, syphilis, fever, bronchitis, various urinary disorder, prolonged rheumatism, and leucorrhea.	Yusuf et al. (1994); Ghani (2003); Siddique et al. (2005); Khan & Rashid (2006); Mukul et al. (2007); Chowdhurya et al. (2009); Rahman et al. (2013); Bhowmik et al. (2014); Hasan et al. (2014); Rahman & Jamila (2016); Ahmmed et al. (2017); Bari et al. (2017); Khatun et al. (2018)
184.	*Talamulika, Talora*	Golden eye-grass, Orchid palm grass, Black Musali	*Curculigoorchioides*	Hypoxidaceae	Root in powdered form is burnt and the smoke is inhaled to get rid of cough and other respiratory infection. Fresh root paste is used as poultice over skin diseases. Root powder with milk is taken against various sexual disorders and weakness.	Yusuf et al. (1994); Ahsan et al. (1997); Dixie et al. (2003); Khan & Rashid (2006); Chowdhurya et al. (2009); Rahmatullah & Mollik (2009); Rahman (2013, 2014); Chhanda et al. (2014); Kabir & Saha (2014); Rahman & Fakir (2015); Sajib & Uddin (2015); Hoque (2019)
185.	*Hathikana*	Large-leaved Leea	*Leea macrophylla*	Vitaceae	Ethnobotanical survey of this plant shows some important therapeutic uses in cancer, dysentery, body-ache, and sexual disability. It has some other traditional uses for tonsillitis, tetanus, nephrolithiasis, rheumatism, arthritis, snake bites, sore, pain, and blood effusion	Yusuf et al. (1994); Ghani (2003); Mukul et al. (2007); Chowdhurya et al. (2009); Rahman et al. (2013); Sajib & Uddin (2013, 2015); Bari et al. (2017); Sarwar (2017); Chowdhury et al. (2019); Hoque (2019)

medicinal plants to sustain their livelihood. For example, a study was conducted covering 30 participant households in five villages of *Natore Sadar upazila*, whose livelihoods were mostly dependent on medicinal plants business. All participant households were asked about the extent to which the production of medicinal plants is helping them to lead a viable livelihood. In reply, around 53% of respondents agreed that their livelihood was better off with income from medicinal plant trading, while 20% replied that it was more or less supportive for their livelihood, and the rest of the respondents replied that it was not enough for their livelihood (Shahidullah, 2007).

7.5.3 HIGH EXPORT POTENTIAL

Like other neighboring countries, Bangladesh can earn foreign currency by producing and exporting medicinal plants. To meet the country's demand and to export, we need to produce medicinal plants on a commercial basis in a sustainable manner. The Indian Union Cabinet recently gave its approval for the Memorandum of Understanding (MoU) for cooperation in the field of traditional methods of medicine and homeopathy between India and Bangladesh. The MoU will provide a designed framework for cooperation between the two countries for the elevation of Indian traditional methods of medicine and homeopathy in Bangladesh. To increase bilateral trade between Dhaka and Delhi, required research, training courses, conferences/meetings will be arranged to raise the trade of AYUSH (Ayurveda, Yoga and Naturopathy, Unani, Siddha, and Homoeopathy) products and also traditional medicine. The total trade of AYUSH products from India in 2012–2013 was Rs. 2167.57 crores, while India's exports of herbal and AYUSH products to Bangladesh for 2013–2014 was Rs. 31.85 crores (Ahmed, 2014). On the other hand, the total export value of medicinal plants from Bangladesh amounted to US $0.39 million in 2014–2015, while the value was US $0.59 million in 2013–2014. The export value of seeds, fruit, and spores for sowing stood at US $0.25 million during 2014–2015. Bangladesh is enjoying duty-free market access to Vietnam, Uganda, Kenya, Somalia, Fiji, Georgia, and other countries for herbal products (Anon., 2016). So, there is a bright export opportunity for Bangladeshi folk medicines and herbal products to India and other countries in the upcoming days, where a large number of people are favoring the Unani and Ayurvedic systems of traditional or folk medicine.

7.5.4 INCREASED SCOPE OF PLANTATION OF MEDICINAL PLANTS

Production of medicinal plants can be increased largely by planting them in some ways which are described in subsections.

7.5.4.1 CULTIVATION IN THE FOREST AND WOODLAND AREAS

In Bangladesh, forests and woodlands cover approximately 16% of the total area. A lot of medicinal plants are grown in these areas. But due to unplanned and overharvesting, the stock of medicinal plants in these areas is decreasing. Therefore, planned planting and judicious harvesting is necessary to raise the stock of medicinal plants in the country.

7.5.4.2 ROADSIDE PLANTATION

According to the Local Government and Engineering Department (LGED), Bangladesh has 271,401 kilometers of road network including 23.47% concreted and 76.53% un-concreted roads. Most of the roadsides are still vacant and there is no plantation on the roadsides. Therefore, medicinal plants may be planted on both sides of the roads. Recently, the government and some NGOs have taken initiative for roadside plantation but very few medicinal plants are included in their program. The roadside landowner may take initiative of planting medicinal plants along the roads. Mass publicity is necessary to create awareness among the roadside owners. Besides, the poor people may be involved for roadside plantation on profit-sharing basis with the concerned government organizations. *Arjun*, *Neem*, *Emblica*, *Bohera*, *Haritaki*, etc., may be grown along the roadside.

7.5.4.3 PLANTATION IN HOMESTEAD AREAS

According to the Bangladesh Bureau of Statistics (BBS), homestead occupies about 0.54 million hectares of land in the country. Presently, some fruits and vegetables are grown in the homestead. A lot of fallow lands are there in every homestead. Therefore, the fallow lands in the homestead area may be planted with different medicinal plants. Approach towards making ‘Medicinal Plant Garden’ in every house may be executed.

7.5.4.4 PLANTATION IN VAST FALLOW LAND

At present, about 402,000 hectares of fallow land exist in Bangladesh. These fallow lands are not suitable for cultivation of crops. Therefore, it is possible to plant medicinal plants in most of the fallow lands as different medicinal plants can be grown in diverse soil conditions. For example, if medicinal plants are grown in sandy or saline soil, then it will save our fertile lands that are presently under crop cultivation. On the other hand, utilization of sandy or saline soils will be possible that are currently fallow. Crops suitable for saline soils include basil, mint, asparagus, *Arjuna*, *Azdirachta*, etc. On the contrary, asparagus, mint, mimosa, and several other cover crops can be grown in sandy soils. But for introducing medicinal crops in the sandy or saline belts, detailed research findings are necessary in our condition.

7.5.5 INCREASED USE IN PHARMACEUTICAL AND COSMETIC INDUSTRY

In Bangladesh, many pharmaceutical and cosmetic industries in the country are manufacturing several herbal medicines and healthcare products. Currently, there are 277 Unani, 201 Ayurvedic, and 32 herbal drug manufacturers in Bangladesh with 6,498 Unani, 4,058 Ayurvedic, and 537 herbal products registered by the Directorate General of Drug Administration (Hoque, 2019). The present annual consumption of medicinal plants in the country is about 20,000 metric tons, from which one-fourth of pharmaceutical drugs are made. To meet the demand, a huge amount of raw materials of medicinal plants is needed. At the moment, herbal raw materials are imported from different countries of the world, which can be produced domestically. In Bangladesh, about 86.7% of drug manufacturing industries are importing raw medicinal plant materials from India; about 53.3% of factories are from Pakistan, and a few are importing from Korea, Nepal, and Iran (Anon., 2016).

7.5.6 INCREASING INTEREST OF THE PEOPLE

Using medicinal plants as remedies for different human ailments has no side effects. Therefore, interest of the people is now increasing towards using medicinal plants and herbal products for various health complications and beauty purposes, respectively. Hence, most people in rural and urban areas are choosing the traditional method of treatment widely.

7.5.7 FAVORABLE CLIMATIC CONDITIONS

By virtue of favorable agro-climate and seasonal variation, Bangladesh is genetically resourceful of medicinal plants. As many as 500 medicinal plant species were identified throughout the country. An organized research program on collection and evaluation of other important medicinal plants from abroad may be undertaken to enrich the genetic pool. Finally, plants that are suitable for growing in this agro-climate may be selected to increase their availability and decrease import.

7.5.8 ECOSYSTEM SERVICES

Tree species of medicinal plants check soil erosion. Mint, mimosa, and other species act as a cover crop. The leguminous species of medicinal plants improve soil health by fixing nitrogen to the soil. Roadside plantation with medicinal trees like *Arjuna, Bohera, Amloki, Horitoki,* and other plants provide shade and help in decreasing global warming. The falling leaves also add organic matter to the soil.

7.6 CONCLUSION

Globally, the use of medicinal plants has increased. To meet the demand, we need a higher amount of raw materials. But due to overextraction, negligence in cultivation, and several other factors, many species of important medicinal plants are now under threat, especially in a country like Bangladesh. Therefore, it is time to take initiative for saving endangered or threatened medicinal plant species. Moreover, as most medicinal plants grow under natural habitat in Bangladesh, we are unconsciously destroying them because many people cannot recognize them or do not know their use and other benefits. There are so many findings available in Bangladesh regarding different aspects of medicinal plants, but these are scattered. An attempt has been undertaken in this chapter to list out endangered medicinal plants of Bangladesh with probable suggestions for saving them. Finally, important medicinal plants have been listed with their possible uses against different human complaints. The information included in this chapter will be helpful to increase the knowledge base of concerned stakeholders.

KEYWORDS

- **Bangladesh**
- **conservation**
- **cultivation**
- **farm level cultivation**
- **high quality planting materials**
- **livelihood**
- **medicinal plants**
- **utilization**

REFERENCES

Ahmed, F., (2014). *Bangladesh Herbal Medicine Export Bright as India to Join Hands.* The Daily Observer.Available from: https://www.business-standard.com/content/b2b-pharma/india-and-bangladesh-to-join-hands-for-traditional-medicines-114090900624_1.html (accessed on 18 December 2023).

Ahmmed, M. R., Ahmed, S., & Sunny, S. M., (2017). A study on diversity of medicinal plant usage by folk medicinal practitioners in different villages of Dhunat Upazila, Bogra district, Bangladesh. *J. Pharmacogn. Phytochem.*, *6*(1), 177–186.

Ahsan, A. T. M. K., Shabuddin, M., Choudhuri, K., Shahriar, M., Huda, A. N., & Chowdhury, Z., (1997). Genetic resources conservation and utilization of medicinal and aromatic plants. Plant genetic resources-Bangladesh perspective. In: *Proc. National Workshop on Plant Genetic Resources.* Bangladesh Agricultural Research Council, Farm Gate, Dhaka.

Akter, R., Uddin, S. J., Grice, I. D., &Tiralongo, E., (2013). Cytotoxic activity screening of Bangladeshi medicinal plant extracts. *J. Nat. Med.*, *68*(1), 1–7. doi: 10.1007/s11418-013-0789-5.

Anonymous, (2009). *Herbal Medicine Sector Seeks Government Support*. Star business report. The Daily Star. Available from: https://www.thedailystar.net/news-detail-88826 (accessed on 18 December 2023).

Anonymous, (2014). *National Botanical Garden, Mirpur, Dhaka. A Booklet* (pp. 1–10). Department of Forest. Ministry of Forest and Environment.

Anonymous, (2016). *A Study on Sector-Based Need Assessment of Business Promotion Council- Herbal Products and Medicinal Plants*. Bangladesh Foreign Trade Institute (BFTI), Kawran Bazar, Dhaka.

Anonymous, (2019a). *Annual Research Progress: 2018–2019*. Bangladesh Forest Research Institute, Chittagong.

Anonymous, (2019b). *Varieties Released.* Research Management Wing, Bangabandhu Sheikh Mujibur Rahman Agricultural University, Gazipur. Available at https://bsmrau.edu.bd/dres/ (accessed on 18 December 2023).

Anonymous, (2019c). Hamdard University. Available at http://www.hamdarduniversity.edu.bd/faculty-of-Unani-and-ayurvedic-medicine/ (accessed on 18 December 2023).

Bari, M. R., Ali, M. A., Miah, M. G., & Abdullah, M. R., (2017). Diversity of medicinal plants in Haluaghat upazila of Mymensingh and their uses. *J. Biosci. Agric. Res., 15*(2), 1266–1273. https://doi.org/10.18801/jbar.150217.156.

Bhowmik, R., Saha, M. R., Rahman, M. A., & Islam, M. A., (2014). Ethno medicinal survey of plants in the southern district Noakhali, Bangladesh. *Bangladesh Pharm. J., 17*(2), 205–214.

Chhanda, S. A., Muslim, T., & Rahman, A., (2014). Phytochemical Studies on *Glinusoppositifolius* (L.) Aug. DC. *Dhaka Univ. J. Sci., 62*(1), 45–48.

Chowdhury, N. S., Farjana, F., Jamaly, S., Begum, M. N., &Zenat, E. K., (2019). Pharmacological values and phytochemical properties of devil's cotton (Ulatkambal): A review. *Bangladesh Pharm. J., 22*(1), 109–116.

Chowdhurya, M. S. H., Koikea, M., & Muhammed, N., (2009). Use of plants in healthcare: A traditional ethno-medicinal practice in rural areas of southeastern Bangladesh. *Int. J. Biodivers. Sci. Manag., 5*(1), 41–51.

Das, U., & Islam, M. S., (2019). A review study on different plants in *Malvaceae* family and their medicinal uses. *Am. J. Biomed. Sci. Res., 3*(2), 94–97.

Dixie, G., Imam, S. A., & Hussain, M. J., (2003). *Medicinal Plant Marketing in Bangladesh* (pp. 8–22). Inter-cooperation Bangladesh, Dhaka.

Ghani, A., (2003). *Medicinal Plants of Bangladesh: Chemical Constituents and Uses*. Asiatic Society of Bangladesh, Dhaka, Bangladesh.

Haque, M. A., Bari, L., Hasan, M. M., Sultana, M. M., & Reza, S. A., (2014). A survey on medicinal plants used by the folk medicinal practitioners in Tangail Sadar Upazilla, Tangail, Bangladesh. *J. Environ. Sci. Nat. Resour*., *7*(1), 35–39.

Hasan, M. M., Hossain, S. A., & Ali, M. A., (2014). Medicinal plant diversity in Chittagong, Bangladesh: A database of 100 medicinal plants. *J. Sci. Innov. Res., 3*(5), 500–514.

Hasan, S., (2014). Pharmacological and medicinal uses of *Achyranthes aspera*. *Int. J. Sci. Environ. Technol., 3*(1), 123–129.

Hoque, M. A., (2019). Education and research on medicinal plants in Bangladesh: Present status and future strategies. In: *Proc. National Seminar on Cultivation, Conservation and Sustainable Utilization of Medicinal Plants for Livelihood Improvement*. Uttar Banga Krishi Viswavidyalaya, Pundibari, West Bengal, India.

Jain, S. K., (2000). Global resurgence of ethnomedico-botany-the Indian scene. *J. Trop. Plants, 1*, 75–81.

Kabir, T., & Saha, S. A., (2014). Study on the indigenous medicinal plants and healing practices of Murong tribe in Khagrachari district (Bangladesh). *Int. J. Pharmacogn., 1*(10), 654–659. http://dx.doi.org/10.13040/IJPSR.0975-8232.IJP.1(10).654-59.

Khan, N. A., & Rashid, A. Z. M. M., (2006). A study on the indigenous medicinal plants and healing practices in Chittagong hill tracts (Bangladesh). *Afr. J. Trad. CAM, 3*(3), 37–47.

Khatun, A., Jannat, K., & Jahan, R., (2018). Some plant-based home remedies used in Narayanganj district, Bangladesh. *J. Med. Plants Stud., 6*(4), 104–106.

Mukul, S. A., Uddin, M. B., & Tito, M. R., (2007). Medicinal plant diversity and local healthcare among the people living in and around a conservation area of northern Bangladesh. *Int. J. For. Usuf. Mngt., 8*(2), 50–63.

Musaddiqa, S., Mustafa, K., & Ahmad, S., (2018). Pharmaceutical, ethnopharmacological, phytochemical and synthetic importance of genus *Aerva*: A review. *Natural Product Communications*, *13*(3), 375–385.

Nawaz, A. H. M. M., Hossain, M., Karim, M., Khan, M., Jahan, R., & Rahmatullah, M., (2009). An ethnobotanical survey of Rajshahi district in Rajshahi division, Bangladesh. *Am-Eurasian J. Sustain. Agric.*, *3*(2), 143–150.

Rahman, A. H. M. M., & Jamila, M., (2016). Angiosperm diversity at Jamtala village of Chapai Nawabganj district, Bangladesh with emphasis on medicinal plants. *Res. Plant Sci.*, *4*(1), 1–9. doi: 10.12691/plant-4-1-1.

Rahman, A. H. M. M., & Keya, M. A., (2015). Traditional medicinal plants used by local people at the village Sabgram under Sadar upazila of Bogra District, Bangladesh. *Res. Plant Sci.*, *3*(2), 31–37. doi: 10.12691/plant-3-2-2.

Rahman, A. H. M. M., (2013). Ethno-medico-botanical studies on cucurbits of Rajshahi division, Bangladesh. *J. Med. Plants Stud.*, *1*(3), 118–125.

Rahman, M. A., Begum, K., Rashid, M. E., & Rashid, M. H., (2012). Medicinal plant diversity in the flora of Bangladesh and their conservation: 2. A report on ten angiosperm families. *Plant Archives*, *12*(2), 1023–1035.

Rahman, M. M., & Fakir, M. S. A., (2015). Biodiversity of medicinal plants in Bangladesh: Prospects and problems of conservation and utilization. *Int. J. Minor Fruits Med. Aromat. Plants*, *1*(1), 1–9.

Rahman, M. M., Masum, G. Z. H., Sharkar, P., & Sima, S. N., (2013). Medicinal plant usage by traditional medical practitioners of rural villages in Chuadanga district, Bangladesh. *Int. J. Biodivers. Sci. Ecosyst. Serv. Managt.*, *9*(4), 330–338. doi: 10.1080/21513732.2013.841757.

Rahman, S., (2014). *CynodonDactylon: Antimicrobial Potential of Crude Extract as Valuable Medicinal Plant*. Department of Mathematics and Natural Sciences, BRAC University, Dhaka.

Rahmatullah, M., Das, A. K., & Mollik, A. H., (2009). An ethnomedicinal survey of Dhamrai sub-district in Dhaka District, Bangladesh. *Amer.-Eurasian J. Sustain. Agric.*, *3*(4), 881–888.

Rashid, M. H., Rashid, M. E., & Rahman, M. A., (2014). Inventory of threatened plants of Bangladesh and their conservation management. *Int. J. Environ.*, *3*(1), 141–167.

Rashid, M. U., &Eakram, M. S., (2009). Identification, prevalence, and use of medicinal plants in Fulbaria upazila of Mymensingh district. *The Agriculturists*, *7*(1, 2), 90–95.

Sajib, N. H., & Uddin, S. B., (2013). Medico-botanical studies of Sandwip Island in Chittagong, Bangladesh. *Bangladesh J. Plant Taxon.*, *20*(1), 39–49.

Sajib, N. H., & Uddin, S. B., (2015). Ethnomedicinal study of plants in Hathazari, Chittagong, Bangladesh. *Pertanika J. Trop. Agric. Sci.*, *38*(2), 197–210.

Sarwar, A. K. M. G., (2017). Ethno-medicinal plant diversity of Bangladesh – their conservation status at the botanical garden of Bangladesh Agricultural University. *J. Plant Pathol. Microbiol.*, *8*, 5. doi: 10.4172/2157-7471-C1-005.

Shahidullah, A. K. M., (2007). *The Role of Medicinal Plants in Livelihood Improvement and Ecological Sustainability in Bangladesh: An Application of a Participatory Approach to Management and Marketing*. MS Thesis. Faculty of Graduate Studies, University of Manitoba, Canada.

Siddique, N. A., Bari, M. A., Pervin, M. M., Nahar, N., Banu, L. A., Paul, K. K., Kabir, M. H., et al., (2005). Screening of endangered medicinal plants species by questionnaire survey in Barind Tract in Bangladesh. *Pak. J. Biol. Sci.*, *8*, 1783–1793.

Tuhin, M. I. H., Asaduzzaman, M., Islam, E., Khatun, Z., & Rahmatullah, M., (2013). Medicinal plants used by folk medicinal herbalists in seven villages of Bhola district, Bangladesh. *Amer.-Eurasian J. Sustain. Agric.*, *7*(3), 210–218.

Yusuf, M., Chowdhury, J. U., Wahab, M. A., & Begum, J., (1994). *Medicinal Plants of Bangladesh* (p. 340). Bangladesh Council of Scientific and Industrial Research (BCSIR), Dhaka.

CHAPTER 8

Some Parasitic and Carnivorous Plants of the Indian Himalayan Region with Ethnomedicinal Use

SUTAPA KUMAR (RAI)

Department of Botany, Vivekananda College, Thakurpukur, Kolkata, West Bengal, India

ABSTRACT

The ethnomedicinal values of two very unique groups of plants, the parasites and carnivores, have been considered here. These plants have very interesting habits, growing in varied and unusual habitats and in some typical ecological niches. The medicinal values, along with the active principles found in some parasitic plants like *Cuscuta reflexa* (Convolvulaceae), *Monotropa uniflora* (Ericaceae), *Dendrophthoe falcata* (Loranthaceae), and *Balanophora polyandra* (Balanophoraceae), are discussed here. The carnivores considered here are *Drosera burmannii* and D. *peltata* (Droseraceae). The study aims at highlighting the valuable medicinal potential of the two fascinating groups of plants. They have long been used in traditional systems of medicine by ethnic tribes and rural people in various parts of India to effectively cure ailments of the lung, stomach, liver, intestine, kidney, brain, and other sense organs. They have been reported to cure even some forms of cancer. Many of them have antioxidant, anti-fertility, and anthelmintic properties. Narrow geographical range and the requirement of specific ecological niches make their distribution very limited. Habitat loss is the major cause for some of them reaching the rare, threatened, or endangered status. Therefore, they call for more active research and urgent conservation measures. In vitro

Bioprospecting Ethnomedicinal Plant Resources: Sustainable Utilization and Restoration.
Gopal Shukla, Jahangeer A. Bhat, A. P. Das, and Sumit Chakravarty (Eds.)

propagation has been successful in *Drosera burmannii* and could be a future method to conserve and propagate these valuable plants.

8.1 INTRODUCTION

Ethnomedicine is the mother of all systems of medicine. It is perhaps as old as mankind itself. The indigenous tribes of India have a sound knowledge of the plants in their surroundings including their economic and medicinal values. Ethnobotanical investigations based on oral traditional knowledge (TK) collected from indigenous people in many parts of the world have been of great benefit, India being no exception. Ethnomedicinal value of the local plant resources in a particular area gained from TK systems have contributed to the development of various pharmaceuticals and nutraceuticals in modern times (Pei, 2001). O'Neill & Rana (2016) have rightly commented on the intimate relationship of the Indian tribals with their environment which has given them insights into the use of natural resources in a sustainable way. India is a hotspot of plant diversity and is home to a number of parasitic and carnivorous plants as well. The ethnomedicinal values of parasitic and carnivorous plants have been considered here. These plants grow in very specific ecological niches in very typical habitats and show unusual habits. They show fascinating adaptive measures for obtaining the essential nutrients they require for their growth. The ecological contribution they make to their ecosystems cannot be ignored. Several parasitic plants belonging to families like Balanophoraceae, Convolvulaceae, Ericaceae, Lauraceae, Loranthaceae, Orobanchaceae, Santalaceae, Scrophulariaceae, and even Rafflesiaceae have been reported from different parts of India (Adhikari et al., 2003; Benniamin et al., 2012; Vijayan et al., 2015; Verma & Yadav, 2018). Ethnomedicinal studies have been conducted in many of them (Deshmukh, 2011; Shanavaskhan et al., 2012; O'Neill & Rana, 2016). The same is true for carnivorous plants as well. Several carnivorous plants of Nepenthaceae, Lentibulariaceae, and Droseraceae have been reported from India and their medicinal uses studied (Kumar et al., 2017).

The present study includes the review of some parasitic or carnivorous plants that were observed during field trips in parts of the hills and adjoining plains of the Indian Himalayan region like the Dooars region at the foothills of the Himalayas, Sikkim in the eastern Himalayas and Kausani and Nainital in the western Himalayas during the period between 2009 and 2016. The review includes a brief description of their distribution, external morphology which would help in their identification, their ethno medicinal uses and some of the important chemical constituents present in them that support their medicinal potentials. The parasitic plants discussed here include *Cuscuta reflexa* (Convolvulaceae), *Monotropa uniflora* (Ericaceae), *Dendrophthoe falcata* (Loranthaceae), and *Balanophora polyandra* (Balanophoraceae) and the two carnivorous plants considered here are *Drosera burmannii* and *D. peltata* (Droseraceae).

8.2 PARASITIC PLANTS AND THEIR ETHNOMEDICINAL USE

Parasites are an interesting group of plants because of their unusual adaptations for procuring their nutrients. They are partially or totally dependent on certain host plants for their nutritional requirements. Holo- or total parasites being non-green are unable to photosynthesize and are completely dependent on their hosts as obligate parasites whereas, hemi- or partial parasites being green do manufacture a part of their food or are facultative parasites (Khwaja et al., 2013). Root-like penetrating structures called haustoria characterize parasitic plants for absorbing nutrients from their hosts. Root parasites connect their haustoria to roots of the host while, stem parasites connect them to stems. Unique among the holoparasites are the mycoparasites or mycoheterotrophs which parasitize mycorrhizal fungi and become associated with many other plants in their surroundings. The distribution, morphology, ethno medicinal uses, and important phytochemicals present in the parasites considered are discussed individually.

8.2.1 CUSCUTAREFLEXAROXB.

The species represents the family Convolvulaceae and is commonly known as dodder in English. It is a very common total parasite. Locally, it is called 'Swarnalata' meaning golden vine. Prolific in growth, this climber parasite is distributed throughout the plains of India. It forms a yellowish-green, rootless, leafless tangled, wiry mass covering the host plant (Plate 8.1(a)). It has small white bell-shaped flowers (Plate 8.1(b)) attached to the nodes. The flowers have a fleshy calyx and reflexed corolla lobes.

PLATE 8.1 (a) *Cuscuta reflexa plant; (b) Cuscuta reflexa flower; (c) Monotropa uniflora; (d) Balanophora polyandra female inflorescence; (e) Balanophora polyandra male inflorescence; (f) Dendrophthoe falcata; (g) Drosera burmannii; and (h) Drosera peltate.*

Cuscutareflexa has been widely used by the indigenous people of India for treating a variety of ailments (Deshmukh, 2011; Shanavaskhan et al., 2012; Khwaja et al., 2013; Lalchand et al., 2017; Verma & Yadav, 2018). A warm paste of the plant is used to treat muscle pain, headache, and rheumatic pain. Plant paste applied on the scalp prevents hair fall. Seeds of the plant have carminative and anthelmintic properties. Fresh juice of the plant along with the juice of sugarcane is used in the treatment of jaundice. Stem decoction is used as a laxative, for liver ailments, and urinary tract infection. Chloroform and ethanol extracts showed antitumor activity against Ehrlich ascites carcinoma tumor in mice. Lupeol, dulcitol, mannitol, sitosterol, carotenoids, and flavonoids isolated from *C. reflexa* are pharmacologically active compounds having antimicrobial, anti-inflammatory, and antitumor activity (Udavant et al., 2012).

8.2.2 MONOTROPAUNIFLORA L.

The species is a member of the family Ericaceae and is commonly known as Indian pipe plant or Ghost plant. In India, it is found in Himachal Pradesh, Jammu Kashmir, Sikkim, Uttarakhand, Mizoram, Nagaland, Manipur, Arunachal Pradesh, and Darjeeling. It is a mycoheterotroph feeding on certain mycorrhizal fungi. Its occurrence is ephemeral, when there is moisture after a dry period and grows in moist, dark environments like the understory of dense forests. The stem is 5–30 cm tall densely clothed with scale-like leaves and bears a single flower with 3–8 translucent tepals, 10–12 stamens, and a single pistil (Plate 8.1(c)). It produces an oval capsule-like fruit. *Monotropa uniflora* has been widely used by the indigenous people of India in their traditional cures. Infusion of the root is used as a remedy for spasms, fainting spells, various nervous conditions, and as a sedative, root infusions given to children suffering from fits, epilepsy, and convulsions. Plant extract alone or combined with rose water has been found to be an excellent application for ophthalmic inflammation, a poultice of the plant if applied to sores effectively heals them, flowers are chewed to bring relief from toothache (Norland & Norland, 2011). One of the main ingredients in the plant is salicylic acid, a naturally occurring phenolic acid with sedative and analgesic properties (Petley, 2013).

8.2.3 BALANOPHORAPOLYANDRA GRIFF.

A member of the family Balanophoraceae, the species is an obligate root parasite. In India, it is found in Darjeeling, Sikkim, Jammu and Kashmir, and Uttarakhand. It is an erect, fleshy, achlorophyllous, holophrastic herb. It forms an underground tuberous structure. The leaves are 3 to 10, distichous, ovate, and connate into a sheath-like whorl. Their unisexual flowers borne in compact inflorescences represent above ground parts during reproduction (Plate 8.1(d) and (e)). The plant is dioecious (Chhetri et al., 2018). The whole plant of *Balanophora polyandra* has hemostatic properties and is traditionally used to treat piles, traumatic injury, and bleeding, it is also reported to relieve rheumatism, and a fresh extract of the plant is used for treatment of irregular menstruation (O'Neill & Rana, 2016). Acetone extract of the fresh plant is reported to possess significant free radical scavenging activities. *Balanophora* contains flavonoids with antioxidant properties, Lupeol, a triterpene with anti-inflammatory properties (Wang, 2012); hydrolysable tannins (HTs) and steroids are also found (Wang, 2013).

8.2.4 DENDROPTHOE FALCATA (L.F.) ETTINGSHSYNONYM LORANTHUS FALCATUS L. F.

The species belongs to Loranthaceae and is commonly called Honeysuckle Mistletoe in English and locally known as '*Banda.*' In India, it is found in Arunachal Pradesh, Himachal Pradesh, Sikkim, Jammu Kashmir, Uttarakhand, Bihar, Uttar Pradesh, West Bengal, Tripura, and Assam. *Dendrophthoe* is an epiphytic partial stem parasite (Photo 8.1(f)) growing on a wide range of host plants. The opposite leaves are leathery, 7-15 cm long. Flowers are borne in racemes and have exerted stamens. The whole plant is used in traditional medicine. It has astringent, narcotic, and diuretic properties. Application of fresh juice of leaf and stem paste topically on the affected area heals wounds and leukoderma, leaf powder with honey is used in treating pulmonary tuberculosis and asthma, leaf juice cures menstrual disorders, urinary troubles, inflammations, and ulcers (Shanavaskhan et al., 2012). Decoction of the plant is used as an antifertility agent and antisyphilitic agent (Khwaja et al., 2013). Leaves contain flavonoids and young shoots contain tannins (Manthri et al., 2011).

8.3 CARNIVOROUS PLANTS AND THEIR USE

Carnivorous plants, better known as insectivores, devour small animals like protozoans, small mammals such as rats, and a variety of arthropods, especially insects, to obtain a part of their nutrition (Mondal & Mondal, 2000; Ghosh, 2003). They thrive on nutrient-poor soils, especially deficient in nitrogen, and have developed different trapping mechanisms to trap, kill, and digest animals to procure the nutrients they require. Some, like *Nepenthes*, have 'Pitfall Traps' with their leaves folded into a pitcher-like structure having a slippery inside and housing digestive enzymes to disintegrate and digest the prey. *Drosera* has a 'Flypaper Trap' with sticky mucilage, in which their prey gets stuck. *Dionaea* or Venus flytrap has a 'Snap Trap' which closes shut as soon as its prey alights. 'Bladder Traps' are found in plants like *Utricularia* with a sucking mechanism that sucks prey within a bladder that generates an internal vacuum. Finally, there are the 'Lobster Traps,' also known as eel traps, in plants like *Genlisea*, which force their prey to move towards the digestive organs with inwardly directed hairs, but there is no exit door for the prey (Gordon, 1992).

In the present scenario, with a dearth of broad-spectrum antifungal drugs, the carnivorous plants are being looked upon as a group which could meet this need. At present, mild to severe forms of athlete's foot or different types of fungal skin infections are not totally curable by medications available in the market. The wide array of secondary metabolites that help the carnivorous plants to survive could in the near future, serve as a new class of antifungal drugs for mankind (Kumar et al., 2017). Besides this, the plants also have digestive, anti-inflammatory, anti-aging, and even antitumor potentials.

8.3.1 DROSERABURMANNII VAHL

A member of the family Droseraceae, the species commonly known as sundew is a small carnivorous plant. In India, it is found in Assam, Meghalaya, West Bengal, Maharashtra, Kerala, Karnataka, and Tamil Nadu. It is a small annual acaulescent herb with radical wedge-shaped leaves in a rosette (Plate 8.1(g)). The leaves contain mucilage-secreting glands which are borne at the end of long and mobile tentacles that can bend 180 degrees in a minute or so, trapping the small insects alighting on it. Flowers are pinkish white borne in terminal racemose inflorescences. The sepals and petals are also glandular. Warm infusion of the whole plant of *Droseraburmannii* has anti-inflammatory activity and is used to cure bronchitis, asthma, and whooping cough. Alcohol extracts are reported to have antifertility and antitumor activities in mice (Raju et al., 2012; Ghate et al., 2015). It contains flavonoids and ascorbic acid which together is strongly anti-inflammatory while, naphthoquinones such as plumbagin are also found as an antifertility agent (Madhavan et al., 2009). Yanthan et al. (2017) have reported an efficient *in vitro* regeneration protocol from shoot tips of *Droseraburmannii*.

8.3.2 DROSERAPELTATATHUNB.

It is a perennial insectivorous herbaceous plant of the family Droseraceae (Plate 8.1(h)). In India, it is found in Arunachal Pradesh, Assam, Mizoram, Meghalaya, Manipur, Nagaland, Darjeeling, Himachal Pradesh, Uttarakhand, Jammu and Kashmir, and Sikkim. It grows in grassy slopes, at elevations from 1,500–3,600 m in the Himalayas. It has an erect or climbing stem about 9–32 cm tall. The plant grows from a subterranean tuber and is only in leaf for a few months each year. *Drosera peltata* has pain relieving, blood purifying and carminative properties. With salt added to the crushed leaves, a poultice is made which has long been used to remove or heal blisters and to bring relief from arthritis and rheumatism. *Drosera peltata* extract is used in the treatment of dental caries and periodontitis (Devi & Ramasubramaniaraja, 2009). Plumbagin, droserone, and droxydroserone are the three naphthoquinones along with the four flavonols quercetin, gossypetin, gossypin, and isogossypitrin that are the active principles found in the plant extract (Nair et al., 1990).

8.4 CONCLUSION

Populations of several parasitic and carnivorous plants are now threatened, chiefly due to overexploitation and habitat degradation. As many members of these two unique groups of plants grow in specific microclimates found in specific ecological niches, their existence is very much under threat. It is also clear from the discussion that they have various medicinal potentials which could tackle the most pressing problems in today's world. They can scavenge free radicals efficiently; others can cure nervous conditions like anxiety, insomnia, and epilepsy. Some are able to improve heart health or cure digestive disorders. Some of them even have antifertility or anticancer properties. It is the need of the hour to conserve this medicinally important group of plants. Immediate measures must be taken for documentation and protection of the indigenous knowledge related to their use. As a recommendation, phytochemical, toxicological, and pharmacological studies should be undertaken to show the efficiency level of these plants or to find out any toxic side effects in their use. Therefore, it is necessary to adopt urgent strategies to protect their habitat. In recent times, *in vitro* techniques are being adopted for their propagation, and this technique has been successfully applied in the case of *Drosera burmannii*. Worldwide research in this direction might yield more fruitful results.

KEYWORDS

- **Carnivorous plants**
- **Ethnomedicinal use**
- **In vitro techniques**
- **Monotropa uniflora**
- **Morphology**
- **Parasitic plants**

REFERENCES

Adhikari, D., Majumder, M., Arunachalam, A., & Samah, R., (2003). A rare root parasitic plant (*Sapria himalayana* Griffith.) in Namdapha National Park, north-eastern India. *Curr. Sci., 85*(12), 1669.

Benniamin, A., Dey, S., Chaturvedi, S. K., & Jamir, M. A., (2012). Supplements to the root parasitic plant in India, a new recorded species. *Christisonia siamensis* Craib. (*Orobanchaceae*). *Taiwania, 57*(2), 217–221.

Chhetri, P. K., O'Neill, A. R., & Chhetri, B., (2018). Root holoparasite *Balanophora polyandra* Griff. (*Balanophoraceae*) in eastern Himalaya (Sikkim, India): Distribution, range, status, and threats. *Journal of Threatened Taxa, 10*(8), 12123–12129.

Deshmukh, P. J., (2011). Some angiospermic parasitic plants from Narnala fort with special references to their ethnobotanical value. *Int. J. Adv. Biotech. Res., 2*(4), 431–436.

Devi, P. B., &Ramasubramaniaraja, R., (2009). Dental caries and medicinal plants: An overview. *J. Pharm. Res., 2*(11), 1669–1675.

Ghate, N. B., Chaudhuri, D., Das, A., Panja, S., & Mandal, M., (2015). An antioxidant extract of the insectivorous plant *Droseraburmannii* Vahl. alleviates iron-induced oxidative stress and hepatic injury in Mice. *PLoS One, 10*(5), e0128221.

Ghosh, D., (2003). Project landscape: 11 hunter plants. *Resonance*, 64–70.

Gordon, C. A., (1992). *Guide to Carnivorous Plants of the World.* Collins Publishers, New York.

Khwaja, S., Gor, S., Visavadia, M., Soni, V., &Tatmia, N., (2013). Ethnobotanical survey of some parasitic plants growing in Girnar forest of Junagadh district of Gujarat, India. *Int. Res. J. Biol. Sci., 2*(4), 59–62.

Kumar, S., Labala, R. K., Thorat, S., & Patra, J. K., (2017). Insectivorous plants of India; source of Yanthan, *in vitro* regeneration of *Droseraburmannii* Vahl.: A carnivorous plant of north-east India. *Biotech.*, *7*, 124.

Lalchand, Gupta, R., Sahu, R., & Rout, O. P., (2017). *Cuscutareflexa* (dodder plant): A critical review on the medicinal plant used in Ayurveda. *International Journal of Research in Ayurveda & Pharmacy, 8*(6), 38.

Madhavan, V., Basnett, H., Gurudeva, M. R., & Yoganarasimhan, S. N., (2009). Pharmacognostical evaluation of *Drosera burmannii* Vahl. (Droseraceae). *IJTK, 8*(3), 326–333.

Manthri, S., Kota, C. S., & Talluri, M., (2011). Pharmacological review of *Dendrophthoe falcata. J. Phytol., 3*(3), 18–25.

Mondal, A. K., & Mondal, S., (2000). Killer plants. *Science Reporter, 37*(3), 9–15.

Nair, A. G. R., Shanmugasundaram, P., & Madhusudhanan, K. P., (1990). Naphthaquinones and flavonols from leaves of *Droserapeltata. Fitoterapia, 61*(1), 85, 86.

Norland, M., & Norland, M., (2011). *Proving: Indian Pipe* (*Monotropa Uniflora*). Orchard Leigh Rodborough Hill Stroud. The School of Homeopathy, Gloucestershire, England.

O'Neill, A. R., & Rana, S. K., (2016). An ethnobotanical analysis of parasitic plants (*Parijibi*) in the Nepal Himalaya. *Journal of Ethnobiology and Ethnomedicine*, *12*, 14.

Pei, S. J., (2001). Ethnobotanical approaches of traditional medicine studies some experiences from Asia. *Int. J. Adv. Biotech. Res.*, *9*, 74–79.

Petley, N. D., (2013). *A Review of Literature Pertaining to the Clinical Use of Monotropauniflora*, a Dissertation. College of Naturopathic Medicine, University of Bridgeport, Bridgeport, CT.

Raju, A., Christina, A. J. M., & Mayakrishnan, A., (2012). Antitumor potential of ethanol and aqueous extracts of *Droseraburmannii* Vahl. against Dalton's ascitic lymphoma bearing mice. *J.Pharmacol. Res.*, *5*, 1418–1423.

Shanavaskhan, A. E., Sivadasan, M., Alfarhan, A. H., & Thomas, J., (2012). Ethnomedicinal aspects of angiospermic epiphytes and parasites of Kerala, India. *IJTK, 11*(2), 250–258.

Udavant, P. B., Satyanarayana, S. V., & Upasani, C. D., (2012). Preliminary screening of *Cuscutareflexa* stems for anti-inflammatory and cytotoxic activity. *Asian Pacific Journal of Tropical Biomedicine*, *2*(3), 1303–1307.

Verma, N., & Yadav, R., (2018). *Cuscutareflexa*: A parasitic medicinal plant. *Plant Archives, 18*(2), 1938–1942.

Vijayan, A., Santhakumar, S., Vivekraj, P., & Kalavathy, S., (2015). Parasitism of host trees by the *Loranthaceae* in the region of Sitheri Hills (Eastern Ghats), Tamil Nadu, India. *Bulletin of Environment, Pharmacology and Life Sciences*, *4*(3), 104–109.

Wang, X., Qiao, W., Liu, Z., & Cheng, R., (2012). Phytochemicals and biological studies of plants from the genus *Balanophora. Chemistry Central Journal*, *6*, 79.

Wang, Y., Yang, J., Wang, A., Ma, J., Tian, J., & Ji, T., (2013). Hydrolyzable tannins from *Balanophora polyandra. Acta Pharmaceutica Sinica B*, *1*, 46–50.

Yanthan, J. S., Kehie, M., Kumaria, S., & Tandon, P., (2017). *In vitro* regeneration of *Drosera burmannii* Vahl.: A carnivorous plant of north-east India. *Biotech*, *7*, 124.

CHAPTER 9

Use and Preference of Medicinal Plants Among Forest Fringe Indigenous Communities in Eastern Sub-Himalayan Region: A Case Study

SUBHRAJYOTI PANDA, YANGLEM LAKSHIMAI DEVI, and
PRABHAT KUMAR PAL

Department of Agricultural Extension, Uttar Banga Krishi Viswavidyalaya, Pundibari, Cooch Behar, West Bengal, India

ABSTRACT

India is famous for its rich heritage and natural resources. From ancient times, Indians used plants for medicinal purposes to cure various diseases. Medicinal plants have a long history in different communities of our country. North Bengal is mainly surrounded by rivers and forests. About 18.94% of the area is covered with forests. The indigenous communities mainly live in and around the forest areas. The tribes living in the North Bengal Forest fringe areas are *Toto, Rabha, Mech, Garo, Oraon*, etc. These communities are considerably dependent on the NTFPs, which also include medicinal plants. These tribes use different types of remedies of plant origin that are only collected from the forest areas. But this is practiced sporadically and not thought of systematically as an entrepreneurial venture. The present writing reviewed different medicinal plants available in the forest fringe communities of North Bengal and assessed the entrepreneurial scope with these plants in the area. The chapter also assessed the economic contribution of NTFP as an occupational option for these communities. Based on these findings, the chapter proposed some strategies to enhance the cultivation of medicinal plants in these areas.

Bioprospecting Ethnomedicinal Plant Resources: Sustainable Utilization and Restoration.
Gopal Shukla, Jahangeer A. Bhat, A. P. Das, and Sumit Chakravarty (Eds.)

9.1 INTRODUCTION

Despite the access to modern healthcare systems worldwide, the majority of the world's common people still depend on traditional plant-based ailment systems for their primary healthcare (Anon., 2002); growing consciousness about the side effects of modern medicines continuously increases such dependency. There are around 250,000 species of plants globally, of which approximately 14–28% are used for medicinal purposes. In India, 6,000–7,500 species out of 15,000–17,000 flowering plants are used for medicinal purposes in different traditional systems of medicine (Rao, 1994). The share of India in the global trade of medicinal and aromatic plants (MAPs) is only 0.5% (Singh & Jha, 2018), although the global market for herbal products is expected to touch the mark of US $5 trillion by the year 2050 (Purohit & Vyas, 2004). In developing countries like India, there is an increasing attempt to incorporate traditional medicines into the local healthcare systems. Although many researchers are involved in exploring the huge potential of ethnobotanical knowledge for treating various diseases (Dutta & Dutta, 2005; Kala, 2005; Jain et al., 2010; Jeyaprakash et al., 2011), these plants are under threat due to deforestation, overgrazing, and reckless utilization. So, the conservation of these biological resources as well as their sustainable usage is an urgent need of today (Payyap-pallimana & Fadeeva, 2013). Among others, the usage of plant products directly as medicines is predominantly found among the indigenous communities and especially the forest fringe dwellers (Bose et al., 2015; Vineeta et al., 2021).

The Terai region of West Bengal is situated in the northern part of West Bengal and characterized by 3Ts, viz. Tea, Timber, and Tourism. This 3T culture shows its uniqueness when it gets intermingled with the livelihood relations of aboriginal ethnic tribes of North Bengal. The forest fringe areas of North Bengal are mainly inhabited by the aboriginal tribals of Duars. They are: Nepali (mostly concentrated in Sikkim–Darjeeling region and consists of Limbu, Gurung, Magar, Rai, Tamang, Mewar, etc.); Bhutia (are the people originally from Tibet who entered Sikkim through Bhutan and settled in higher altitudes); Mech (under the broader Boro group of Eastern India); Lepcha (a tribal group living mainly in mountain valleys in Sikkim, western Bhutan, and parts of Nepal and West Bengal); Rava (the tribes of North Bengal); Drukpa (the people of Bhutan are called Drukpas); Sherpa (Sherpas of Darjeeling have attained worldwide fame as tough porters and expert guides in high altitudes); Toto (an isolated tribal group of North Bengal), etc. (Anon., 2001; Bhattacharya, 2004; Das et al., 2003, 2007). They make good use of all the possible natural resources for food, shelter, clothes, weaving, handicrafts, etc. They use medicinal plants of local origin for the treatment of different diseases and ailments by local medicine-men. These men do not disclose their secret TK on medicinal properties of many local medicinal plants, which they have gained through their age-old trial and error methods (Jain & Mudgal, 1999). In this backdrop, the present study was undertaken to document the medicinal plants and their utilization by the forest fringe communities of North Bengal for their livelihood along with a comparative assessment of these medicinal plants in comparison to conventional treatment (CT) methods. The study also attempted to advocate extension strategies to enhance cultivation of medicinal plants in the forest fringe communities of North Bengal.

9.2 METHODOLOGY

The present study was conducted in the forest fringe tribal communities of North Bengal. Purposive and simple random sampling procedures were followed in the present study. The districts Cooch Behar and Alipurduar were selected purposively due to the selected tribes for the study being found in these areas. The three tribes, namely Toto, Rabha, and Mech, were selected purposively from the selected districts. The Toto tribes are found in the Totopara village of Alipurduar district. In Cooch Behar district, Rabha tribes live in the villages of Rasikbill and Paglir Kuthi. Rajabhat Khawa, Chilapata, and Uttar Dhalkaur villages in Alipurduar district were found to be specific for both the Rabha and Mech tribes. The villages were therefore purposively selected for the present study. From the six villages, an exhaustive list of tribal people was selected with the help of the Block officials, Panchayat officials, and local tourist guides. From this exhaustive list, 200 proactive tribal people were selected to constitute the respondents of the present study for final data collection. To assess the contribution of NTFP on livelihood, a schedule was used to collect data through personal interview method. Information was also collected through exploring literature and group discussion of key informants. The type of vocation found in the respondents' families, like agriculture, animal husbandry, poultry, fisheries, business, service, NTFP related, and others. From the occupations, their annual income was calculated. Focus group discussions in each area were conducted to compare the performance of medicinal plants as a component of treatment. Matrix ranking, as advocated by Chambers (1988), was done to quantify the preferences of medicinal plants as an ailment over the conventional method of treatment (allopathic treatment). The number of indicators, as prescribed by the key informants, was considered for computation of this matrix for comparison among the ailments.

9.3 RESULTS AND DISCUSSION

9.3.1 *ETHNOMEDICINAL PLANT*

The medicinal plants are used by the tribes in North Bengal from time immemorial. The knowledge was transmitted from generation to generation. From literature survey, it is found that many plants were used for medicinal values by the indigenous communities of North Bengal (Sarkar & Das, 2010; Bose et al., 2015). Triangulation was done for confirmation of the findings of previous works in the area during primary survey. Table 9.1 presents the name of common diseases along with the name of plants (with scientific names) which are used against these diseases by the indigenous communities of the study area.

TABLE 9.1 Common Diseases and Associated Ethnomedicinal Plants in the Study Area

Sl. No	Diseases/Ailments	Ethnomedicinal Plant Species	
		VN	SN
1.	Liver tonic	*Kalmegh; Chirata*	*Andrographis paniculata* (Brum. f.) Wall. Ex Nees.
2.	Anemia	*Kulekhara*	*Hygrophila schulli* (Buch. Ham) M. Ret. et. S.M. Almeida.
3.	Chronic bronchitis, cough, and cold.	*Basak*	*Justicia adhatoda* L.
4.	Snake bite	*Githa*	*Thunbergia grandiflora* Roxb.
5.	Sex diseases	*Apang;Chirchiti*	*Achyranthes aspera* L.
6.	Jaundice	*Chhorachurisag*	*Deeringiaam aranthoides* (Lamarck) Merrill.
7.	Jaundice and diarrhea	*Astachatur*	*Crinum amoenum* Roxb. Ex Ker. Gawler.
8.	Digestive stimulant and anti-vomiting.	*Dhoney*	*Coriandrum sativum* L.
9.	Diarrhea and dysentery.	*Thankuni*	*Centella asiatica* (L.) Urban.
10.	Acidity and child constipation.	*Chatim*	*Alstoniascholaris* (L.) R.Br.
11.	Muscular pain	*Kukursonga*	*Blumealacera* (Burm. f.) DC
12.	Skin disease	*Kesuti*	*Eclipta prostrate* (L.) L.
13.	Migraine pain	*Jonglidopati*	*Impatiens trilobata* Colebrook.
14.	Tuberculosis	*Puinsak*	*Basella alba* Stewart.
15.	Stomach pain	*Sona*	*Oroxylum indicum* Vent.
16	Bone fracture and blood dysentery, amoebiasis.	*Simul*	*Bombax ceiba* L.
17.	Eye treatment.	*Hatisur*	*Heliotropium indicum* L.

TABLE 9.1 *(Continued)*

Sl. No	Diseases/Ailments	Ethnomedicinal Plant Species	
		VN	SN
18.	Scurvy	*Anaras*	*Ananas comosus* (L.) Merr.
19.	Cough and cold, and pain.	*Chimjera*	*Drymaria diandra* (Blume.) Duke
20.	Blood in urine, diabetes	*Betlahari*	*Costusspeciosus* (Koen. Ex retz) Smith.
21.	Rheumatism	*Mohaboli*	*Trichosanthes cordata* Roxb.
22.	Asthma and snakebite.	*Ban-alu*	*Dioscorea bulbifera* L.
23.	Insect bite/sting	*Nagmoni*	*Dracaena angustifolia*
24.	Pain killer	*Bherenda*	*Ricinus communis* L.
25.	Cough and cold	*Tulsi*	*Ocimum sanctum* L.
26.	Running nose	*Pachroli*	*Pogostemon plectranthoides* Desf.
27.	Redness of urine and other urinary problem.	*Bhut raj Musinto*	*Lygodium pinnatifidum* (L.) Sw.
28.	Burning sensation, fatigue, skin diseases, cough, and fever, blood dysentery.	*Jaba*	*Hibiscus rosa-sinensis* L.
29.	Fever and acidity.	*Neem*	*Azadirachta indica* A. Juss.
30.	Headache	*Kurdadra*	*Parabaena sagittata* Miers.
31.	Urinary diseases	*Nimuka Tambarki*	*Stephania hernandifolia* Walp.
32.	Insomnia	*Akundi*	*Stephania japonica* (Thumb.) Miers.
33.	Blood sugar	*Kuchuli*	*Ficus hispida* Linn. f.
34.	Fungal infection of feet.	*Surangi*	*Morinda citrifolia* L.
35.	Rheumatism, leucorrhea, dysentery, and blood dysentery, night blindness.	*Gandhalpata*	*Paederia scandens (Lour.) Merr.*
36.	Fever, liver disorder.	*Ashshewra*	*Glycosmis arborea* (Roxb.) DC.
37.	Stomach problem	*Bel*	*Aegle marmelos* (L.) Corr ex. Roxb.
38.	Boils and tumors, pneumonia.	*Mithapata*	*Scoparia dulcis* Roxb.
39.	Dog bite	*Dhatura*	*Datura stramonium* L.
40.	Toothache	*Kantikari*	*Solanum khasianum*
41.	Gonorrhea	*Jingsingphang*	*Clerodendron fragrans* Vent.
42.	Vomiting and diarrhea.	*Gamari*	*Gmelina arborea* Roxb.
43.	Irregular menstrual cycle.	*Segoon*	*Tectona grandis* L.f.
44.	Whitening of hair and memory booster.	*Nishinda*	*Vitex negundo* L.
45.	Skin diseases and inflammation.	*Halud*	*Curcuma longa* L.

Abbreviations: VN: Vernacular name; and SN: scientific name.

9.3.2 ETHNOMEDICINAL PLANTS FOR LIVELIHOOD

NTFPs provide livelihood to a large proportion of the global population. This is more frequent in the case of people living in forest-fringe communities. These communities are dependent on NTFPs in terms of their day-to-day needs of food, medicine, employment, income (Shukla & Chakravarty, 2012; Verma & Paul, 2016; Lepcha et al., 2018, 2020). NTFPs play a role as a safety net for reducing poverty and scarcity and are considered as a flagship of their socio-cultural and economic development (Angelsen & Wunder, 2003; Suleiman et al., 2017). However, the communities under study at present have undergone a huge transformation and reaped the advantage of Government schemes for mainstreaming them. Table 9.2 presents the primary occupations found in tribal families of forest fringe communities of the study area, whereas Figure 9.1 depicts a comparative picture of different occupations and their contribution to their livelihood.

TABLE 9.2 Primary Occupations of the Indigenous Communities in the Study Area

SL. No.	Occupation	Percent of Families	Annual Income (INR ₹)
1.	NTFP	4.0	36,300.00
2.	Agriculture	35.0	42,667.00
3.	Animal husbandry	4.0	41,375.00
4.	Business	16.0	1,15,800.00
5.	Service	35.0	1,37,689.00
6.	Others	6.0	61,629.00

Source: Primary Survey.

From Table 9.2, it is seen that most of the indigenous families were extracting their primary livelihood from agriculture (35% with an annual income of INR ₹42,667.00 on average) and service (35% with an annual income of INR ₹1,37,689.00 on average). As many as 16% of the families were also practicing business as their primary livelihood with an average income of INR ₹ 1,15,800.00 per annum, whereas only 4% of families were living with the collection of NTFP as their primary occupation with a lowest annual income of INR ₹36,300.00 from this source. Some of the families were also extracting their primary livelihood from animal husbandry (4%) and other sources (6%) like generic occupation, artisan, and van rickshaw pulling. However, Figure 9.1 depicts that although NTFP provides the least income as a primary occupation (average income for only 4% of families who only undertake NTFP as a primary occupation), it contributes 11.10% of the total annual income for all the families as a primary or secondary

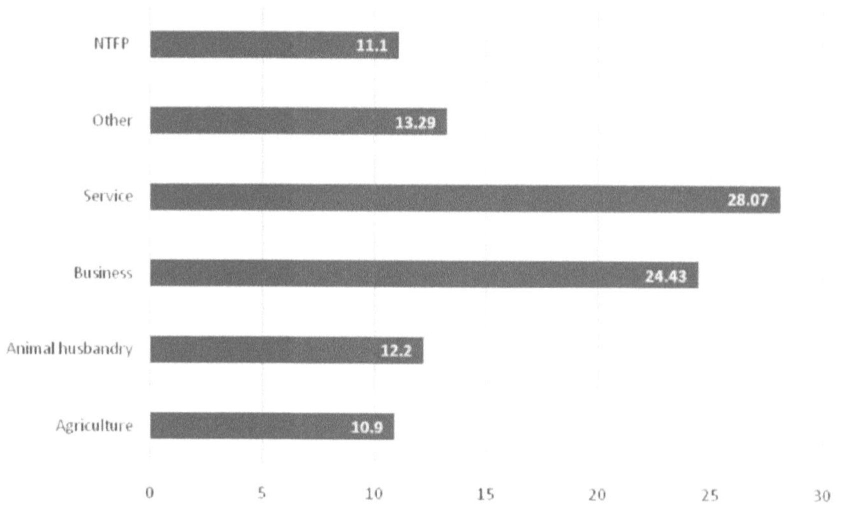

FIGURE 9.1 Comparative contribution of NTFP on livelihood.

Source: Primary Survey.

source. Considering all sources, here again, service contributes the highest proportion (28.07%) of the total annual income, and agriculture contributes the least (10.90%).

9.3.3 PREFERENCES OF MEDICINAL PLANTS OVER CONVENTIONAL TREATMENT (CT)

A preference matrix ranking with scorecard is presented in Table 9.3 to represent the comparison between conventional treatment methods (TMs) versus treatment with local medicinal plants.

It is seen that from the eight parameters, locally available (LA), low cost (LC), alternative usage, having knowledge and less adverse effect had got more average score (AS) (with values of 7.40, 8.40, 8.40, 5.00, and 8.80, respectively) in case of treatment with local medicinal plant than the conventional treatment method. The indigenous communities collect medicinal plants directly from the forest and prepare the medicine in their home, so the cost is very less than the conventional treatment method. In case of effectiveness and market demand (MD), conventional treatment method was more effective than the treatment with ethnomedicinal plant because the conventional method was more specific and gives instant cure to the disease in comparison to the local treatment method. Considering preference of the

TABLE 9.3 Comparison between Conventional and Ethnomedicinal Plant Treatment Method

TM	RG	LA	Ef	LC	AU	Pr	HK	MD	LAE	TS
CM	Gr-1	6	8	1	2	4	1	8	5	35
	Gr-2	5	9	1	3	6	2	8	4	38
	Gr-3	8	9	4	2	3	0	9	3	38
	Gr-4	6	8	2	4	6	2	9	6	43
	Gr-5	7	8	2	1	7	0	9	4	38
AS		**6.40**	**8.40**	**2.00**	**2.40**	**5.20**	**1.00**	**8.60**	**4.40**	**38.40**
ET	Gr-1	7	5	8	8	7	6	3	9	53
	Gr-2	7	6	9	8	5	5	3	8	51
	Gr-3	8	4	9	8	6	3	5	8	51
	Gr-4	9	6	8	8	5	5	4	10	55
	Gr-5	6	3	8	10	4	6	6	9	52
AS		**7.40**	**4.80**	**8.40**	**8.40**	**5.40**	**5.00**	**4.20**	**8.80**	**52.40**

Note: Mann-Whitney U-value = 21.00 (p = 0.24).
Abbreviations: TM: Treatment methods (CT: conventional treatment; ET: ethnomedicinal treatment); AS: average score; RG: ranking groups; LA: locally available; Ef: effectiveness; LC: low cost; AU: alternative utilization; Pr: preference; HK: having knowledge; MD: market demand; LAE: less adverse effects; and TS: total score.

local people, both get almost similar score viz. 5.20 and 5.40, respectively. The overall score of all the parameters showed that treatment with ethnomedicinal plant was more popular than the conventional treatment method in the forest fringe areas of North Bengal. These findings are in consistent with the findings of different previous works (Engebretson, 2002; Conboy et al., 2007; Rishton, 2008; Schmidt et al., 2008). These studies concluded that the traditional people perceived the folk treatment method through ethnomedicinal plants serves a holistic ailment rather than treating only specific problem. Traditional Chinese medicines which also uses indigenous TMs based on traditional herbs is also an example of such holistic treatment and is being used over more than 3,000 years (Xutian et al., 2009). The statistical value indicates that there is no significant difference between the two methods of treatment.

9.4 CONCLUSION

Based on the previous discussion, it can be concluded that medicinal plants as a component of NTFP have a significant contribution to the economic as well as the daily livelihood of the indigenous communities living in and around forest fringes of North Bengal. These plants are similarly preferred as a conventional treatment method in these communities. In the forest fringe areas, there is potential to establish an entrepreneurship on medicinal plants among the youths for their economic development. So, the youths of these communities can be employed by staying in their homestead instead of going outside for livelihood. Following extension strategies may enhance cultivation of these plants among these communities: Economical recognition of medicinal plants as a potential occupation of the forest fringe communities. The government can develop expertise among the tribes in using medicinal plants for human health. The extension officer can create awareness among the tribes to cultivate medicinal plants in their home. Providing improved package and practices for medicinal plant cultivation. Motivate the tribes towards more medicinal plant cultivation. Providing them the information about demand of the medicinal plants about its cultivation and market will improve their livelihood with better healthcare as well.

KEYWORDS

- **cultivation**
- **extension strategy**
- **forest fringe communities**
- **livelihood**
- **market demand**
- **medicinal plants**

REFERENCES

Angelsen, A., & Wunder, S., (2003). *Exploring the Forest Poverty Link: Key Concepts, Issues, and Research Implications*. CIFOR Bogor, Indonesia.

Anonymous, (2001). *Statistical Data for West Bengal*. Government of West Bengal, Calcutta.

Anonymous, (2002). *Traditional Medicine Strategy 2002–2005*. World Health Organization Rome. Available at: www.who.int/medicines/library/trm/trm_start_eng.pdf (accessed on 18 December 2023).

Bhattacharya, K., (2004). *B for Bengal, T for Tribes* (pp. 45–48). Papyrus, Calcutta.

Bhattacharyya, U. C., (1997). *Flora of West Bengal* (Vol. I). Botanical Survey of India, Kolkata.

Bose, D., Ghosh, R. J., Das, M. S., Datta, T., Das, M. S., & Biswas, H., (2015). Medicinal plants used by tribals in Jalpaiguri district, West Bengal, India. *Journal of Medicinal Plants Studies*, *3*, 15–21.

Chambers, R., (1988). *RRA Notes*. IIED, London.

Conboy, L., Kaptchuk, T. J., Eisenberg, D. M., Gottlieb, B., & Acevedo-Garcia, D., (2007). The relationship between social factors and attitudes toward conventional and CAM practitioners. *Comple. Ther. Clin. Pract.*, *13*, 146–157.

Das, A. P., Ghosh, C., Sarkar, A., & Biswas, R., (2007). Ethnobotanical studies in India with notes on Terai-Duars and Hills of Darjeeling and Sikkim. *NBU J. Pl Sci.*, *1*, 67–83.

Das, S. N., Janardhanan, K. P., & Roy, S. C., (1983). Some observations on the ethnobotany of the tribes of Totopara and adjoining areas in Jalpaiguri Districts of West Bengal. *Journal of Economic and Taxonomic Botany*, *4*(2), 453–474.

Dutta, B. K., & Dutta, P. K., (2005). Potential of ethnobotanical studies in North East India: An overview. *Indian Journal of Traditional Knowledge*, *4*(1), 7–14.

Engebretson, J., (2002). Culture and complementary therapies. *Complement Ther. Nurs. Midwifery*, *8*, 177–184.

Jain, D. L., Baheti, A. M., Jain, S. R., & Khandelwal, K. R., (2010). Use of medicinal plants among tribes in Satpuda region of Dhule and Jalgaon districts of Maharashtra: an ethnobotanical survey. *Indian Journal of Traditional Knowledge*, *9*(1), 152–157.

Jain, S. K., & Mudgal, V. A., (1999). *Hand Book of Ethnobotany*. Bishen Singh Mahendra Pal Singh, Dehra Dun, India.

Jeyaprakash, K., Ayyanar, M., Geetha, K. N., & Sekar, T., (2011). Traditional uses of medicinal plants among the tribal people in Theni districts (Western Ghats), Southern India. *Asian Pacific Journal of Tropical Biomedicine*, 20–25.

Kala, C. P., (2005). Current status of medicinal plants used by traditional Vaidyas in Uttaranchal State of India. *Ethnobotany Research & Application*, *3*, 267–278.

Lepcha, L. D., Shukla, G., Pala, N. A., Vineeta, Pal, P. K., & Chakravarty, S., (2018). Contribution of NTFPs on livelihood of forest fringe communities in Jaldapara National Park, India. *Journal of Sustainable Forestry.* https://doi.org/10.1080/10549811.2018.1528158.

Lepcha, L. D., Vineeta, Shukla, G., & Chakravarty, S., (2020). Livelihood dependency on NTFPs among Forest dependent communities. *Indian Forester, 146*, 603–612.

Payyappallimana, U., & Fadeeva, Z., (2013). *Traditional Knowledge and Biodiversity*. United Nations University-IAS, Yokohama.

Purohit, S. S., & Vyas, S. P., (2004). *Marketing of Medicinal and Aromatic Plants in Rajasthan*. National Consultative Workshop on Medicinal and Artomatic Plants, held at GBPUAT, Pantnagar.

Rao, R. R., (1994). *Biodiversity in India (Floristic Aspects)*. Bishen Singh Mahendra Pal Singh, Dehra Dun.

Rishton, G. M., (2008). Natural products as a robust source of new drugs and drug leads: Past successes and present-day issues. *American Journal of Cardiology*, *101*, 43–49.

Sarkar, A., & Das, A. P., (2010). Ethnobotanical formulations for the treatment of jaundice by the Mech tribe in Duars of West Bengal. *Indian Journal of Traditional Knowledge*, *9*(1), 134–136.

Schmidt, B., Ribnicky, D. M., Poulev, A., Logendra, S., Cefalu, W. T., & Raskin, I., (2008). A natural history of botanical therapeutics. *Metabolism*, *57*, S3–S9.

Shukla, G., & Chakravarty, S., (2012). Ethnobotanical plant use of Chilapatta reserved forest in West Bengal. *Indian Forester*, *138*, 1116–1124.

Singh, K. M., & Jha, A. K., (2018). *Medicinal and Aromatic Plants Cultivation in Bihar, India: Economic Potential and Condition for Adoption*. http://dx.doi.org/10.2139/ssrn.2213308.

Suleiman, M. S., Wasonga, V. O., Mbau, J. S., Suleiman, A., & Elhadi, Y. A., (2017). Non-timber forest products and their contribution to household's income around Falgore game reserve in Kano, Nigeria. *Ecological Processes*, *6*, 23.

Verma, S. K., & Paul, S. K., (2016). Sustaining the non-timber forest products (NTFPs) based rural livelihood of tribal's in Jharkhand: Issues and challenges. *Jharkhand Journal of Development and Management Studies*, *14*, 6865–6883.

Vineeta, Shukla, G., Bhat, J. A., & Chakravarty, S., (2021). Species richness and folk therapeutic uses of ethnomedicinal plants in West Bengal, India: A meta-analysis. *Phyto Plus*. https://doi.org/10.1016/j.phyplu.2021.100158.

Xutian, S., Zhang, J., & Louise, W., (2009). New exploration and understanding of traditional Chinese medicine. *American Journal of Chinese Medicine*, *37*, 411–426.

CHAPTER 10

A Directory of Ethnomedicinal Plants of Reang Ethnic Community in Northeast India

NAZIMUR RAHMAN TALUKDAR,[1,2] ROFIK AHMED BARBHUIYA,[1,2] PARTHANKAR CHOUDHURY,[1,2] and FIROZ AHMED[3]

[1]*Wildlife Conservation Research Center, Department of Ecology and Environmental Sciences, Assam University, Silchar, Assam, India*

[2]*Center for Biodiversity and Climate Change Research, Udayan, Hailakandi, Assam, India*

[3]*Vindhyan Ecology and Natural History Foundation, Mirzapur, Uttar Pradesh, India*

ABSTRACT

A literature review was carried out to prepare a checklist of ethnomedicinal plant species used by the Reang tribe of Northeast India. A total of 230 plant species, represented by 182 genera and 83 families, were listed along with their vernacular name, botanical name, family name, and mode of preparation for diseases. The prominent plant families were Lamiaceae (12 species), Euphorbiaceae (11 species), and Apocynaceae, Asteraceae, Fabaceae, and Zingiberaceae (nine species each), Amaranthaceae (eight species), Rubiaceae (seven species), Acanthaceae, Leguminosae, Malvaceae, and Orchidaceae (six species each), and Combretaceae and Liliaceae (five species each). The leaf was the dominant plant part (39.12%) used for the treatment of diseases, followed by the root (19.24%), bark (10.09%), and fruit (5.68%). Among the disease categories, dermatological diseases were more predominant, followed by pain and inflammation, gastrointestinal disorders, general health problems,

Bioprospecting Ethnomedicinal Plant Resources: Sustainable Utilization and Restoration.
Gopal Shukla, Jahangeer A. Bhat, A. P. Das, and Sumit Chakravarty (Eds.)

and urinogenital disorders. The ethnomedicinal plant species are now threatened due to deforestation and are losing their importance to the new generation. A good number of plants were found to be used for multiple diseases. Such ethnic uses need to be validated through ethno pharmacological studies.

10.1 INTRODUCTION

Numerous plants and plant products are harvested from the forest for a variety of purposes like food, fiber, fodder, ethnomedicine, agriculture, domestic, and industrial uses (Dattagupta & Gupta, 2014; Talukdar & Gupta, 2018; Hazarika & Marak, 2019; Pachuau & Dutta, 2020). The NTFPs are used for subsistence and also the source of livelihood for many people including primary healthcare (Sarmah et al., 2008; Talukdar et al., 2021). This is especially common where basic health facilities and communication are not available. Traditional ethnomedicinal knowledge is passed from generation to generation. Numerous NTFPs are found in Northeast India as two-thirds of the region is covered with forest (Anon., 2015). It is also the home of large numbers of indigenous communities, and they are dependent on many forest products for their basic healthcare (Kushwaha, 2006). One such indigenous community is Reang and they are inhabitants of Tripura and Mizoram. In Tripura, Reang is the second largest indigenous community after Tripuri and were considered as the primitive community in the state of Tripura with a total population of about two lakhs speaking a dialect *Kaubru* closely related to Tibeto-Burman (https://trci.tripura.gov.in/tribal_population; Shil et al., 2014). The community practices slash and burn type of shifting cultivation. They have traditional cultural values and use a variety of plants for primary healthcare which is passed from generation to generation through verbal communication.

It is estimated that about 80% of the population of the world living in developing countries uses ethnomedicine for their healthcare (Anon., 2001). These plants are easily available, LC, effectiveness, and with no side effects (Odhav et al., 2010). Ethnobotanical studies focus on documenting traditional medicinal knowledge of plants and their ways of practicing for particular diseases. Many of such plants have already established pharmaceutical importance and provided a gateway for modern drugs (Newman et al., 2000; Kunle et al., 2012). However, much of the TK is fast disappearing with the loss of local cultural values (Reyes-Garcíaa et al., 2013; Kaunga, 2017). Therefore, priorities are being given to document such plants and integration of knowledge towards conservation of the species (Heneidy et al., 2017). Many studies were undertaken in Northeast India to document the ethnomedicinal values of plants (Bhardwaj & Ghakar, 2005; Dutta & Dutta, 2005; Barbhuiya et al.,

2009; Talukdar & Gupta, 2018). Many earlier studies had documented the ethnomedicinal uses of the Reang community (Das et al., 2008, 2009; Shil & Choudhury, 2009a, b; Banik et al., 2010; Sen et al., 2011; Debnath et al., 2014; Shil et al., 2014). The present study was an attempt to gather all the NTFPs used by the Reang tribe of Northeast India in their primary healthcare practices. The cumulative information will help in pharmaceutical tests and to create greater consciousness for the conservation of species.

10.2 MATERIAL AND METHODS

A systematic review was carried out to prepare the checklist of NTFPs used for ethnomedicinal purposes by the Reang ethnic community of Northeast India. Relevant published literature in the form of technical reports and scientific articles available on the internet were searched, downloaded, and reviewed. Various keywords such as Reang tribe, ethnobotany, ethnomedicine, ethno-pharmacology, and NTFPs were used in search engines to find the published literature. The gathered information was analyzed, and the checklist was prepared, including the botanical name, their family, vernacular name, and method of using particular parts for diseases. The diseases were categorized into major groups like dermatological disorders, pain and inflammation, gastrointestinal disorders, general health problems, urinogenital disorders, fever, diabetes, respiratory disorders, and hepatic disorders.

10.3 RESULTS AND DISCUSSION

Review of earlier studies on ethnomedicinal plant uses by Reang ethnic community of Northeast India resulted in a directory of 230 plants of ethnomedicinal importance, which were represented by 182 genera under 83 families (Table 10.1). Families represented by more than two species are illustrated in Figure 10.1. The genera *Crotalaria*, *Alternanthera*, *Cassia*, *Cissus*, *Curcuma*, *Ficus*, *Phyllanthus*, and *Terminalia* were represented by three or more species, while others were represented by less than three species each. The majority of the documented species were used against a single ailment, while a few of them were used against multiple diseases (Table 10.2). Only a few plant species were commonly enlisted by the majority of the studies, while the majority of the plants were commonly documented by only a few studies. For example, the ethnomedicinal importance of *Streblus asper* was mentioned in five reports, *Andrographis paniculata*, *Plumbago zeylanica*, *Sterculia villosa* were mentioned in four reports, 17 species were mentioned in three reports, and 59 species were mentioned in two reports, and the remaining species were each mentioned in a single report.

TABLE 10.1 Ethnomedicinal Plant Species Used by the Reang Ethnic Community

SL. No.	S; F	VN	PU	D/A	T
1.	*Abrus precatorius* L.; Papilionaceae	*Sonakaich, takharichum*	L, Sd, R	Joint, muscular, and abdominal pain, poisonous bite.	Paste applied externally.
2.	*Acanthus leucostachys* Wallich; Acanthaceae	*Mussali*	L	Blood clotting, antiseptic.	Paste applied externally.
3.	*Achyranthes aspera* L.; Amaranthaceae	*Apang, Samtai, Ulta lengra*	Wp	Bleeding in time of delivery, kidney, urinary disorders.	Decoction taken orally.
4.	*Achyranthes bidentata* Blume; Amaranthaceae	*Ultalingra*	L	Water born foot disease.	Paste applied externally.
5.	*Adhatoda vasica* L.; Acanthaceae	*Basak, Basakblai*	L	Cough, cold, dysentery, fresh wounds.	Decoction taken orally for cough, cold, dysentery, and applied externally for blood clotting.
6.	*Ageratum conyzoides* L.; Asteraceae	*Shyamnama, Uchunti, Dandotpal*	L	Blood clotting, eye trouble, jaundice, stomach disorders, menstruation pain, wound.	Extract applied externally in fresh cuts, wounds, and in eye trouble and consumed orally once daily to treat jaundice, stomach disorder and menstruation pain.
7.	*Aloe indica* L.; Liliaceae	*Ghritakumari, Ghrita kanchan*	L	Hair fall, stomach disorders.	Juice prepared from leaf is taken orally for stomach problem and used externally on hair for reducing hair fall.
8.	*Alpinia allughas* Rosc.; Zingiberaceae	*Taru*	Rh	Bronchitis, rheumatic pain.	Decoction prepared by boiling 10 g each of stem and rhizome, 10 g of powdered Moong pulse and small amount of asafetida mixed (8–9 cups) with water and taken twice a day to get relief.
9.	*Alpinia nigra* (Gaertn.) Burtt.; Zingiberaceae	*Thri*	Rh	Cut and wounds.	Extract applied externally.

TABLE 10.1 *(Continued)*

SL. No.	S; F	VN	PU	D/A	T
10.	*Alstonia scholaris* R.Br.; Apocynaceae	*Chatim, Chaituang*	B, Lx, L	Headache, stomach problem, allergy, abscesses.	Leaf is used against headache; latex is applied in allergy and abscesses; bark decoction of bark is used against stomach problem.
11.	*Alternanthera brasiliana* Linn.; Amaranthaceae	*Lalagunsita (Reang)*	Rh	Menstrual problem	Extract mixed with water is consumed.
12.	*Alternanthera paronychioides* A. St.-Hil.; Amaranthaceae	*Khurmurmu*	L	Cut and wounds	Extract is applied externally.
13.	*Alternanthera philoxeroides* Griseb.; Amaranthaceae	*Helencha, Alancha*	Wp	Stomach disorder, diarrhea.	Decoction is taken orally.
14.	*Amaranthus gracilis* Desf.; Amaranthaceae	*Khentamaira*	Wp	Allergenic swelling	Paste is applied externally.
15.	*Amaranthus spinosus* L.; Amaranthaceae	*Khentyimyra*	L, St	Dysentery	Extract taken orally.
16.	*Amischotolype hookeri* (Hassk.) Hara; Commelinaceae	*Msautotra*	L	Naval pain	Decoction is consumed with rice.
17.	*Ampelocissus divaricata* (Wall. ex Laws.) Planch.; Vitaceae	*Yosrem*	L	Blood clotting, antiseptic	Leaves of the plant are crushed with leaves of China rose to prepare paste and applied.
18.	*Anacolosa ilicoides* Mast.; Olacaceae	*Musafama*	L	Pain killer, antiseptic	Paste applied orally.
19.	*Andrographis paniculata* (Burm. f.) Wall. ex Nees.; Acanthaceae	*Kairata, Chorpai, Kalamegh, Kalmegh balai*	Wp	Small worms, stomach troubles, dysentery, fever.	Extract consumed orally.
20.	*Angiopteris evecta* (Forst.) Hoffm.; Angiopteridaceae	*Skemamuidui*	Rh	Bone fracture, carbuncle.	Paste applied externally and dressed.

TABLE 10.1 *(Continued)*

SL. No.	S; F	VN	PU	D/A	T
21.	*Aphanamixis polystachya* (Wall.); Meliaceae	*Pitra*	B, L	Stomach pain, skin disease.	Paste applied externally against skin disease; bark and leaf decoction are applied against stomach pain.
22.	*Ardisia paniculata* Roxb.; Myrsinaceae	*Siphufu*	B	Burn	Paste applied externally.
23.	*Argereia nervosa* (Burm. f.) Hort.; Convolvulaceae	*Konglabli*	R	Sprain	Paste applied externally.
24.	*Asclepias curassavica* L.; Asclepiadaceae	*Parija*	R	Blood coagulation.	Paste applied externally.
25.	*Asparagus racemosus* Willd.; Liliaceae	*Shatamuli*	L, R	Diarrhea, fever, measles, stomach pain.	Leaf and root decoction are eaten.
26.	*Azadirachta indica* A. Juss.; Meliaceae	*Neem, Ing*	L, B	Leprosy, skin infections, chicken pox, diabetes, respiratory disorders.	Leaf paste is applied for leprosy, skin infections and chicken pox; leaf and bark decoction are consumed for diabetes and respiratory disorders.
27.	*Baccaurea ramiflora* Lour.; Phyllanthaceae	*Samai*	R	Insect sting	Paste is applied externally.
28.	*Barleria prionitis* Linn.; Acanthaceae	*Ziodi*	L	Pruritis, rubefacient, blotch.	Pills prepared by mixing the leaves and fruits of Terminalia chebula and sulfur-dioxide (*Gandhak*) and 1–2 pills are mixed with coconut oil for body massage.
29.	*Bauhinia purpurea* L.; Caesalpiniaceae	*Sitakaia*	B	Leucorrhea, menstruation problem.	Extract is mixed with honey and taken orally.
30.	*Bauhinia variegata* Linn.; Leguminosae	*Blongsitakaya*	L	Bone fracture	Paste is applied externally.

TABLE 10.1 *(Continued)*

SL. No.	S; F	VN	PU	D/A	T
31.	*Begonia thomsonii* A. DC.; Begoniaceae	*Thaichengmokoi*	R	Diarrhea	Roots boiled in water and supernatant is taken orally thrice a day.
32.	*Blechnum orientale* L.; Blechnaceae	*Sikiomamoidu*	P	Antiseptic	Decoction is applied externally.
33.	*Blumea lanceolaria* (Roxb.) Druce; Asteraceae	*Muisingha*	L	Vaginal protrusion.	Cooked and consumed.
34.	*Boehmeria macrophylla* D. Don; Utricaceae	*Muithlimsu*	R	Leucorrhea	Paste mixed with water consumed orally.
35.	*Boerhaavia diffusa* L.; Nyctaginaceae	*Punurnava, Twisa jolap*	L, R, Sd	Asthma, urinary disorders, blood impurities.	Decoction consumed orally.
36.	*Bombax ceiba* L.; Malvaceae	*Shimul, Borchuk, Bochu*	L, R	Head, tooth, ear, and body pain, bone fracture.	Decoction is on head, tooth, ear, and body; leaf paste applied externally on fractured bone.
37.	*Bougainvillea glabra* Choisy; Nyctaginaceae	*Khumlekha*	F	Diabetes	Decoction is consumed orally.
38.	*Brunella vulgaris* L.; Lamiaceae	*Lamaku*	B	Toothache	Fresh bark extract is applied.
39.	*Bryophyllum calycinum* Salisb.; Crassulaceae	*Ginghor (Reang)*	L	Diarrhea, indigestion.	Extract is consumed orally.
40.	*Butea monosperma* (Lamk.) Taub.; Fabaceae	*Palas*	B, F	Dysentery, toothache, and wound.	Decoction is taken orally for dysentery; bark or leaf paste is applied externally.
41.	*Callicarpa arborea* Roxb.; Lamiaceae	*Chamathuichla*	B, R	Leucorrhea, gastritis	Pills or decoction is consumed orally.
42.	*Calotropis gigantea* LR Br.; Asclepiadaceae	*Akanda, Angon*	F, L, R	Asthma, cough, fever, gum infection.	Decoction is taken orally.

TABLE 10.1 *(Continued)*

SL. No.	S; F	VN	PU	D/A	T
43.	*Calotropis procera* (Aiton) W.T. Aiton; Apocynaceae	*Shepalimphang, Hangra pata*	L	Rheumatism, eczema, and other skin diseases.	Poultice is applied externally.
44.	*Cassia alata* L.; Fabaceae	*Khaspoiblai, Dadmari, Dangduraja*	L	Skin infection, ringworm infection.	Leaf is pasted with *Alliumsativum* leaf and applied externally.
45.	*Cassia fistula* L.; Fabaceae	*Sonali, Bandarlati, Sundal, Skthuri*	B, Fr, L, Sd	Constipation, flatulence, inflammation, tapeworm, wounds.	Decoction is consumed orally.
46.	*Cassia tora* L.; Caesalpiniaceae	*Luthamphang, Edanchi, Lotho, Chakunda*	L, Sd	Worm infection, vision problem, liver disease, leprosy.	Paste is applied externally on the skin to control ringworm and leprosy; Leaf extract is applied to treat eye problems and taken orally against liver problems.
47.	*Catharanthus roseus* (L.) G. Don.; Apocynaceae	*Khumboiragi*	F, L, R	Diabetic, insect bite, wound.	Fresh leaves consumed in the morning to regulate blood sugar, leaf, and root paste applied topically on stings and wound.
48.	*Celosia argentia* L.; Amaranthaceae	*Khumchakorma*	L	Fever	Extract taken orally.
49.	*Centella asiatica* (L.) Urba; Apiaceae	*Thankuni, Masundwi, Samsota*	Wp	Asthma, stomach disorder, dysentery, leprosy.	The plant is eaten raw.
50.	*Cheilanthes tenuifolia* (Burm. f.) Sw.; Pteridaceae	*Dalamkhundruj*	Fd	Antiseptic	Paste is applied externally.
51.	*Chenopodium ambrosioides* Linn.; Chenopodiaceae	*Batto*	L, St	Scabies	Decoction applied externally.

TABLE 10.1 *(Continued)*

SL. No.	S; F	VN	PU	D/A	T
52.	*Chonemorpha fragrans* (Moon) Alston; Apocynaceae	*Yiangma*	L	Blood coagulation	Paste is applied externally.
53.	*Chonemorpha griffithii* Hook. f.; Apocynaceae	*Yiangma*	L	Blood clotting, antiseptic.	Paste is applied externally.
54.	*Cinnamomum camphora* (Linn.) Sieb.; Lauraceae	*Ducaphur*	B	Cardiac problem	Decoction is consumed orally.
55.	*Cinnamomum tamala* (Buch. Ham.); Lauraceae	*Tejpata*	B, L	Stimulant, spice	Culinary
56.	*Cissampelos pareira* L.; Menispermaceae	*Tousabachoriaha*	L	Fracture	Paste applied on fracture and dressed.
57.	*Cissus adnata* Roxb; Vitaceae	*Bhaitya lata, Khumthaili*	B, L	Fracture, pain, carbuncle.	Paste applied externally.
58.	*Cissus assamica* (Laws.) Craib.; Vitaceae	*Dudebra*	R	Carbuncle	Paste applied externally.
59.	*Cissus quadrangularis* L.; Vitaceae	*Harbhanga*	Wp	Bone fracture, indigestion.	Paste applied externally on fracture and decoction consumed orally for indigestion.
60.	*Clausena heptaphylla* (Roxb.) Wt. & Arn.; Rutaceae	*Sataukura*	Fr	Cough, asthma	Juice mixed with rock salt consumed orally.
61.	*Clerodendrum indicum* L.; Verbenaceae	*Brahma jasthi, Asubahang*	L, R	Asthma, cough, rheumatism, tooth decay.	Extract taken orally for cough and asthma; decoction applied externally for rheumatism and tooth decay.
62.	*Clerodendrum japonicum* (Thunb.); Verbenaceae	*Moishebak* (*Reang*)	L	Rheumatism	Decoction is taken orally.

TABLE 10.1 *(Continued)*

SL. No.	S; F	VN	PU	D/A	T
63.	*Clerodendrum paniculatum* L.; Lamiaceae	*Boiec*	R	Typhoid	Roots of this plant, *Tamarindus indica* and *Ananas comosus* are decocted together and consumed orally twice a day.
64.	*Clerodendrum philippinum* (Thumb); Lamiaceae	*Maishebak*	R	Rheumatism	Extract is consumed orally.
65.	*Clerodendrum viscosum* Vent.; Lamiaceae	*Araokhuinaima, Baitgach, Chayakkhui bufang, Ghetu*	L, R	Cough, dysentery, inflammation, skin disease	Extract is consumed orally.
66.	*Clitoria ternatea* L.; Fabaceae	*Aparajita, Khumalali*	L, R, Sd	Constipation, mental disorder, memory booster.	Decoction is taken orally.
67.	*Coccinia grandis* L.; Cucurbitaceae	*Telakuchi pata, Takhahaichumu*	Wp	Blood sugar; diabetes, scabies, skin infection.	Leaves are eaten to regulate blood sugar; paste applied externally against scabies and skin infection.
68.	*Coccinia indica* (L.) Voigt; Cucurbitaceae	*Taokhaspoitha*	L	Diabetes, rheumatism	Extract is taken orally for rheumatism and fresh leaves eaten to regulate blood sugar.
69.	*Cochlospermum religiosum* (L.) Alst.; Bixaceae	*Kenkechi*	R	Urinary infection.	Decoction is taken orally.
70.	*Codariocalyx motorius* (Houtt.) H. Ohashi; Leguminosae	*Turkimondon*	L	Health tonic	Decoction is taken orally.
71.	*Combretum decundrum* Roxb.; Combretaceae	*Kechi*	R	Malarial fever	Decoction is taken orally.
72.	*Combretum latifolium* Blume.; Combretaceae	*Kaechiblai (Reang)*	L	Dysentery, goiter	Extract used consumed taken orally.

TABLE 10.1 *(Continued)*

SL. No.	S; F	VN	PU	D/A	T
73.	*Cordyline fruticosa* (L.) A. Chev.; Asparagaceae	*Chanthi*	R	Malarial fever	Pills prepared from root paste taken twice a day.
74.	*Costus speciosus* (Koenig) Sm.; Costaceae	*Keu, Mailuma kothoma*	R	Constipation, inflammation, worm infection.	Decoction is taken orally.
75.	*Crotalaria albida* Heyne ex Roth.; Papilionaceae	*Ban atasi, Zhunzhuni, Bishut*	L, R	Indigestion, swelling.	Decoction is taken orally.
76.	*Crotalaria pallida* Ait.; Fabaceae	*Karamakakho, Skamyouhma kakho*	L	Ringworm	Paste is applied externally.
77.	*Crotalaria spectabilis* Roth.; Leguminosae	*Skamayomakakho*	L, R	Health tonic, ringworm.	Decoction consumed orally as tonic and paste applied on infected parts.
78.	*Crotalaria verrucosa* L.; Leguminosae	*Skemakakho*	R	Rheumatism	Pills prepared from root paste are taken orally.
79.	*Croton caudatus* Geis.; Euphorbiaceae	*Jiran*	R	Constipation	Decoction is consumed orally.
80.	*Curculigo capitulata* (Lour.) O. Kuntze; Hypoxidaceae	*Lairdam*	Lb	Cut and wounds.	Paste is applied externally.
81.	*Curculigo orchioides* Gaertn.; Hypoxidaceae	*Lairudamsla*	Rh	Diarrhea, jaundice, piles, sex stimulant.	Rhizome is consumed orally.
82.	*Curcuma aromatica* Salisb.; Zingiberaceae	*Bloungkarma*	Rh	Stomach ailments, skin infection.	About 20 gm of fresh rhizome is taken with molasses in empty stomach for about a week to treat stomach problems and paste is applied externally for dermal problems.
83.	*Curcuma caesia* Roxb.; Zingiberaceae	*Karmoksom*	Rh	Malarial fever	Pills prepared from rhizome are taken.

TABLE 10.1 *(Continued)*

SL. No.	S; F	VN	PU	D/A	T
84.	*Curcuma domestica* Valeton; Zingiberaceae	*Karma*	L, Rh	Diabetes	Decoction is consumed orally.
85.	*Cuscuta reflexa* Roxb.; Convolvulaceae	*Banalata, Tarulata*	Wp	Cough, jaundice, kidneys, and Uterine problems, viral infection, itchy skin.	Decoction is used externally.
86.	*Cyathea contaminans* (Wall. ex Hook.) Copel.; Cyatheaceae	*Bongreng*	C	Anticoagulant, antiseptic.	Paste is prepared from apical soft portion of the caudex cut to apply topically.
87.	*Cyathea henryi* (Bak.) Copel.; Cyatheaceae	*Bongreng*	St	Anticoagulant	Paste prepared from apical portion of the trunk is applied externally.
88.	*Cymbidium aloifolium* (L.) Sw.; Orchidaceae	*Khelang*	R	Bone fracture	Aerial roots are pasted and applied as poultice with bamboo splints as support.
89.	*Cymbopogon citratus* (DC.) Stapf.; Poaceae	*Chaimandar*	L	Cough and cold.	Decoction is taken orally.
90.	*Cynodon dactylon* (L.) Pers.; Poaceae	*Dinedurba*	Wp	Blood clotting, cough, dysentery, dermatitis, painful menstruation.	Extract is taken orally to cure dysentery, applied externally to get relief from headache, stop bleeding and also mixed with flower extract of *Hibiscus rosa sinensis* with sugar to consume orally for a week once a day for overcoming menstruation problem.
91.	*Cyperus rotundus* L.; Cyperaceae	*Mutha*	Bl, L	Skin disease, urinary disorders.	Paste is consumed orally to treat urinary disorders and applied topically to cure skin disease.
92.	*Datura stramonium* Linn.; Solanaceae	*Kalodutra*	L, Sd	Anxiety, body pain, rheuma-tism, dermal problems.	Paste is used as poultice to treat rheumatism and body pain; seeds are taken orally once a day against dermal problems and anxiety.

TABLE 10.1 *(Continued)*

SL. No.	S; F	VN	PU	D/A	T
93.	*Desmodium motorium* (Houtt.) Merr.; Papilionaceae	*Turkimondon*	L	Improve health	The leaves are cooked with meat for healthy body.
94.	*Desmos longiflorus* (Roxb.) Safford; Annonaceae	*Khorjun*	B	Toothache	Extract prepared by boiling in water and used as mouth wash.
95.	*Dicarnopteris lineris* (Burm. f.) Underwood; Gleicheniaceae	*Muikandochla*	Fd	Throat pain	Decoction is taken internally.
96.	*Dillenia indica* L.; Dilleniaceae	*Chailta, Thaiplok*	Fr, L	Cough, fever, and wound.	Extract is consumed orally.
97.	*Dioscorea bulbifera* L.; Dioscoreaceae	*Rangrang*	T	Health tonic.	Extract is consumed orally.
98.	*Diplazium esculentum* (Retz.) Sw.; Athyriaceae	*Sikiomamoidu, Maikhando*	Fd	Improve digestion and health.	Fronds cooked as vegetable.
99.	*Drymoglossum heterophyllum* (Linn.) Trimen; Polypodiaceae	*Sikitang*	P	Fracture	Paste applied on affected bones and supported with bamboo splints.
100.	*Drynaria quercifolia* (Linn) J. Smith.; Polypodiaceae	*Bandartala*	Rh	Blood clotting, fracture, cardiac problem.	Paste mixed with molasses are consumed orally during cardiac problem and applied as poultice on fractured area with bamboo splints as support and also to stop bleeding.
101.	*Drypetes assamica* (Hook. f.) Pax & Hoffm.; Euphorbiaceae	*Lamaku*	B	Toothache	Fresh bark decoction of is applied on gums.
102.	*Eclipta prostrata* (L.) L. Mant.; Asteraceae	*Profit*	L	Cut and wounds.	Paste is applied topically.

TABLE 10.1 *(Continued)*

SL. No.	S; F	VN	PU	D/A	T
103.	*Ehretia acuminata* R. Br.; Boraginaceae	*Dusukchaoma*	L	Fracture	Paste is applied on affected parts.
104.	*Entada pursaetha* DC.; Mimosaceae	*Kung*	Sd	Body pain	Paste is used for body massage.
105.	*Entada rheedii* Spreng.; Leguminosae	*Kung*	Sd	Body pain	Paste is used for body massage.
106.	*Eria pubescens* (Hook.) Lindl. ex Stend.; Orchidaceae	*Khelangpachi*	Sd	Ear sore	Seeds of ripe fruits are powdered and about 1 gm of it is applied twice daily.
107.	*Euphorbia hirta* L.; Euphorbiaceae	*Dara dudhi, Shyamkhai, Khemychu*	F, L, St	Cough, gonorrhea, pimples, skin disease, throat pain.	Decoction is applied externally.
108.	*Eusteralis stellata* (Lour.) Panig.; Lamiaceae	*Mulomandar*	L	Night blindness	Paste of the leaves and kidney of goat (10 g) is taken orally twice a day for a week.
109.	*Ficus hispida* L.; Moraceae	*Thaichuwathia*	H	Cough	Paste is consumed orally.
110.	*Ficus pumila* L.; Moraceae	*Dusalua*	L	Anti-lice	Extract is applied externally.
111.	*Ficus rigida* Jack.; Moraceae	*Jola*	R	Laxative	One pill prepared from paste is consumed at bedtime.
112.	*Flacourtia indica* (Burm. f.) Merr.; Flacourtiaceae	*Lukluki*	L	Carbuncle	Paste is applied topically.
113.	*Galinsoga parviflora* Cav; Asteraceae	*Gangaful, Garingburani sam*	Wp	Fever, hepatic problem, inflammation, insect bite, toothache, wound.	Paste (only aerial parts) is applied externally to treat wound, inflammation, toothache, and insect bite and extract is consumed orally to get relief from fever and liver problem.
114.	*Gelonium multiflorum* A. Juss.; Euphorbiaceae	*Thairungtu*	B	Gastritis	Pills prepared by pounding bark pieces are taken orally.

TABLE 10.1 *(Continued)*

SL. No.	S; F	VN	PU	D/A	T
115.	*Globba multiflora* Wall. ex Baker; Zingiberaceae	*Hiching*	Rh	Cough	Half a cup of hot decoction is taken orally twice a day.
116.	*Goniothalamus sesquipedalis* (Wall.) Hook. f.; Annonaceae	*Damnagra*	R	Leucorrhea	Half a cup of supernatant from fresh root paste solution is taken orally thrice a day till it is cured.
117.	*Grewia nervosa* (Lour.) Panigr.; Malvaceae	*Lamaku*	B	Toothache	Extract is applied externally.
118.	*Hedyotis auricularia* L.; Rubiaceae	*Multia lata, Shyamkajal*	L	Abdominal disorder.	Pills prepared from fresh leaves are taken orally.
119.	*Hippeastrum puniceum* (Lam.) Urb.; Amaryllidaceae	*Kalaful*	L	Swelling, skin infection.	Paste is applied externally.
120.	*Holarrhena antidysenterica* L.; Apocynaceae	*Kurchi, Pandhra kuda, Kuichama*	B, R	Cold, dysentery, fever, piles.	Decoction is taken consumed orally.
121.	*Holarrhena pubescens* (Buch. Ham.) Wall ex G. Don.; Apocynaceae	*Kuichama*	B	Dog bite	Pills prepared by pounding bark pieces are consumed.
122.	*Homalomena aromatica* (Spreng.) Schott.; Araceae	*Kamaitru*	Rh	White spot in eye.	Decoction is taken internally.
123.	*Ichnocarpus frutescens* R.Br.; Apocynaceae	*Dugdha lata, Soyalata*	F, L, R	Bleeding in gum, skin disease, fever, kidney disorder.	Paste is applied topically to treat diseases and hardening of gum; decoction is taken orally for fever and kidney disease.
124.	*Jasminum laurifolium* Roxb.; Oleaceae	*Mali*	L	Anti-vomiting	Half a cup leaf decoction mix of this species and *Aeglemarmelos* is taken orally.
125.	*Justicia adhatoda* L.; Acanthaceae	*Asuamfang*	L, R	Pneumonia, cough	Decoction is consumed orally to treat pneumonia and extract is consumed orally for cough.

TABLE 10.1 *(Continued)*

SL. No.	S; F	VN	PU	D/A	T
126.	*Kaempferia rotunda* L.; Zingiberaceae	*Khuntowia*	T	Chest pain	Paste is applied externally.
127.	*Kalanchoe pinnata* (Lam.) Pers.; Crassulaceae	*Zihor*	L	Fever, gallbladder, kidney stone, jaundice, blood clotting, stomach disorder, wound.	Extract is taken orally to cure gallbladder and kidney stone, paste applied topically to stop bleeding and also as antiseptic; pills prepared from leaves are taken orally to treat stomach disorder, fever, and jaundice.
128.	*Lantana camara* L.; Verbenaceae	*Khangsiniha*	L	High blood pressure	Extract is taken orally.
129.	*Leea asiatica* L.; Liliaceae	*Banchalita, Koknal*	L, R	Diabetes, liver disorders, worm infection.	Extract is taken orally.
130.	*Leea indica* (Burm. F.) Merr.; Liliaceae	*Kukur jibba*	R	Diarrhea, stomach pain.	Extract is taken orally.
131.	*Lepisanthes senegalensis* (Poir) Leenh. Sapindaceae	*Khikrakhidang*	B, R	Dog bite, stomach pain	Bark decoction is taken orally to get relief from stomach pain and root paste is applied topically on dog bite.
132.	*Leucas lavandulaefolia* Rees; Labiatae	*Halkusa, Dondo kolos*	L, R, St	Anxiety, skin infection, wound.	Paste is applied topically.
133.	*Leucus aspara* Spreng.; Lamiaceae	*Dron*	F, L	Liver ailments, headache.	Decoction is consumed orally.
134.	*Litsea glutinosa* (Lour.) C. B. Rob.; Lauraceae	*Musafaoma*	L	Gonorrhea	Paste is taken orally.

TABLE 10.1 *(Continued)*

SL. No.	S; F	VN	PU	D/A	T
135.	*Litsea glutinosa* (Lour.) C. B. Rob.; Lauraceae	*Balmaimy*	B, L	Fracture, injury, muscle pain.	Bark powder mixed with goat milk is consumed orally to ease chest pain; bark and stem paste is applied topically and dressed to heal wound and fracture.
136.	*Lycopodium cernuum* Linn.; Lycopodiaceae	*Shibjonta*	Wp	Swelling, itching.	Paste is applied topically.
137.	*Lygodium flexuosum* (L.) Sw.; Schizaeaceae	*Duoreng*	Ra	Headache	Rachis is tied over forehead.
138.	*Mallotus philippensis* (Lamk.) Muell.; Euphorbiaceae	*Boltauthu*	Lb	Leucorrhea	Cooked and consumed.
139.	*Mallotus tetracoccus* (Roxb.) Kurz.; Euphorbiaceae	*Laichoblai*	L	Skin diseases, ring worm.	Paste is applied topically.
140.	*Melastoma malabathricum* L.; Melastomaceae	*Loloti*	Fr, St	Impotency, tooth ache.	Fruit is consumed ray and stem paste applied for toothache.
141.	*Melothria heterophylla* (Lour.) Cogn.; Cucurbitaceae	*Dupoitha*	R	Urinary troubles and blockages.	Decoction is taken orally thrice a day for a week.
142.	*Mentha arvensis* L.; Lamiaceae	*Khumbawbaw*	L	Asthma	Decoction mix prepared from its leaves and *Piper nigrum* seeds are taken internally thrice a day.
143.	*Meriandra benghalensis* Cav.; Labiatae	*Bangal sage*	Wp	Diabetes, blood pressure, skin disease, tonsillitis.	Paste is applied topically over skin infection; decoction consumed orally regulates blood pressure, diabetes, and tonsillitis.
144.	*Merremia umbellata* (Linn.) Hall. f.; Convolvulaceae	*Dungbaomdu, Goiramrang*	L	Bone fracture, dysentery.	Decoction is taken internally against dysentery and paste is applied topically on bone fracture.
145.	*Mesua ferrea* L.; Guttiferae	*Khersai*	Sd	Pain killer	Paste is used for body message.

TABLE 10.1 *(Continued)*

SL. No.	S; F	VN	PU	D/A	T
146.	*Meyna spinosa* Roxb.; Rubiaceae	*Man kata, Gaittara*	Fr	Diabetes, skin disease, hepatic disorder, peptic ulcer.	Immature fruit paste is applied topically to treat skin disease and fruits are consumed fresh to get cure from peptic ulcer, diabetes, and hepatic disorder.
147.	*Microcos paniculata* L.; Malvaceae	*Kushnanaliya, Ashar bufang*	Fr, L	Bone fracture, cold, typhoid fever.	Leaf paste is applied externally on fracture and fruit is consumed to cure cold and typhoid fever.
148.	*Microsorium superficiale* (Bl.) Ching.; Polypodiaceae	*Hulowukto*	Rh	Cough and cold	Paste of its rhizome prepared from rhizome mixed with *Piper nigrum* seeds is consumed orally thrice a day to cure cough and cold.
149.	*Mikania cordata* (Burm. f.) Robinson.; Asteraceae	*Deshmara*	L	Blood clotting	Paste is applied on wound.
150.	*Mikania micrantha* Kunth.; Asteraceae	*Refugi*	L	Blood clotting, antiseptic, diarrhea, eye trouble, headache.	Extract is applied on wound to stop bleeding and as an antiseptic. It is also applied on eyes and over forehead and also consumed orally to control diarrhea.
151.	*Mimosa pudica* L.; Fabaceae	*Shyamsati*	L, R	Burning sensation, fever, leprosy, tooth ache	Leaf extract is used externally on piles and boils; leaf paste is applied topically on tumor and on bone injuries as pain killer. Root extract mixed with *Dillenia indica* and a few drops of honey is taken orally once daily to cure dysentery. Dried root powder is taken orally to cure general fevers and fresh root is used during toothache.
152.	*Mimusops elengi* L.; Sapotaceae	*Bakul*	B, Fr, Sd	Dental problem, tonsillitis, fever.	Decoction is taken orally.
153.	*Mitragyna rotundifolia* (Roxb.); Rubiaceae	*Kalomphang* (*Reang*)	Lx	Tumor	Leaf latex is applied externally.

TABLE 10.1 *(Continued)*

SL. No.	S; F	VN	PU	D/A	T
154.	*Moringa oleifera* Lam.; Moringaceae	*Signahbhaong, Sedan*	B, Fr, L, R	Cough and cold, throat infection, tooth ache, urinary tract infection.	Bark, young fruits and stem are boiled in water and consumed orally against hookworm, cough, and cold; leaf and root paste are applied in toothache topically to cure urinary tract and throat infection.
155.	*Murraya koenigii* (L.) Spreng; Rutaceae	*Santaukhi*	R	Leucorrhea	Pills are taken against leucorrhea.
156.	*Murraya paniculata* L.; Rutaceae	*Kamini, Kamini ghum*	B, R	Tooth decay, swollen gums, sex stimulant.	Decoction is taken orally.
157.	*Musa acuminate* L.; Musaceae	*Ram kala, Thailik*	F, Fr, L	Allergy infection, bronchitis, dysentery.	Decoction is taken orally.
158.	*Mussaenda roxburghii* Hook. f.; Rubiaceae	*Khurmumu*	B	Diarrhea	About 20 ml fresh bark decoction is administered orally thrice in a day.
159.	*Nyctanthes arbortristis* L.; Oleaceae	*Siuli, Hengra*	F, L	Bronchitis, asthma, inflammation, worm infection.	Decoction is taken orally.
160.	*Ocimum basilicum* L.; Lamiaceae	*Tulasisom*	L	Cough	Decoction is taken orally.
161.	*Ocimum tenuiflorum* L.; Lamiaceae	*Tulasi*	L	Cough	Decoction is taken orally.
162.	*Ophiorrhiza nutans* Clarke; Rubiaceae	*Salbuaio*	L	Antiseptic	Paste is used as an antiseptic.
163.	*Oroxylum indicum* (Linn.) Vent.; Bignoniaceae	*Ttaokharung*	B	Eczema, jaundice	Extract is administered orally.
164.	*Oxalis corniculata* L.; Oxalidaceae	*Amlati, Amchukai*	Wp	Dysentery, rheumatism, stomach disorders, tooth ache.	Decoction is taken orally.

TABLE 10.1 *(Continued)*

SL. No.	S; F	VN	PU	D/A	T
165.	*Pandanus foetidus* Roxb.; Pandanaceae	*Tawkha phauhkhoi*	Fr, L, R	Amoebiasis, tooth ache, stomach pain.	Decoction is taken orally.
166.	*Pandanus odoratissimus* Roxb.; Pandanaceae	*Keora, Ketaki*	R	Poisonous bite, red urine disorder.	Decoction is taken orally.
167.	*Papilionanthes teres* (Roxb.) Schult.; Orchidaceae	*Khelang*	R	Fracture	Paste is applied externally.
168.	*Passiflora foetida* Linn.; Passifloraceae	*Ban kamala, Fok gula*	Fruit and leaf	Blood purifier, burns, scabies.	Leaf ash mixed with ghee is applied on burns and scabies and ripe fruits are eaten as a blood purifier.
169.	*Pelatantheria insectifer* (Reichb. F.) Ridl.; Orchidaceae	*Uamaifry*	Leaf	Bone fracture, pain killer.	Its leaves and aerial roots of *Papilionanthe teres* are pasted together and applied on fractured bones and also used as pain killer.
170.	*Peperomia pellucida* (L.) H.B.K.; Piperaceae	*Charkouma*	L	Wounds	Paste is applied externally.
171.	*Phlogacanthus thyrsiflorus* (Roxb.) Ness.; Acanthaceae	*Ficong*	L	Ear pain	Extract is applied externally.
172.	*Phragmites karka* (Retz.) Trin. ex Steud.; Poaceae	*Nusui*	Lb	Sprain	Paste is applied externally.
173.	*Phyllanthus acidus* L.; Euphorbiaceae	*Harbarai, Leur, Iheri*	Fruit and leaf	Hepatic disorders, rheumatism, smallpox, stomach disorder.	Decoction is taken orally.
174.	*Phyllanthus emblica* Linn.; Euphorbiaceae	*Amloi*	B, Fr	Conjunctivitis, diabetes, hair fall, urinary problem.	Paste is smeared on head about an hour before shower to reduce hair fall and dandruff; pills are administered 2–3 times a day to cure conjunctivitis and urinary problem.
175.	*Phyllanthus fraternus* Webster; Euphorbiaceae	*Bhuiamla, Samuk gach*	L, R	Diabetes, dysentery, menstrual problem.	Decoction is taken internally.

TABLE 10.1 *(Continued)*

SL. No.	S; F	VN	PU	D/A	T
176.	*Pithecellobium heterophyllum* (Roxb.) Macbr.; Mimosaceae	*Samtawkhi*	L	Jaundice	A few drops of the decoction are administered empty stomach orally.
177.	*Pleomele spicata* (Roxb.) N.E. Brown; Agavaceae	*Chanthi*	R	Stomach trouble.	Soft inner portion of the root is eaten raw.
178.	*Plumbago zeylanica* Linn. (Chitrak); Plumbaginaceae	*Agunishita, Chitogachh*	B, R, Sd	Jaundice, snakebite, digestive, to induce abortion, rheumatism.	Half a cup of root decoction is taken once for three days to induce abortion. A few drops of decoction are taken after snakebite and also to cure rheumatism and leaf extract is applied on forehead to treat jaundice.
179.	*Polygonum barbatum* L.; Polygonaceae	*Neetfhgang*	St	Gastric trouble, ulcers.	Stem is consumed orally.
180.	*Polygonum orientale* L.; Polygonaceae	*Biskatali*	L	Asthma and cough.	Extract is taken internally.
181.	*Pronephrium nudatum* (Roxb.) Holttum; Thelypteridaceae	*Uabamthu*	P	Pyorrhea	Cold decoction is used to wash mouth 2–3 times a day.
182.	*Prunella vulgaris* L.; Lamiaceae	*Lamaku*	B	Tooth ache	Decoction is applied externally.
183.	*Prunus arborea* (Bl.) Kalkman; Rosaceae	*Lutaukha*	L	Appetizer	Pills are administered orally twice a day.
184.	*Psidium guineense* Swartz; Myrtaceae	*Bangayam, Jarbogoyam*	L, Tw	Dendrites, scurvy.	Leaves and twigs are boiled in water and solution is used as mouth wash to get relief from scurvy. Twigs are used as toothbrush to make gums strong.
185.	*Pteris ensiformis* Burm.; Pteridaceae	*Jobamfang*	Fd	Joint swelling	Paste is applied topically twice a day.

TABLE 10.1 *(Continued)*

SL. No.	S; F	VN	PU	D/A	T
186.	*Pteris semipinnata* L.; Pteridaceae	*Skaiumamoidu*	Fd	Carbuncle	Paste is applied topically.
187.	*Pyrrosia adnascens* (Forst.) Ching; Polypodiaceae	*Bormondi*	Rh	Cough and cold	Cold decoction mixed with a little powdered *Piper nigrum* seeds are administered orally for a week.
188.	*Rauwolfia serpentina* L.; Euphorbiaceae	*Sarpagandha, Lairusich*	L, R	Anxiety, bitter, hypertension, snake bite.	Decoction is administered orally.
189.	*Rhus chinensis* Miller; Anacardiaceae	*Dounilbly*	L	Blood clotting	Fresh leaf paste is applied externally as blood coagulant and antiseptic.
190.	*Rhynchostylis retusa* (L.) Blume; Orchidaceae	*Sibraimyfry*	L	Pyorrhea	Paste is applied externally.
191.	*Ricinus communis* Linn.; Euphorbiaceae	*Urtoukchock*	L	Asthma, cough, leprosy, tooth ache.	Extract is consumed orally to cure cough and piles and paste is consumed orally to treat ulcer, leprosy, and toothache.
192.	*Saurauia roxburghii* Wall.; Saurauiaceae	*Dubaikang*	R	Leucorrhea	Decoction is taken internally.
193.	*Scindapsus officinalis* (Roxb.) Schott.; Araceae	*Duleiro*	L	Leprosy	Extract is applied externally.
194.	*Scoparia dulcis* L.; Scrophulariaceae	*Burahanganja*	L	Jaundice, menstruation problem.	Extract is consumed orally to treat jaundice and paste is consumed orally to cure irregular menstruation.
195.	*Senna occidentalis* L.; Fabaceae	*Kalkashunda, Masinga*	L, R, Sd	Constipation, cough, skin infection, urinary disorder.	Decoction is taken orally.
196.	*Senna alata* (L.) Roxb.; Fabaceae	*Khaspoi*	L	Ringworm	Paste is applied externally.

TABLE 10.1 *(Continued)*

SL. No.	S; F	VN	PU	D/A	T
197.	*Sesbania grandiflora* L.; Fabaceae	*Bakful*	B, F, L	Cough, smallpox	Decoction is consumed orally.
198.	*Sida rhombifolia* L.; Malvaceae	*Bonisuo*	L	Blood coagulation	Paste is applied externally.
199.	*Smilax zeylanica* L.; Smilacaceae	*Komarialata, Koyarma*	L, R	Hepatitis, skin, and urinary disease.	Decoction is consumed orally to cure hepatitis and urinary disease while, leaf paste is applied topically to treat skin diseases.
200.	*Solanam xanthocarpum* Chard & Wendl.; Solanaceae	*Khangkhastey (Reang)*	R	Detoxication	Extract is administered orally to detoxicate inebriation.
201.	*Sonchus brachyotus* DC. Prodr.; Asteraceae	*Blonghamchen*	L	Rheumatism	Fresh leaf extract is taken orally.
202.	*Spermacoce hispida* L.; Rubiaceae	*Rifugilata*	L	Blood clotting, wound.	Leaves are used for blood clotting and as antiseptic.
203.	*Spilanthes paniculata* Wall.; Asteraceae	*Ushnui*	F, L, St	Deworming, constipation, liver trouble, tooth ache, as bait for fishing.	Leaf and young stem decoction are administered orally to deworm and treat constipation and liver trouble. Fresh flower is applied during toothache. Leaf paste is used as bait.
204.	*Spondias pinnata* (L. f.) Kurz.; Anacardiaceae	*Amra*	B, Fr, L	Dysentery, skin disease.	Paste is topically applied on infected skin and bark decoction, fruit, and leaf is taken orally to treat dysentery.
205.	*Stephania glandulifera* Miers.; Menispermaceae	*Thandamanik*	Co	Gastritis	Decoction is taken orally.
206.	*Stephania japonica* (Thunb.) Miers.; Menispermaceae	*Tousabachoiyama*	R	Rheumatism, abdominal pain, flatulence, urinary disorder.	Pills prepared from root paste are orally administered thrice a day.

TABLE 10.1 *(Continued)*

SL. No.	S; F	VN	PU	D/A	T
207.	*Sterculia villosa* Roxb. ex Masters; Sterculiaceae	*Fathi*	R	Blood dysentery, tonic.	Extract is consumed orally as tonic and treat blood dysentery.
208.	*Streblus asper* Lour.; Moraceae	*Pathar sauna, Doi Shang, Salua, Mukhuichiechai*	L, R	Tooth ache, diabetes, fracture, dysentery, hair loss, hair dye.	Warm crushed fresh leaves are administered topically on fractured portion. Extract is used as hair dye, to prevent hair loss, applied on tooth ache and taken orally to cure dysentery.
209.	*Tabernaemontana divaricata* (L.) R. Br.; Apocynaceae	*Khuntautai*	B, R	Tooth ache	Extract is applied externally.
210.	*Tamarindus indica* L.; Caesalpiniaceae	*Taintai*	L, Sd	Rheumatism, snakebite.	Seeds and leaf extract is administered orally.
211.	*Teramnus labialis* (L. f.) Spreng.; Leguminosae	*Kircha*	L	Tonic	Extract is taken orally.
212.	*Terminalia arjuna* (Roxb.) Wight. & Arn.; Combretaceae	*Arjun, Arjunfang*	B	Earache, heart, and stomach problem.	About 10 gm bark powder boiled in milk is consumed orally.
213.	*Terminalia bellirica* (Gaertn.) Roxb.; Combretaceae	*Bhoera*	Fr	Cough, stomach problem.	Fruit is eaten.
214.	*Terminalia chebula* Retz.; Combretaceae	*Bakhla, Haritaki*	Fr	Cough, diabetes, smallpox, stomach pain, vomiting.	Fruit is eaten.
215.	*Thysanolaena latifolia* (Roxb.) Kuntz.; Poaceae	*Nusui blai*	L	Blood clotting	Paste is used externally.
216.	*Tinospora cordifolia* (DC.) Miers.; Menispermaceae	*Mourgiada, Gulancha, Duksa sungdari, Duboei*	Wp	Abdominal pain, arthritis, diabetes, immunity booster, urinary disorders.	Stem soup is served.
217.	*Tournefortia montana* Lour.; Boraginaceae	*Pamomlenima*	L	Antiseptic	Extract is administered externally.

TABLE 10.1 *(Continued)*

SL. No.	S; F	VN	PU	D/A	T
218.	*Triumfetta rhomboidea* Lindl.; Malvaceae	*Dumso*	R	Carbuncle	Paste is applied externally.
219.	*Typhonium trilobatum* Schott; Araceae	*Khargan*	L	Bowl disease, body pain, constipation.	Extract is taken orally.
220.	*Urena lobata* Linn.; Malvaceae	*Yak khunjur; Santhai*	L, R	Fever, wound.	Paste is applied topically and dressed to heal wounds and root decoction is taken orally against fever.
221.	*Urginea indica* Kunth.; Liliaceae	*Sitri*	Bl	Stomach disorders of cattle.	Decoction administered orally.
222.	*Vanda roxburghii* R.Br.; Orchidaceae	*Khelangmba, Slan tawgba*	L	Cough, epilepsy	Extract is consumed orally.
223.	*Vernonia cinerea* (L.) Less.; Asteraceae	*Khumplaksom, Siyalmutra, Debpari*	L, R	Abdominal pain, diarrhea, malarial fever, dermal infection.	Freshly prepared leaf extract is taken to treat abdominal pain, diarrhea, and malarial fever and root paste is used externally on dermal infection.
224.	*Vitex heterophylla* Roxb.; Lamiaceae	*Prung*	L	Fracture	Paste is applied topically.
225.	*Vitex negundo* L.; Verbenaceae	*Nishinda*	Wp	Black fever, joint pain, malaria, vomiting, weakness.	Decoction is taken orally.
226.	*Xeromphis spinosa* (Thunb.); Rubiaceae	*Mainaphal, Monkata*	Fr, L	Inflammation, pain, vomiting.	Decoction is taken orally.
227.	*Zanthoxylum armatum* DC.; Rutaceae	*Chaingslong*	R	Gastric pain	Paste prepared from root is eaten as pills thrice a day.
228.	*Zingiber cassumunar* Roxb.; Zingiberaceae	*Hiching*	Rh	Cough and cold	Decoction is taken orally.

TABLE 10.1 *(Continued)*

SL. No.	S; F	VN	PU	D/A	T
229.	*Zingiber rubens* Roxb.; Zingiberaceae	*Brintaukru*	Rh	Urinary infection	Paste is applied externally.
230.	*Ziziphus mauritiana* Lam.; Rhamnaceae	*Baroi, Baray*	Fr	Constipation, tooth pain, tonic.	Fruit is eaten fresh.

Abbreviations: SN: Serial number; S; F: Species; Family; VN: vernacular name (Kaobru); PU: parts used (L: leaf; Sd: seed; R: root; Wp: whole plant; Rh: rhizome; B: bark; Lx: latex; St: stem; P: pinna; F: flower; Fr: fruit; Fd: frond; Lb: leaf bud; C: caudex; Bl: bulb; T: tuber; H: hypanthodia; Ra: rachis; Tw: twig; Co: corm); D/A: disease/ailment; and T: therapy.

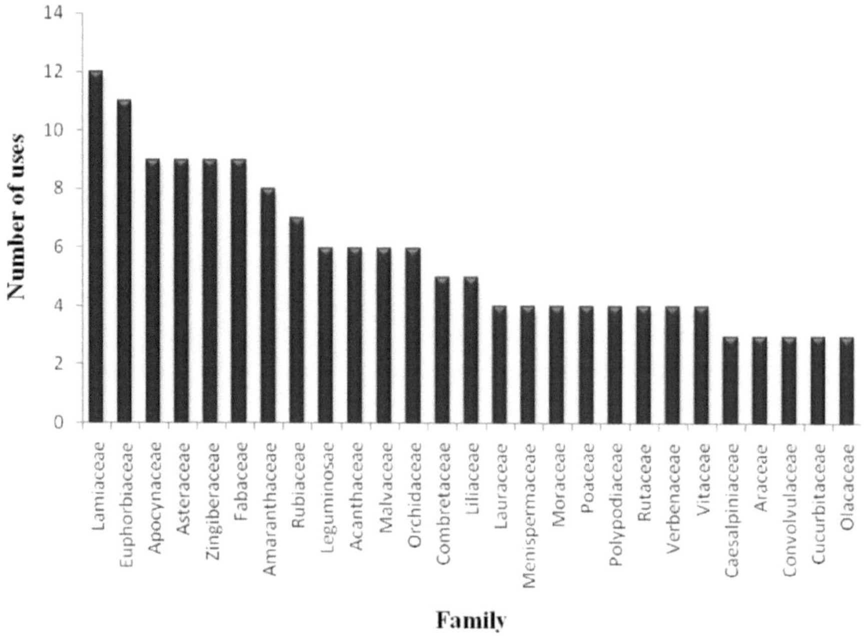

FIGURE 10.1 Prominent ethnomedicinal plant families used by the Reang community.

TABLE 10.2 Ethnomedicinal Plant Species Used by the Reang Community

Disease Category	Number of Uses
Dermatological disorders (abscess/carbuncle, allergenic swelling, antiseptic/cut and wounds, burns, eczema, leprosy, ring worm, and water born foot disease).	91
Pain and inflammation (bone fracture, body pain, ear pain, headache, naval pain, pyorrhea, rheumatism, sprain, throat pain, toothache).	78
Gastrointestinal disorders (diarrhea, dysentery, gallbladder stone, indigestion, small, and tape worm, and stomach disorder).	72
General health problems (anti-lice, blood coagulation, cough, and cold, dog bite, eye disease, health tonic, insect sting, night blindness, and vomiting).	57
Urinogenital disorder (gonorrhea, leucorrhea, urinary infection, and vaginal protrusion).	33
Fever	27
Diabetes	14
Respiratory disorders	13
Hepatic disorders	9

The Reang community has a very strong knowledge of medicinal important plants. Plants were used mostly for dermatological disorders (91 species) which are followed by pain and inflammation (78 species), gastro-intestinal (72 species), general health problems (57 species), urinogenital disorder (33 species), fever (27 species), diabetes (14 species), respiratory disorders (13 species) and hepatic disorders (nine species). Plant taxa used for different disease categories are shown in Table 10.1. Many of the plants have one or more uses reported by one healer, while others have informed other uses indicating the importance of such review. Thus, compiling data reveals that many of the species have multiple uses. For example, species like *Cuscuta reflexa*, *Kalanchoe pinnata*, *Ageratum conyzoides*, *Galinsoga parviflora*, *Streblus asper*, *Tinospora cordifolia*, *Azadirachta indica*, *Bombax ceiba*, *Cassia fistula*, *Cynodon dactylon*, *Euphorbia hirta*, *Plumbago zeylanica*, *Spilanthes paniculata*, and *Terminalia chebula* were used to cure five or more diseases.

Among the plant parts used, the leaf was dominant (39.12%), followed by root (19.24%), bark (10.09%), fruit (5.68%), rhizome (4.42%), whole plant (4.42%), and flower (4.1%; Figure 10.2). The plant parts were used in different forms like paste, juice, extract, infusion, pills, curry (cooking) and decoction and used either externally or internally for management of

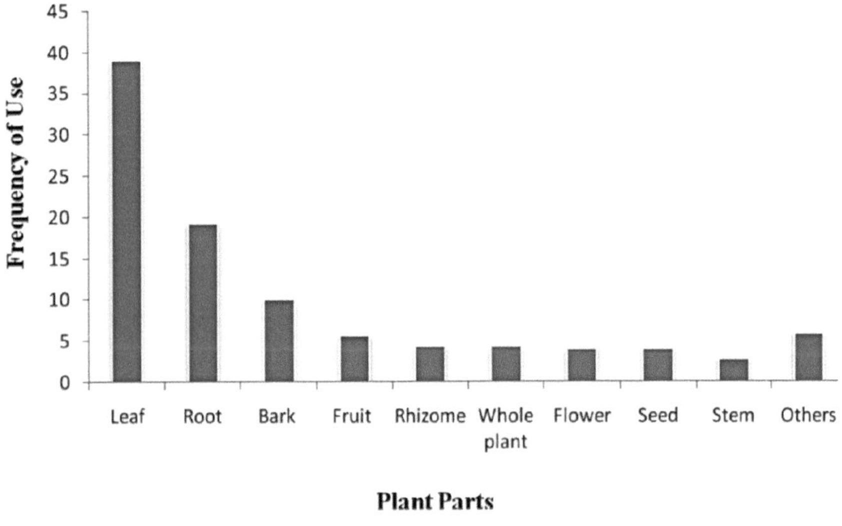

FIGURE 10.2 Plant parts of ethnomedicinal plant species used by the Reang community.

various ailments. The dosage of particular medicine is used based on the age of patients and time of recovery. Children were given less amount of dosage and many of the treatments were continued for several days until the recovery of a disease. Whenever the primary health practice by using this folk medicine does not show any sign of recovery, they consult with doctors in the modern healthcare center. It has been observed that many of the plants have more than one vernacular name reported from different workers. Many of the similar plants were called by the same name because of their similarity in structure and habitat (Sen et al., 2011; Shil et al., 2014). Many reported the use of the particular part for a particular disease while others reported the use of other parts for the same ailments indicating the need for pharmacological analysis of the plants.

10.4 CONCLUSION

Northeast India is endowed with forests and forest products. People living in and around the forest are using forest products in their daily lives for a variety of purposes. The present review provides an updated checklist of ethnomedicinal plant resources used by the Reang ethnic community of Northeast India for the treatment of different diseases. The Reangs have a rich source of traditional knowledge on ethnomedicinal uses of a wide diversity of plants. A total of 230 plant species were reported with single and multiple uses for ailments. Their traditional knowledge of primary healthcare based on ethnomedicinal plants is passed verbally from generation to generation. The rapid growth of industrialization, encroachment, loss of habitats, and fragmentation are the major factors causing the degradation of biodiversity of ethnomedicinal plant resources. Furthermore, the implication of modern tools and healthcare in the lives of tribal people has caused the deterioration of such traditional knowledge and thereby the importance of non-timber forest products (NTFPs). Efforts on the preservation of traditional knowledge and NTFPs should be given due priority as many of the species are still unexplored and therefore, there is an urgent need for pharmacological analysis of the plants used by the Reangs and all other ethnic communities.

KEYWORDS

- **conservation**
- **Cynodon dactylon**
- **ethnomedicinal plants**
- **northeast India**
- **pharmacological analysis**
- **Reang ethnic community**

REFERENCES

Anonymous, (2001). *General Guidelines for Methodologies on Research and Evaluation of Traditional Medicine*. World Health Organization, Geneva, Switzerland.

Anonymous, (2015). *State of Forest Report*. Forest Survey of India, https://www.fsi.nic.in/forest-report-2015 (accessed on 18 December 2023).

Banik, G., Bawari, M., Choudhury, M. D., Choudhury, S., & Sharma, G. D., (2010). Some anti-diabetic plants of southern Assam. *Assam University Journal of Science & Technology: Biological and Environmental Sciences*, *5*, 114–119.

Barbhuiya, A., Sharma, G. D., Arunachalam, A., & Deb, S., (2009). Diversity and conservation of medicinal plants in Barak valley, South Assam, northeast India. *Ind. J. Trad. Know.*, *8*, 169–175.

Bhardwaj, S., & Ghakar, S. K., (2005). Ethnomedicinal plants used by the tribals of Mizoram to cure cut and wound. *Ind. J. Trad. Know.*, *4*, 75–80.

Das, A. K., Dutta, B. K., & Sharma, G. D., (2008). Medicinal plants used by different tribes of Cachar district, Assam. *Ind. J. Trad. Know.*, *7*, 446–454.

Das, H. B., Majumdar, K., Datta, B. K., & Ray, D., (2009). Ethnobotanical uses of some plants by *Tripuri* and *Reang* tribes of Tripura. *Natural Product Radiance*, *8*, 172–180.

Dattagupta, S., & Gupta, A., (2014). Traditional processing of non-timber forest products in Cachar, Assam, India, traditional processing of non-timber forest products in Cachar, Assam, India. *Ind. J. Trad. Know.*, *13*, 427–433.

Debnath, B., Debnath, A., Shilsharma, S., & Paul, C., (2014). Ethnomedicinal knowledge of Mog and Reang communities of south district of Tripura, India. *Indian Journal of Advances in Plant Research*, *1*, 49–54.

Dutta, B. K., & Dutta, P. K., (2005). Potential of ethnobotanical studies in north east India: An overview. *Ind. J. Trad. Know.*, *4*, 7–14.

Hazarika, T. K., & Marak, S., (2019). Wild edible fruits of Meghalaya, North-East India: An unexplored potential for nutritional security and economic prosperity. *Acta Hort.*, *1241*, 717–727. doi: 10.17660/ActaHortic.2019.1241.104.

Heneidy, S. Z., Halmy, M. W. A., & Bidak, L. M., (2017). The ethnobotanical importance and conservation value of native plants in eastern Arabian Peninsula. *Feddes Repertorium, 128*, 105–128. https://doi.org/10.1002/fedr.201600024.

Kaunga, J. M. O., (2017). The use of indigenous traditional knowledge for ecological and bio-diverse resource management by the Laikipia Maasai and the Samburu. In: Roue, M., Oteng-Yeboah, A., Kariuki, P., & Adou, Y., (eds.), UNESCO *Knowing our Lands and Resources: Indigenous and Local Knowledge of Biodiversity and Ecosystem Services in Africa. Knowledges of Nature* (pp. 6–17). Paris.

Kunle, O. F., Egharevba, H. O., & Ahmadu, P. O., (2012). Standardization of herbal medicines: A review. *Int. J. Biodivers. Conserv.*, *4*, 101–112. https://doi.org/10.5897/IJBC11.163.

Kushwaha, S. P. S., (2006). Land use/ cover changes in the North-Eastern Hills region. In: Ramakrishnan, P. S., Saxena, K. G., & Rao, K. S., (eds.), *Shifting Agriculture and Sustainable Development of North-Eastern India- Tradition and Transition* (pp. 3–15). Oxford & IBH Publishing Pvt. Ltd., New Delhi, UNESCO (MAB).

Newman, D. J. G., Cragg, M., & Snader, K. M., (2000). The influence of natural products upon drug discovery. *Nat. Prod. Res.*, *17*, 215–234.

Odhav, B., Kandasamy, T., Khumalo, N., & Baijnath, H., (2010). Screening of African traditional vegetables for their alpha-amylase inhibitory effect. *J. Med. Plants Res.*, *4*, 1502–1507.

Pachuau, L., & Dutta, R. S., (2020). Wild edible fruits of northeast India: Medicinal values and traditional practices. In: Sen, S., & Chakraborty, R., (eds.), *Herbal Medicine in India*. Springer, Singapore, https://doi.org/10.1007/978-981-13-7248-3_27.

Reyes-Garcíaa, V., Guèzeb, M., Luzb, A. C., Paneque-Gálvezb, J., Macíac, M. J., Orta-Martínezb, M., Pinod, J., & Rubio-Campilloe, X., (2013). Evidence of traditional knowledge loss among a contemporary indigenous society. *Evol. Hum. Behav.*, *34*, 249–257. https://doi.org/10.1016/j.evolhumbehav.2013.03.002.

Sarmah, R., Adhikari, D., Majumder, M., & Arunachalam, A., (2008). Traditional medico-botany of Chakma community residing in the north-western periphery of Namdapha National Park in Arunachal Pradesh. *Ind. J. Trad. Know.*, *7*, 587–593.

Sen, S., Chakraborty, R., De, B., & Devanna, N., (2011). An ethnobotanical survey of medicinal plants used by ethnic people in West and South district of Tripura, India. *Journal of Forestry Research, 22*, 417–426. http://dx.doi.org/10.1007/s11676-011-0184-6.

Shil, S., & Choudhury, M. D., (2009a). Ethnomedicinal importance of Pteridophytes used by Reang tribe of Tripura, North East India. *Ethnobotanical Leaflets, 13*, 634–643.

Shil, S., & Choudhury, M. D., (2009b). Indigenous knowledge on healthcare practices by the Reang Tribe of Dhalai District of Tripura, North East India. *Ethnobotanical Leaflets, 13*, 775–790.

Shil, S., Choudhury, M. D., & Das, S., (2014). Indigenous knowledge of medicinal plants used by the Reang tribe of Tripura state of India. *Journal of Ethnopharmacology, 152*, 135–141. http://dx.doi.org/10.1016/j.jep.2013.12.037.

Talukdar, N. R., Choudhury, P., Barbhuiya, R. A., & Singh, B., (2021). Importance of non-timber forest products (NTFPs) in rural livelihood: A study in Patharia Hills Reserve Forest, northeast India. *Trees, Forests and People*, *3*, 100042. https://doi.org/10.1016/j.tfp.2020.100042.

Talukdar, S., & Gupta, A., (2018). Ethnomedicinal knowledge of the Garo community of two villages in Western Assam, India. *J. Herb. Med.*, *20*, 100229. https://doi.org/10.1016/j.hermed.2018.08.004.

CHAPTER 11

Sacred Groves: In Situ Conservation of Ethnomedicinal Plants

KUNTAL NARAYAN CHAUDHURI

Department of Botany, Vivekananda College, Thakurpukur, Kolkata, West Bengal, India

ABSTRACT

Sacred groves are sacred natural sites in patches of virgin forests that are protected by autochthonous communities as abodes of local guardian deities. Although found throughout the hills and plains of India, sacred groves are most commonly located in the biodiversity-rich ethnic areas of the Eastern Himalayas and the Western Ghats. Sacred groves share certain common features that include a small or large tree cover. These culturally protected natural habitats can effectively safeguard the local biodiversity. These informal institutions predate the existence and supplement the functions of the formal ones. The chapter outlines the role of sacred groves of India in *in-situ* conservation of ethnomedicinal plants. There are numerous reports on the wealth of ethnomedicinal plant species found in the sacred groves across India. These relic forests function as natural resource bases for the local communities by harboring key biological resources including ethnomedicinal plants. These medicinal plant resources are vital to the well-being of remote communities who still depend on traditional herbal remedies for their primary healthcare. Sacred groves have little or no human interference due to the customary restrictions on their access and biomass extraction activities. The smaller groves are virtually inaccessible, but the larger ones allow limited harvesting through sustainable resource management practices. This is commonly in the form of natural resource-related social taboos

Bioprospecting Ethnomedicinal Plant Resources: Sustainable Utilization and Restoration.
Gopal Shukla, Jahangeer A. Bhat, A. P. Das, and Sumit Chakravarty (Eds.)

of the local communities which prevent overharvesting of plant resources including ethnomedicinal plants. The community-based traditional management practice in these sacred groves also checks degradation of natural habitats of medicinal plants and supports a rich plant spectrum of which they are an important part. These traditional repositories of local medicinal plants also function as the last refuge for ecologically important and fragile plant species with medicinal value. These ancient ethnoecological institutions involving the stewardship of the local communities provide a decentralized, inclusive, and participatory model for the *in-situ* conservation of the wealth of indigenous medicinal plant species of India.

11.1 INTRODUCTION

The plant kingdom is a veritable treasure trove of myriad natural products bequeathed with the amazing power of healing. Since the dawn of civilization, humanity has used the powerful effects of these plant substances on the human body and mind to alleviate their suffering from pain and diseases. In addition to the traditional crude drugs, a plethora of fine chemicals obtained from plants constitute or perform the role of precursors and prototypes for many modern pharmaceutical drugs (Newman et al., 2003). According to the WHO, an estimated 80% of the global population still depends on traditional herbal remedies for their primary healthcare (Bodeker et al., 2005). However, this could be as high as 95% in the developing world (Tilburt & Kaptchuk, 2008). Even in the developed world, there is a rising trend in the popularity of herbal medicine (Khan & Ahmad, 2019). India's rich heritage in traditional medicine comprises not only a handful of codified (written) classical systems such as Ayurveda, Siddha, Unani, and Amchi (Pushpangadan, 2018), but also a plethora of non-codified (oral) folk systems of the rural and tribal communities (Singh, 2001). However, biological diversity, which is fundamental to human welfare, is being lost at an alarming rate across the globe (Cardinale et al., 2012). With their growing demand but limited cultivation, the increasing pressure on the wild populations of indigenous medicinal plants results from factors including habitat destruction and overharvesting (Dhyani, 2016). According to the IUCN and the WWF, among the 50,000–80,000 medicinal plant species found worldwide, about 15,000 are on the brink of extinction (Bentley, 2010). The objective of CBD (1992) includes the preservation of biodiversity, sustainable use of its components, and fair and equitable sharing of the benefits. Therefore, environmentalists have often emphasized that the policies for biodiversity conservation need a human dimension in order to be more effective (Lubchenco, 1998). Social institutions can play an important role in this respect (Berkes & Folke, 1998). This chapter reviews the important role played by the sacred groves of India in the *in-situ* conservation of indigenous medicinal plants.

11.2 SACRED GROVES OF INDIA

The term 'biocultural diversity' was coined as part of the Declaration of Belem (1988) to articulate the idea of an inextricable link that bridges the disparate worlds of biological and cultural diversities. Maffi (2007)defined biocultural diversity as "the diversity of life in all its manifestations: biological, cultural, and linguistic – which are interrelated (and possibly coevolved) within a complex socio-ecological adaptive system." Human beings, since the prehistoric times, have closely interacted with nature and they became increasingly dependent upon its bounty (Maffi, 2001). Transmitted across the generations through the millennia, the community-based protection and management of local natural habitats that function as key natural resource bases is deeply entrenched in the cultural beliefs and social practices of all autochthonous communities. This is also an integral part of their TK or TEK (Lévi-Strauss, 1966). Their culturally protected sacred natural spaces constitute the oldest form of *in situ* conservation. These traditional (informal) ones predate the existence and supplement the functions of the modern (formal) ones such as reserve forests, wildlife sanctuaries, national parks, and biosphere reserves (Pretty et al., 2009).

Sacred natural spaces, areas or sites are an integral part of cultural landscapes in the form of islands of untouched natural ecosystems that hold special spiritual significance for autochthonous communities as sanctuaries for practicing their animistic beliefs. Sacred (or spirit) groves – the most common form of sacred natural sites – are small or large patches of virgin forests or wetlands along with their vegetation and wildlife that are protected in a natural state by the local communities in their common belief that these natural landscapes are the sacred dwelling grounds of the local guardian deities or spirits (Hughes & Chandran, 1998). These relic forests are the last remains of a primeval world that epitomizes the reverence to nature as a key survival strategy for indigenous cultures.

The sacred groves of India represent a living, albeit declining, ancient tradition (Chaudhuri & Chaudhuri, 2014). Although once common throughout rural India due to the traditional slash-and-burn (*Jhum*) farming, these had declined during the colonial period when shifting cultivation was prohibited and the community-owned forests were usurped as state property by the British administration (Gadgil & Vartak, 1976). However, it was also during this period that the scientific survey of sacred groves was initiated by the colonial authorities in the 19th century, probably for the documentation of their bioresources for future profit (Chaudhuri & Chaudhuri, 2014). The extant sacred groves are perhaps the only ecosystems in India in near-virgin conditions (Gadgil & Vartak, 1976). Although reported from across India, both in the hills as well as the plains, and even in the deserts, the best-preserved ones exist only in the remote ethnic belts located in the biodiversity-rich regions of the Eastern Himalayas and the Western Ghats. These sacred groves encompass extremely diverse cultural and ecological dimensions. Godbole et al. (1998) recognized three types of groves. The informal groves in remote locations are spirit sanctuaries and the entire vegetation is deemed sacred. The formal groves located near settlements are dedicated to village deities with shrines and idols. The memorial groves found at cremation or burial sites are dedicated to the clan ancestors or heroes. In India, although the sacred groves can range from a small cluster of few trees to large tracts of forests, however, they share certain common features such as an arboreal cover, a strategic position near a perennial water source, a local guardian deity, and regular ritualized worship (Manikandan et al., 2011; Plate 11.1(a)–(d)). These ancient ethnoecological institutions encompass all the forms of heritage-intangible and cultural (Ahmad, 2006), physical and cultural (Lowenthal, 2005). These sylvan sacred spaces are known by diverse local names in the different regions of India (Table 11.1). Malhotra et al. (2000) had reported that a total number of 13,750 sacred groves were recorded across India. However, the estimated figure of sacred groves existing in the country could be as high as 1,50,000 (Malhotra, 1998). Thus, most of them still have not been studied and documented.

11.3 MANAGEMENT OF ETHNOMEDICINAL PLANT RESOURCES IN SACRED GROVES

The term taboo, from the Polynesian word *tapu* meaning forbidden, is a tacit prohibition against an utterance or behavior based on cultural sense that holds something as profane (impure) or sacred (pure). Social taboos are voluntary codes of conduct in the form of proscriptions that have been

PLATE 11.1 Few sacred groves from the tribal-dominated lateritic south-western (Rarh) region of West Bengal, India. (a) Thutaraiburi sacred grove in Mangalda, Raghunathpur, Purulia; (b) Khandarani sacred grove in Deulbhira, Taldangra, Bankura; (c) Madnathakur sacred grove in Chhandar, Barjora, Bankura; and (d) Kanakdurga sacred grove in Chilkigarh, Jamboni, Jhargram.

TABLE 11.1 Local Names for Sacred Groves in Different Regions of India

State	Local Name (Language)	State	Local Name (Language)
Assam	*Than* (Assamese)	Maharashtra	*Deovani* (Marathi)
Andhra Pradesh	*Pavithravanam* (Telegu)	Manipur	*Lai* (Manipuri)
Bihar	*Sarna* (Santali)	Meghalaya	*Law* (Khasi)
Chhattisgarh	*Devgudi* (Mundari)	Mizoram	*Ngawpui* (Mizo)
Himachal Pradesh	*Deorali* (Pahari)	Odisha	*Thana* (Odia)
Jharkhand	*Sarna* (Santali)	Rajasthan	*Oran* (*Rajasthani*)
Karnataka	*Kan* (Kannada)	Tamil Nadu	*Kovilkadu* (Tamil)
Kerala	*Kavu* (Malayalam)	Uttarakhand	*Devvan* (Garhwali)
Madhya Pradesh	*Sarna* (Santali)	West Bengal	*Than* (Bengali)

handed down through the generations as customary laws that are designed to regulate human behavior within a traditional society (Colding & Folke, 2001). Taboos are self-enforced as the concept of automatic sanctions, i.e., the belief is that the violator's action carries its own penalty, and any act of sacrilege invokes the wrath of the guardian deities and spirits (Posner & Rasmusen, 1999). Social taboos exist in all traditional cultures as informal institutions and contrast with the formal ones, i.e., the rules and laws framed and enforced by external political authorities (Colding & Folke, 2000). These taboos may be highly adaptive from an ecological viewpoint, and they contribute to the conservation of the local bioresource pools. Therefore, social taboos guide human conduct towards the physical environment which is treated as 'sacred' and held with care and respect in most traditional cultures (Posey, 2000).

The sacred groves of India are protected by diverse traditional beliefs and practices of the locals. This is based on their age-old empirical and concrete understanding of the local ecosystems – traditional ecological knowledge (TEK). This includes a cache of social taboos that operate as 'invisible' systems for the conservation of the local biodiversity and the management of the local bioresources (Colding & Folke, 2001). In India, natural resource-related taboos associated with the sacred groves play a critical role in restricting all forms of human interference to preserve these natural habitats in a near-virgin condition and prevent over-exploitation of their bioresource pools. These taboos allow little or no access to these ecosystems and restrict all forms of biomass extraction activities such as hunting, gathering, felling, fishing, farming, and grazing. These taboos not only protect the sacred groves as natural habitats of indigenous medicinal plants, but they also prevent the over-harvesting of the wild populations of ethnomedicinal plants present therein. Both the women and men of the Kurichiyas of Kannur, Kerala, are not only involved in conserving their local medicinal plants in their *Kavus* (sacred groves), but they also sustainably use and manage this vital natural resource by growing them (*ex situ*) in their homegardens (Rajith & Ramachandran, 2010). Chandran & Gadgil (1993) described two types of sacred groves as far as bioresources extraction is concerned. The smaller ones such as the Malleeswaran Mudi sacred grove in Palakkad, Kerala are rarely accessed by the locals and are completely inaccessible for any material needs. In contrast, the larger ones such as the Jatlatada grove of Srikakulam, Andhra Pradesh can sustain local communities by allowing limited volume of biomass harvesting. The ritualized collection of medicinal herbs from these sacred groves is usually performed after soliciting the permission of the cadastral deities with ritual offerings made by the shaman-priests. They are variously called *Baigas* (Chhattisgarh), *Bhagats* (Maharashtra), *Gaudas* (Karnataka), *Lyngdohs* (Meghalaya) or *Maibas* (Manipur). Based on the classification of natural resource-related taboos by Colding & Folke (2001), a review of the literature revealed the following four categories of social taboos that play important roles in the sustainable use and management of local ethnomedicinal plants in the sacred groves of India (Table 11.2).

TABLE 11.2 Natural Resource-Related Taboos Associated with the Sacred Groves of India

Category (Restriction)	Natural Resource Management Principle	Example of Sacred Grove (References)
Habitat taboo (Where not to collect?)	Restricts access and use of bioresources in a specific (sacred) habitat.	Malleeswaran Mudi sacred grove of the Mudugars in Palakkad, Kerala (Nadanakunjidarn, 2000).
Species taboo (What not to collect?)	Protects a specific (sacred) species of plants or animals as potential bioresources.	Mahabali sacred grove of the Meiteis in Imphal, Manipur (Khumbongmayum et al., 2006).
Temporal taboo (When not to collect?)	Regulates the access and use of bioresources during specific time periods.	Sacred groves (*Kavus*) of the Kurumbas of Palakkad, Kerala (Tharakan, 2007).
Segment taboo (Who cannot collect?)	Regulates the access and use of bioresources by specific social segments.	Jatlatada grove of the Jatapus in Srikakulam, Andhra Pradesh (Das et al., 2004).

11.3.1 HABITAT TABOOS

This type of taboos of cultural groups are in general a management strategy that bars the access to, and the extraction of local bioresources from specific resource-rich local habitats or landscapes that they considered to be sacred as the abodes of deities and spirits (Colding & Folke, 2001). These taboos protect entire natural habitats in the form of sacred groves (Chaudhuri & Chaudhuri, 2015) and other sacred natural spaces (Chaudhuri & Chaudhuri, 2018). The Malleeswaran Mudi (Hill) sacred grove in Palakkad, Kerala, is seldom entered by anyone except the local Mudugar people who would go there only for the annual hunting festival. Otherwise trespassing and no form of biomass harvesting including the collection of medicinal herbs is permitted. Not even rotten wood, leaf litter or animal carcass can be removed from this sacred habitat. Even carrying any form of hunting weapons and agricultural tools is strictly forbidden in this remote spirit sanctuary (Nadanakunjidarn, 2000). However, many sacred groves, especially those near human dwellings, often perform the key function of local resource bases. These allow some form of restricted access and biomass extraction within the sacred habitats. However, this is strictly regulated by some categories of natural resource-related taboos are described in subsections.

11.3.2 SPECIES TABOOS

This sort of taboos of cultural groups are in general another management strategy that bans the slaughter or any other harmful use of key local bioresources in the form of particular species of plants (or animals) that they considered to be sacred as totems, symbols or memorials (Colding & Folke, 2001). These taboos often involve keystone or endemic species that play important ecological roles in the local ecosystems (Colding & Folke, 1997). In the Mahabali sacred grove in Imphal, Manipur, the dominant *Heibong* trees, *Ficus racemosa* L. (= *F. glomerata* Roxb.) (Moraceae) is protected by the local Meiteis as a sacred tree. This is a keystone species that supports *frugivorous* insects, birds, and mammals that are pollinators of wild plants including the medicinal ones (Khumbongmayum et al., 2006). This arborescent species is mentioned in Ayurveda as *Udumbara* whose bark can cure dysentery, ulcers, diabetes, urinary problems and other conditions (Deep et al., 2013). This is also true for the Akoli sacred grove of Jalore, Rajasthan.

11.3.3 TEMPORAL TABOOS

This category of taboos of cultural groups is in general yet another management strategy that prohibits the extraction of vital local bioresources during certain time intervals with frequencies that could be sporadic, daily, weekly, monthly, seasonal or annual (Colding & Folke, 2001). However, these social embargoes are commonly imposed on certain months or seasons of the year. Referred to as 'closed seasons,' these time periods usually coincide with their productive seasons of important wild species. In the *Kavus* (sacred groves) of Palakkad, Kerala (and the nearby Nilgiris, Tamil Nadu) the local Kurumba people collect the honey of giant rock bees (*Apis dorsata* Fabricius), a pollinator of many wild plants including medicinal ones. This bioextraction activity is limited from April to June in the form of an annual honey hunting ritual. This starts only after the breeding season in spring, coincides with the peak forest blooming time in May, and ends before the migration season during the monsoons (Tharakan, 2007). Before scaling the dangerous cliffs in the sacred groves and adjoining forests to reach the honeycombs, the hunters seek the permission and protection of the bee spirit. The brood comb is left unharmed. Some cliffs in these relic forests are also worshipped as 'honey rocks' and the colonies of this important forest pollinator species on these sacred rocks are left intact (Roy, 2002).

11.3.4 SEGMENTAL TABOOS

This kind of taboos of cultural groups are in general still another management strategy that excludes the use of critical local bioresources for definite social segments based on their age, gender, profession or social status (Colding & Folke, 2001). These proscriptions may function as strategic responses to avoid resource depletion due to over-extraction. Over-harvesting from the wild is a widespread problem for most medicinal plant species whose collection is still largely unregulated. However, these taboos practiced in the sacred groves regulate bioresource withdrawal through a 'quota system' or 'bag limit' to effectively regulate the harvest volume of medicinally important plant species. In the Jatlatada grove of Srikakulam, Andhra Pradesh, although felling and hunting are strictly prohibited, only the male elders of the local Jatapus, who as folk healers and well-versed with the local flora, are permitted to collect the local medicinal herbs. This is done after making ritual offerings to obtain the permission of the cadastral deity (Das et al., 2004). This is also true of the *Maibas* as the healer-priests of the local Meitei people in the sacred grove of Manipur (Khumbongmayum et al., 2006).

11.4 *IN SITU* CONSERVATION OF ETHNOMEDICINAL PLANTS IN SACRED GROVES

Sacred groves in India, especially those in the remote areas, have largely survived the global onslaught of anthropogenic intrusions during the recent past. These sacred natural habitats along with the biodiversity therein have survived the widespread degradation and fragmentation of natural ecosystems along with vegetation and wildlife, thus preserving them in a near-pristine state. For the autochthonous communities, this timeless ingenuity defines their relationship with these local ecosystems, as manifested in their reverence to nature. Biodiversity conservation is one of the many important ecological services provided by the sacred groves of India (Gadgil & Chandran, 1992). Social taboos practiced in the sacred groves are unique sustainable resource management systems that effectively protect, maintain, and even enhance the local biodiversity (Gadgil et al., 1993). The key role played by the array of natural resources-related taboos has resulted in the sustainable management of the local bioresource pools of the sacred groves, including their wealth of ethnomedicinal plants (Table 11.3). Therefore, these resilient patches of virgin forests harbor a wide range of plant species including medicinal ones, and function as the last refuge for an immense diversity of ethnomedicinal plants that often play important ecological roles or are themselves ecologically sensitive.

TABLE 11.3 Some Medicinal Plant Resources Reported from the Sacred Groves of India

Medicinal Plant Species (Family)	Medicinal Use	Sacred Grove (State)	References
Bergenia pacumbis (Saxifragaceae)	Roots used to treat kidney stones.	Mawphlang (Meghalaya)	Kapur (1983)
Canarium strictum (Burseraceae)	Resin used to treat skin problems.	Deovani (Maharashtra)	Gadgil & Vartak (1976)
Dipterocarpus indicus (Dipterocarpaceae)	Oleoresin used to cure rheumatism.	Kathalekan and Karikan (Karnataka)	Chandran & Gadgil (1998)
Elaeocarpus serratus (Elaeocarpaceae)	Seeds useful in asthma, epilepsy, and hypertension.	Mawphlang (Meghalaya)	Mishra & Jeeva (2012)
Entada rheedei (Leguminosae)	Bark used to treat cattle against snake bite.	Tunbad (Maharashtra)	Gadgil & Vartak (1976)
Ficus benghalensis (Moraceae)	Stem latex applied on cracked heels.	Mahendergarh (Haryana)	Yadav et al. (2010)
Ficus racemosa (Moraceae)	Bark used for diabetes, dysentery, ulcers, and urinary problems.	Mahabali (Manipur)	Khumbongmayum et al. (2006)
Ficus religiosa (Moraceae)	Bark useful in respiratory disorders.	Mahendergarh (Haryana)	Yadav et al. (2010)
Flacourtia jangomas (Salicaceae)	Fruits and leaves used to treat diarrhea.	Mahabali (Manipur)	Malhotra et al. (2001)
Justicia betonica (Acanthaceae)	Leaves are used to treat abscesses and swellings.	Kavu (Kerala)	Santhosh Kumar et al. (2002)
Kunstleria keralensis (Leguminosae)	Bark used to treat body ache.	Kavu (Kerala)	Mohanan & Nair (1981)
Madhuca bourdillonii (Sapotaceae)	Bark used to treat rheu- matic conditions.	Ankola (Karnataka)	Chandran et al. (2008)

TABLE 11.3 *(Continued)*

Medicinal Plant Species (Family)	Medicinal Use	Sacred Grove (State)	References
Marsdenia sylvestris (Aslepiadaceae)	Leaves useful in the treatment of diabetes.	Kanakdurga (West Bengal)	Bhakat (2015)
Mesua ferrea (Calophyllaceae)	Stamens used for leprosy and skin diseases, bark used for itches and scabies.	Malleeswaran Mudi (Kerala)	Sreekumar et al. (2014)
Prosopis cineraria (Leguminosae)	Bark used to treat leprosy and skin diseases.	Akoli (Rajasthan)	Meena & Singh (2012)
Santalum album (Santalaceae)	Heartwood oil useful in anxiety, depression, and stress, and bark and wood used for headaches.	Malleeswaran Mudi (Kerala)	Sreekumar et al. (2014)
Semecarpus kathalekanensis (Anacardiaceae)	Seeds used to treat asthma, bronchitis, and fever.	Kathalekan (Karnataka)	Dasappa & Swaminath (2000)
Strychnos nux-vomica (Loganiaceae)	Seeds used for treatment of arthritis.	Kanakdurga (West Bengal)	Bhakat (2015)
Syzygium travancoricum (Myrtaceae)	Bark used for treatment of arthritis.	Aikad (Kerala)	Nair & Mohanan (1981)
Tylophora indica (Aslepiadaceae)	Leaves are used to treat asthma.	Kanakdurga (West Bengal)	Bhakat (2015)
Vateria indica (Dipterocarpaceae)	Resin used to treat intestinal worms and wounds.	Yelakundlikan (Karnataka)	Chandran &Gadgil (1998)

The crucial role played by the sacred groves of India in the *in situ* conservation of indigenous medicinal plants has been illustrated by a large number of studies across India (Vartak et al., 1987; Chandran et al., 1998; Pushpangadan et al., 1998; Bhakat & Pandit, 2003, 2004; Bhandary & Chandrasekhar, 2003; Khumbongmayum et al., 2005; Bhakat & Sen, 2008; Ganesan et al., 2009; Bhakat, 2015; Gadgil, 2018). In a small number of pioneering studies, the presence of a large number of important ethnomedicinal plants was reported from the *Kavus* of Kerala by Pushpangadan et al. (1998) and from the adjoining *Kans* of Uttara Kannada (Karnataka), by Chandran et al. (1998). Later, numerous reports started to appear from across India with the turn of the century which includes, but is not limited to 120 medicinal plant species that were documented in four *Lais* (sacred groves) of Imphal, Manipur, by Khumbongmayum et al. (2005) and 105 medicinal species from eight *Kovilkadus* of Madurai (Tamil Nadu) by Ganesan et al. (2009). Recently, Bhakat (2015) also reported 105 species of medicinal trees, lianas, shrubs, and herbs in the large than (sacred grove) of the local Lodha people in Chilkigarh, Midnapore (now Jhargram), West Bengal (Plate 11.1(d)). However, the medicinal plant wealth of many smaller ones in this tribal-dominated southwest (Rarh) region of West Bengal still remains unreported (Plate 11.1(a)–(c)). These reports clearly point out that the sacred groves of India as culturally protected natural habitats stand out as 'islands' or oases of climax forests harboring a rich plant spectrum that function as treasure troves of ethnomedicinal plant wealth. These ancient ethnobotanical institutions are the last strongholds of global biodiversity. They serve as the refugia for keystone species, as well as threatened, endemic, and rare ones with immense medicinal values and even new plant taxa with medicinal potential.

11.4.1 MEDICINAL PLANTS IN ISLANDS OF CLIMAX FORESTS

Studies across India have recorded that on account of their cultural protection, most sacred groves have survived the onslaught of unchecked deforestation over the centuries. Therefore, they appear now as isolated patches of climax forests. These untouched natural habitats harbor rich biodiversity, including indigenous medicinal plants. The now lost virgin forests surrounding them have been transformed by anthropogenic activities into denuded landscapes such as agro-ecosystems, wetlands, or meadows. A typical example of this is the Mawphlang sacred grove of the local Khasi people in the East Khasi Hills, Meghalaya. The dense thickets of this *Law* (sacred grove) spread across 0.8 $km^2$are in stark contrast with the open grasslands surrounding this green 'island.' Mishra &Jeeva (2012) reported that this forest is dominated by the trees of *Castanopsis lanceifolia* (Oerst.) Hickel & A.Camus [= *C. kurzii* (Hance) S.N.Biswas] (Fagaceae) along with 80 other woody angiosperm species. These include *Elaeocarpus serratus* L. (= *E. Ganitrus* Roxb. ex G.Don) (Elaeocarpaceae) or *Rudraksha* of Ayurveda whose seeds are used to treat hypertension, asthma, epilepsy, among many others (Hardainiyan et al., 2015). Kapur (1983) documented the medicinal herbs found in this forest which prominently includes *Bergenia pacumbis* (Buch.-Ham. ex D.Don) C.Y.Wu & J.T.Pan (= *B. ligulata* Engl.; Saxifragaceae; Plate 11.2(a)) or *Prashanbheda* of Ayurveda whose roots are primarily used to treat kidney stones (Kumar & Tyagi, 2013).

PLATE 11.2 Few ethnomedicinal plants in flowering or fruiting stages are commonly found in the sacred groves in India. (a) *Bergenia pacumbis* (Buch.-Ham. ex D.Don) C. Y. Wu & J. T. Pan (Saxifragaceae); (b) *Mesua ferrea* L. (Calophyllaceae); (c) *Santalum album* L. (Santalaceae); (d) *Tylophora indica* (Burm.f.) Merr. (Apocynaceae); (e) *Strychnos nux-vomica* L. (Loganiaceae); and (f) *Prosopis cineraria* (L.) Druce (Leguminosae).

11.4.2 MEDICINAL PLANTS AMONG THE RICH PLANT SPECTRA

Surveys across India have also documented the rich plant spectra present in these informal conservation areas. These were closely comparable to, if not better than, those of the adjoining formal conservation sites. A classic example is a study in Palakkad, Kerala, by Pushpangadan et al. (1998) who showed that a typical *Kavu* (sacred grove) such as the Malleeswaran Mudi sacred grove with a relatively small area of only 1.5 km2 recorded 722 angiosperm species. In comparison, the adjoining Silent Valley NP with a much larger area (90 km2) had a record of only 960 species. Later, Sreekumar et al. (2014) found that most (60%) of the tree species documented in this sacred grove were indigenous medicinal plants. This includes the two prominent species, i.e., *Mesua ferrea* L. (Calophyllaceae; Plate 11.2(b)) or *Nagkesar* and *Santalum album* L. (Santalaceae; Plate 11.2(c)) or *Chandana*.

The stamens of *M. ferrea* are used to treat skin diseases and leprosy in Ayurvedic system (Asif et al., 2017) while the bark is used in folk remedies to cure scabies and itches (Sreekumar et al., 2014). *S. album* is a vulnerable (VU) species. In Ayurvedic system its heartwood oil is used to treat stress, anxiety, and depression (Kumar et al., 2015) while, bark, and wood is used in folk remedies for headaches (Sreekumar et al., 2014).

11.4.3 TRADITIONAL REPOSITORIES OF MEDICINAL PLANTS

Studies across India have also highlighted that with the gradual disappearance of the surrounding natural vegetation, the sacred groves survived as tracts of 'relic' forests that function as repositories of a wide diversity of ethno-medicinal plants. These are protected as near-virgin forest ecosystems and sustainably used by the local communities (Chandran et al., 1998). Kanak-durga sacred grove in West Medinipur, West Bengal was documented with 388 vascular plants including 105 indigenous medicinal plant species (Plate 11.1(d); Bhakat, 2015). This largest sacred grove in the state is spread across 0.2 km^2in the heart of the tribal-dominated *lateritic* south-western (Rarh) *region*. Many of these ethnomedicinal plants have become increasingly rare in the surrounding wilderness, agricultural fields and rural settlements. The list includes *Marsdenia sylvestris* (Retz.) P.I.Forst. [= *Gymnema sylvestre* (Retz.) Schult.] (Aslepiadaceae) or *Meshashringi* which is a common constituent of Ayurvedic formulation prepared from leaves to treat diabetes (Thakur et al., 2012). Another herb used in Ayurvedic system is *Tylophora indica* (Burm.f.) Merr. (= *T. Asthmatica*; Apocynaceae; Plate 11.2(d)) or *Antamula* whose leaves are used to treat asthma (Butani et al., 2007). Similarly, seeds of *Strychnos nux-vomica* L. (Loganiaceae; Plate 11.2(e)) or *Kapilu* are used in Ayurvedic system to treat arthritis (Maji & Banerji, 2017). Boraiah et al. (2003) working in the central Western Ghats reported the resilience of ethnomedicinal plant resources in these repositories. About 60% of the regenerating plant species were medicinally important. The density of this was twice as that of the state-managed reserve forests in the same region.

11.4.4 HABITATS FOR MEDICINAL PLANTS AS KEYSTONE SPECIES

Analogous to the keystone in an arch, a keystone species has an exceptional effect on its natural environment. It dominates an ecosystem, maintains the structure of an ecological community, and influences the diversity of other species in the habitat. Without it, the ecosystem would drastically change or even collapse (Paine, 1995). Sacred groves across India have been documented to harbor these ecologically important keystone species, often with medicinal value, and are protected by the locals as sacred plants. A typical model for this phenomenon is the *Orans* (sacred groves) of Rajasthan. In the Akoli sacred grove of the local Bishnoi people in Jalore, Rajasthan, the sacred *Khejri* tree *Prosopis cineraria* (L.) Druce (= *P. spicigera* L.; Leguminosae; Plate 11.2(f)) is a keystone species of the desert which provides food and shelter to the near-threatened blackbucks (also sacred to the Bishnois), as well as livestock and even humans (Meena & Singh, 2012). This tree is one of the most common medicinal plants of the Indian desert and the bark used in folk medicine to treat skin diseases and leprosy (Girase et al., 2016). Other similar instances include species of the genus *Ficus* (Moraceae), such as the sacred *Pipal* (*F. religiosa* L.) and *Barh (F. benghalensis* L.) trees as reported in the miniature sacred groves of Mahendergarh, Haryana (Yadav et al., 2010) and the sacred *Heibong* (*Ficus racemosa* L.), in the Mahabali sacred grove, Imphal, Manipur. In Ayurvedic system, the bark of *F. religiosa* is used to treat respiratory disorders, and the stem latex of *F. benghalensis* is applied on cracked heels (Sharma et al., 2020).

11.4.5 REFUGES FOR RARE, ENDANGERED, AND THREATENED ETHNOMEDICINAL PLANTS

Rare species have small populations that are at risk but not presently endangered. Threatened species are likely to become endangered soon. Endangered species face the immediate danger of extinction (Nayar & Sastry, 1990). Sacred groves in India have been documented to serve as refuges for these ecologically fragile species that are often ethnomedicinal plants. The natural populations of *Vateria indica* L. and *Dipterocarpus indicus* Bedd. (Dipterocarpaceae), both critically endangered (CR) trees endemic to the southern Western Ghats, were virtually confined to Karnataka's Yelakundlikan sacred grove in Shimoga, and adjoining Kathalekan and Karikan sacred groves in Dakshina Kannada, respectively. This also happened to be the northern limit of their geographical range (Chandran & Gadgil, 1998). The resin from the first taxon has folkloric use in treating wounds and intestinal worms (Shrijani et al., 2018). The oleoresin from the second taxon, called *Gurjun* in Ayurveda, is used to cure rheumatism (Kapoor, 2000). Similarly, in the relict tropical ecosystems of the Myristica swamps found along the Konkan coast which are believed to be millions of years old, *Virola surinamensis* (Rol. ex Rottb.) Warb (= *Myristica fatua* Houtt.) (Myristicaceae), an endangered (EN) mangrove tree that is also endemic to the southern Western Ghats, was reported from the Katlekan sacred grove in Dakshina Kannada, Karnataka (Chandran et al., 1998). Its leaf vapor is used in folk medicine to treat malaria (Lim, 2012).

11.4.6 EXTENDED DISTRIBUTION RANGES FOR ENDEMIC MEDICINAL PLANTS

Endemic plants have limited geographical distribution that could range from an island to a continent (Morrone, 1994). Endemism is an important criterion for a geographical region to qualify as a biodiversity hotspot. In India, several medicinally important endemics have been reported from the sacred groves of the biodiversity hotspots in the Western Ghats and also in the Indo-Burma (Northeast India). Interestingly enough, some of these endemic plants have been documented from sacred groves located well beyond their native ranges. This data suggests that some of the populations of these paleoendemics, which were formerly widespread, have lingered on in these culturally protected relic forests that are now located well beyond the current geographical range of these taxa, which is restricted to a much smaller area. A good example of this is from the Northeast, in the Mahabali sacred grove of Imphal, Manipur. A rare and thriving natural population of coffee plum *Flacourtia jangomas* (Lour.) Raeusch. (Salicaceae), a tree endemic to Southeast Asia, was reported by Malhotra et al. (2001) from this relic forest. This plant has been domesticated and is now widely cultivated in the Terai region of Bihar and Uttar Pradesh. Its fruits and leaves are used to treat diarrhea (Sasi et al., 2018). Boraiah et al. (2003) made a comparative study in the central Western Ghats and found that 40% of medicinally important species were endemics to the sacred groves, and in contrast, only 11% of them were endemics in the state-managed reserve forests.

11.4.7 SHELTERS FOR FEW SPECIMENS OF RARE ETHNOMEDICINAL PLANTS

Sacred groves are forest patches of the indigenous communities that are customarily revered, protected, and conserved due to their immense value as local bioresource pools. In India, some of these relic habitats, perhaps as the last vestiges of the now lost virgin forests, are the final shelters for the natural populations or few specimens of rare medicinal plants. In extreme cases, there could even be only a solitary specimen. Possibly the best example of this was reported by Gadgil & Vartak (1976) who had found a single, but giant, specimen of the rare liana *Entada rheedei* Spreng. (= *E. Pursaetha* DC.) (Leguminosae) in the Tunbad sacred grove of Kolkaba (now Raigad), Maharashtra. Villagers from a wide area of 40 km radius came here to collect its bark with ethnoveterinary use in the treatment of cattle against snake bite. They also reported a pair of fine specimens of *Canarium strictum* Roxb. (Burseraceae), a vulnerable (VU) tree endemic to Uttara Kannada, Karnataka, in a *Devani* (sacred grove) of Maharashtra. Its resin, called *Sambrani* in Siddha, is used to treat dermal problems (Meena et al., 2012).

11.4.8 REDISCOVERY OF LOST SPECIES OF ETHNOMEDICINAL PLANTS

Sacred groves not only provide refuge to numerous rare and endemic medicinal plants but also some medicinal plant species which are believed to be extinct in the wild. These were later rediscovered from the sacred groves in whose culturally protected environments they remained protected from extinction (Chandran et al., 2008; Nair & Mohanan, 1981). There are many reports of medicinal tree species endemic to the southern Western Ghats from *Kavus* and *Kans* (sacred groves). In the Ankola sacred grove of Uttara Kannada, Karnataka, Chandran et al. (2008) rediscovered a small population of 13 trees of *Madhuca bourdillonii* (Gamble) H.J. Lam (Sapotaceae), an endangered (EN) species which was till then thought to be extinct. Its bark was used in folk medicine for treatment of rheumatic conditions (Amrithalingam, 2000). Similarly, in the Aikad sacred grove of Quilon, Kerala, Nair & Mohanan (1981) rediscovered a population of 35 trees of *Syzygium travancoricum* Gamble (Myrtaceae), a critically endangered (CR) species which too was till then widely considered to be extinct. Its bark was used for the folkloric treatment of arthritis (Rajalakshmi, 2016). At the regional level, in Kerala, the floristic analysis of the sacred groves has revealed the presence of many medicinal plant species that were new records for the state, such as *Justicia betonica* L. (Acanthaceae) among many (Santhosh Kumar et al., 2002). Its leaves are used in folk medicine to treat abscesses and swellings (Eapen et al., 2019).

11.4.9 DISCOVERY OF NEW SPECIES OF MEDICINAL PLANTS

Sacred groves as culturally protected pristine habitats are the ultimate refuge for plant life. In India, new species of plants with potential medicinal value have been discovered in the secluded environments of the sacred groves. A new species of the leguminous climber *Kunstleria keralensis* C. N. Mohanan & N. C. Nair (Leguminosae) was reported from a *Kavu* (sacred grove) in Kerala (Mohanan & Nair, 1981). The plant is endemic to the southern Western Ghats. Its bark was used in folk medicine to treat body ache (Binu, 2011). Another *Semecarpus kathalekanensis* Dasappa and M.H. Swaminath (Anacardiaceae) was reported later by Dasappa & Swaminath (2000) from the Kathalekan sacred grove Uttara Kannada, Karnataka. The species is critically endangered (CR) and endemic to the Myristica swamps of the southern Western Ghats. Its allied species are indigenous medicinal plants whose seeds are used in folk medicine to treat fever, bronchitis, and asthma (Gupta et al., 2013). Boraiah et al. (2003) reported that nearly equal proportions of 'non-medicinal' plant species were unique to the sacred groves and to state-managed reserve forests in the central Western Ghats. This suggests that the informal management system has not only conserved medicinally useful plant species but also many other taxa whose usefulness remains to be discovered in the future.

11.5 CONCLUSION

Rendering a sacred dimension to the natural space where the physical and spiritual meet is a characteristic feature of the ecocentric worldviews of autochthonous cultures. Living in close proximity to the environment, these indigenous communities have developed across generations a profound respect for the spirit world that is manifested in nature. This reverence to nature is part of a timeless, yet dynamic, way of life that harvests local natural resources in an unbroken cycle for food, fodder, fiber, fuel, and medicine. However, this harvesting is done with a profound realization of the laws and limits of nature. Their sacred natural spaces, and the beliefs, practices, and wisdom associated with them, epitomize their intimate relationship with the local environment. Sacred groves of India, as culturally protected natural habitats, have faced minimum anthropogenic interference due to a plethora of social taboos. These protected habitats still function as vital resource pools for remote communities. The medicinal plant resources of these ancient ethnoecological institutions are sustainably managed through an array of interconnected natural resource-related taboos which protect these natural habitats and their diversity of medicinal plants. Therefore, these isolated remains of climax forest ecosystems with rich plant spectra and function as traditional repositories of a rich diversity of indigenous medicinal plant species. These also harbor keystone species, serve as the final refuge for rare, threatened, endangered as well as endemic species, and even provide shelter to solitary specimens of rare medicinal plants. The sacred groves of India and the sustainable use of their medicinal plant resources by the local communities point out to the notion that the world's cultural diversity is part of its biodiversity that needs to be protected and preserved. This community-based approach not only prevents overharvesting of medicinal plants but also protects climax vegetations harboring these plants and checks degradation and fragmentation of their natural habitats. These informal institutions must be preserved, revived, and even newly created as these can supplement the formal ones in an effective manner for the *in-situ* conservation of the fast-disappearing wild medicinal plant resources across India. In recent years, radical 'bottom-up' conservation strategies have gained ground worldwide. Sacred groves are an excellent model for a decentralized, inclusive, and participatory approach for the sustainable use and conservation of ethnomedicinal plants where the locals as key stakeholders are involved not as 'owners' but rather as 'stewards' of this invaluable natural and cultural heritage.

ACKNOWLEDGMENTS

I am indebted to Dr. Shreyashi Chaudhuri, Lecturer, Department of History, Calcutta Girls' College, Kolkata, for her help in understanding the cultural dimensions of the sacred groves of India.

KEYWORDS

- **climax forests**
- **endemic medicinal plants**
- **in situ conservation**
- **medicinal plants**
- **sacred groves**
- **segmental taboos**
- **social taboos**

REFERENCES

Ahmad, Y., (2006). The scope and definitions of heritage: From tangible to intangible. *International Journal of Heritage Studies, 12*, 292–300.

Amrithalingam, M., (2000). *Indian Butter Tree: Ethnobotanical and Environmental Aspects* (Vol. 10, pp. 9, 10). MFP News.

Asif, M., Jafari, S. F., Iqbal, Z., Revadigar, V., Oon, C. E., Majid, A. S. A., & Majid, A. M. S. A., (2017). Ethnobotanical and phytopharmacological attributes of *Mesua ferrea*: A mini review. *Journal of Applied Pharmaceutical Science, 7*, 242–251.

Bentley, R., (2010). *Medicinal Plants*. Domville-Fife Press, London.

Berkes, F., & Folke, C., (1998). *Linking Social and Ecological Systems*. Cambridge University Press, Cambridge.

Bhakat, R. K., & Pandit, P. K., (2003). Role of a sacred grove in conservation of medicinal plants. *Indian Forester, 129*,224–232.

Bhakat, R. K., & Pandit, P. K., (2004). An inventory of medicinal plants of some sacred groves of Purulia District, West Bengal. *Indian Forester, 130*,37–43.

Bhakat, R. K., & Sen, U. K., (2008). Ethnomedicinal plant conservation through sacred groves. *Tribes and Tribals, 2*,55–58.

Bhakat, R. K., (2015). Biodiversity conservation through a sacred grove. *Indian Journal of Biological Sciences, 21*,59–62.

Bhandary, M. J., & Chandrasekhar, K. R., (2003). Sacred groves of Dakhina Kanada and Udupi Districts of Karnataka. *Current Science, 85*, 1655, 1656.

Binu, S., (2011). Medicinal plants used for treating body pain by the tribals in Pathanamthitta District, Kerala, India. *Indian Journal of Traditional Knowledge, 10*, 547–549.

Bodeker, G., Ong, C. K., Grundy, C.,Burford, G., & Shein, K., (2005). *WHO Global Atlas of Traditional, Complementary and Alternative Medicine.* WHO Press, Geneva.

Boraiah, K. T., Vasudeva, R., Shonil, A., Bhagwat, & Kushalappa, C. G., (2003). Do informally managed sacred groves have higher richness and regeneration of medicinal plants than state-managed reserve forests? *Current Science, 84*, 804–808.

Butani, A. K., Panchal, S. R., Vyas, N. Y., Khan, Y., & Kurnar, V., (2007). *Tylophora indica* an ancient anti-asthmatic medicinal plant: A review. *International Journal of Green Pharmacy, 1*,2–6.

Cardinale, B. J., Duffy, J. E., Gonzalez, A., Hooper, D. U., Perrings, C., Venail, P., Narwani, A., et al., (2012). Biodiversity loss and its impact on humanity. *Nature,456*, 59–68.

Chandran, M. D. S., & Gadgil, M., (1993). Kans-safety forests of Uttara Kannada. In: Brandl, M., (ed.), *Proceedings of the IUFRO Forest History Group Meeting on Peasant Forestry, 1991, Freiburg, Germany,* (pp. 49–57). Forstliche Versuchs und Forschungsanstalt, Freiburg.

Chandran, M. D. S., & Gadgil, M., (1998). Sacred groves and sacred trees of Uttara Kannada. In: Saraswati, B., (ed.), *Life Style and Ecology* (pp. 85–138). Indira Gandhi National Centre for the Arts and D.K. Print world, New Delhi.

Chandran, M. D. S., Mesta, D. K., Rao, G. R., Ali, S., Gururaja, K. V., & Ramachandra, T. V., (2008). Discovery of two critically endangered tree species and issues related to relic forests of the Western Ghats. *The Open Conservation Biology Journal, 2*, 1–8.

Chandran, M. D., Gadgil, M., & Hughes, J. D., (1998). Sacred groves of the Western Ghats of India. In: Ramakrishnan, P. S., Saxena, K. G., & Chandrashekara, U. M., (eds.), *Conserving*

the Sacred for Biodiversity Management (pp. 211–231).Oxford-IBH Publishing, New Delhi.
Chaudhuri, K. N., & Chaudhuri, S., (2015). The sacred groves of India: An overview of their role in the conservation of biological diversity. In: Jha, P., (ed.), *Biodiversity, Conservation and Sustainable Development: Issues and Approaches* (Vol. I., pp. 46–58). New Academic Publishers, New Delhi.
Chaudhuri, S., & Chaudhuri, K. N., (2014). Sacred groves of India: Exploring the cultural dimensions of an ancient ethnoecological institution. In: Ghosh, L., & Ghosh, S., (eds.), *Proceedings of the Sixth Biennial International Conference, IAAPS, 2012, Kolkata* (pp. 36–46). Maha Bodhi Book Agency, Kolkata.
Chaudhuri, S., & Chaudhuri, K. N., (2018). Sacred natural sites of the BCIM region: Exploring the role of intangible cultural heritage in protecting tangible natural heritage. In: Majumdar, A. J., & Banerjee, S., (eds.), *Sub-Regionalism and Connectivity: The BCIM Corridor* (pp. 150–163). Setu Publication, Kolkata.
Colding, J., & Folke, C., (1997). The relations among threatened species, their protection, and taboos. *Conservation Ecology*, *1*, 1–6.
Colding, J., & Folke, C., (2000). The taboo system: Lessons about informal institutions for nature management. *The Georgetown International Environmental Review*, *12*,413–445.
Colding, J., & Folke, C., (2001). Social taboos: 'Invisible' systems of local resource management and biological conservation. *Ecological Applications*, *11*,584–600.
Das, S. K., Rao, P. S., & Rao, D. B. S., (2004). Sacred and protected groves of Andhra Pradesh. In: Kannan, C. K., & Singh, B. G., (eds.), *Strategy for Conservation of Sacred Groves* (pp. 1–10). Institute of Forests Genetics and Tree Breeding, Coimbatore.
Dasappa, & Swaminath, M. H., (2000). A new species of *Semecarpus* (*Anacardiaceae*) from the Myristica Swamps of North Kanara, Karnataka, India. *Indian Forester,126*,78–82.
Deep, P., Singh, A. K., Ansari, M. T., & Raghav, P., (2013). Pharmacological potentials of *Ficus racemosa*: A review. *International Journal of Pharmaceutical SciencesReviewandResearch,* 22,29–34.
Dhyani, A., (2016). Critically endangered Indian medicinal plants. *Heritage Amruth, 11*, 42–45.
Eapen, J., Ansary, P. Y., Shahul, H. A., Indulekha, V. C., & Resny, A. R., (2019). Pharmacognosy of leaf, stem, and root of *Justicia betonica* Linn. *International Journal of Research in Ayurveda and Pharmacy, 10*,44–48.
Gadgil, M., & Chandran, M. D. S., (1992). Sacred groves. Indigenous vision: Peoples of India, attitudes to the environment. *India International Centre Quarterly, 19*, 184–187.
Gadgil, M., & Vartak, V. D., (1976). The sacred groves of Western Ghats in India. *Economic Botany, 30*, 152–160.
Gadgil, M., (2018). Sacred groves. *Scientific American, 319*, 48–57.
Gadgil, M., Berkes, F., & Folke, C., (1993). Indigenous knowledge for biodiversity conservation. *Ambio, 2*,151–156.
Ganesan, S., Ponnuchamy, M., Kesavan, L., & Selvaraj, A., (2009). Floristic composition and practices on the selected sacred groves of Pallapatty village (Reserved Forest), Tamil Nadu. *Indian Journal of Traditional Knowledge, 8*, 154–162.
Girase, M. V., Jadhav, M. L., & Jain, A. S., (2016). *Prosopis spicigera*: A nature's gift. *International Journal of Pharmaceutical Chemistry and Analysis, 3*,49–52.
Godbole, A., Watve, A., Prabhu, S., & Sarnik, J., (1998). Role of sacred groves in biodiversity conservation with local people's participation: A case study from Ratnagiri

District, Maharashtra. In: Ramakrishnan, P. S., Saxena, K. G., & Chandrashekara, U. M., (eds.), *Conserving the Sacred for Biodiversity Management* (pp. 233–246). Oxford-IBH Publishing, New Delhi.

Gupta, A. K., Bindal, M. C., Gupta, S. K., & Vedpal, D. P., (2013). Aphrodisiac activity of *Semecarpus anacardium* nut. *International Research Journal of Pharmacy, 3*,202–204.

Hardainiyan, S., Nandy, B. C., & Kumar, K., (2015). *Elaeocarpus ganitrus* (*Rudraksha*): A reservoir plant with their pharmacological effects. *International Journal of Pharmaceutical SciencesReview* and *Research, 34*,55–64.

Hughes, D. J., & Chandran, S. M. D., (1998). Sacred grove around the earth: An overview. In: Ramakrishnan, P. S., Saxena, K. G., & Chandrashekara, U. M., (eds.), *Conserving the Sacred for Biodiversity Management* (pp. 69–86). Oxford-IBH Publishing, New Delhi.

*Kapoor, L. D., (2000). CRC Handbook of Ayurvedic Medicinal Plants.*CRC Press, New Delhi.

Kapur, S. K., (1983). Medico-botanic survey of medicinal and aromatic plants of Mawphlang (Shillong). *Indian Drugs, 21*, 1–5.

Khan, M. S. A., & Ahmad, I., (2019). Herbal medicine: Current trends and future prospects. In: Khan, M. S. A., Ahmad, I., & Chattopadhyay, D., (eds.), *Advancements in Herbal Products as Novel Drug Leads* (pp. 3–13). Academic Press, Cambridge.

Khumbongmayum, A. D., Khan, M. L., & Tripathi, R. S., (2005). An ethnobotanical study of medicinal plants in the sacred groves of Manipur, Northeast India. *Journal of Traditional Ecological Knowledge, 4*, 33–38.

Khumbongmayum, A. D., Khan, M. L., & Tripathi, R. S., (2006). Biodiversity conservation in sacred groves of Manipur, Northeast India: Population structure and regeneration status of woody species. *Biodiversity and Conservation, 15*, 2439–2456.

Kumar, R., Anjum, N., & Tripathi, Y. C., (2015). Phytochemistry and pharmacology of *Santalum album* L.: A review. *World Journal of Pharmaceutical Research, 4*,1842–1876.

Kumar, V., & Tyagi, D., (2013). Review on phytochemical, ethnomedical and biological studies of medically useful genus *Bergenia. International Journal of Current Microbiology and Applied Sciences, 2*,328–334.

Lévi-Strauss, C., (1966). *The Savage Mind.* University of Chicago Press, Illinois.

Lim, T. K., (2012). *Myristica fatua.* In: Lim, T. K., (ed.), *Edible Medicinal and Non-Medicinal Plants* (pp. 572–574). Springer Dordrecht.

Lowenthal, D., (2005). Natural and cultural heritage. *International Journal of Heritage Studies, 11*,81–92.

Lubchenco, J., (1998). Entering the century of the environment: A new social contract for science. *Science, 279*,491–496.

Maffi, L., (2001). *On Biocultural Diversity*. Smithsonian Institution Press, Washington, DC.

Maffi, L., (2007). Biocultural diversity and sustainability. In: *Pretty, J., Ball,* A.,*Benton,* T.,*Guivant,* J.,*Lee,* D. R.,*Orr,* D.,*Pfeffer,* M., *& Ward, H., (eds.), The Sage Handbook of Environment and Society (*pp 267–277*).* Sage, London.

Maji, A. K., & Banerji, P., (2017). *Strychnos nux-vomica*: A poisonous plant with various aspects of therapeutic significance. *Journal of Basic and Clinical Pharmacy, 8*,87–103.

Malhotra, K. C., (1998). Anthropological dimensions of sacred groves in India: An overview. In: Ramakrishnan, P. S., Saxena, K. G., & Chandrashekara, U. M., (eds.), *Conserving the Sacred for Biodiversity Management* (pp. 423–438). Oxford-IBH Publishing, New Delhi.

Malhotra, K. C., Gokhale, Y., Chatterjee, S., & Srivastava, S., (2001). *Cultural and Ecological Dimensions of Sacred Groves in India.* Indira Gandhi Rashtriya Manav Sangrahalaya, Bhopal.

Malhotra, K. C., Gokhale, Y., Chatterjee, S., & Srivastava, S., (2000). *Sacred Groves in India: An Overview*. Indira Gandhi Rashtriya Manav Sangrahalaya, Bhopal.

Manikandan, P., Venkatesh, D. R., & Muthuchelian, K., (2011). Conservation and management of sacred groves in Theni District, Tamil Nadu, India. *Journal of Bioscience Research, 2*, 76–80.

Meena, D., & Singh, A., (2012). Oran of Rohida: An endangered tree species of Rajasthan. *Current Science, 103*, 25.

Meena, D., Binaibabu, N., & Doss, J., (2012). Future prospects for the critically endangered medicinally important species, *Canarium strictum* Roxb. a review. *International Journal of Conservation Science, 3*,231–237.

Mishra, B. P., & Jeeva, S., (2012). Plant diversity and community attributes of woody plants in two climax subtropical humid forests of Meghalaya. *Applied Ecology and Environmental Research, 10*,417–436.

Mohanan, C. N., & Nair, N. C., (1981). *Kunstleriaprain*: A new genus record for India and a new species in the genus. *Proceedings of the Indian Academy of Science, 90*, 207–210.

Morrone, J. J., (1994). On the identification of areas of endemism. *Systematic Biology, 43*, 438–441.

Nadanakunjidarn, S., (2000). *Ethnobotanical Studies of Atiapadi Hills of Western Ghats, India*. Ph. D. Dissertation. Pondicherry University, Puducherry.

Nair, N. C., & Mohanan, C. N., (1981). On the rediscovery of four threatened species from the sacred groves of Kerala. *Journal of Economic and Taxonomic Botany, 2*,233–235.

Nayar, M. P., & Sastry, A. R. K., (1990). *Red Data Book of Indian Plants* (Vol. I–III). Botanical Survey of India, Calcutta.

Newman, D. J., Cragg, G. M., & Snader, K. M., (2003). Natural products as sources of new drugs over the period 1981–2002. *Journal of Natural Products, 66*, 1022–1037.

Paine, R. T., (1995). A conversation on refining the concept of keystone species. *Conservation Biology, 9*, 962–964.

Posey, D. A., (2000). Introduction: Culture and nature, the inextricable link. In: Posey, D. A., (ed.), *Cultural and Spiritual Values of Biodiversity* (pp. 3–16). Intermediate Technology Publications, London.

Posner, R. A., & Rasmusen, E. B., (1999). Creating and enforcing norms, with special reference to sanctions. *International Review of Law and Economics, 19*, 369–382.

Pretty, J., Adams, B., Berkes, F., De Athayde, S. F., Dudley, N., Hunn, E., Maffi, L., et al., (2009). The intersections of biological diversity and cultural diversity: Towards integration. *Conservation and Society, 7*, 100–112.

Pushpangadan, P. A., (2018). Comparison of codified traditional systems of medicine and the non-codified systems of medicine like folk remedies and tribal medicine. *Advances in Complementary & Alternative Medicine, 2*,117–118.

Pushpangadan, P., Rajendraprasad, M., & Krishnan, R. N., (1998). Sacred groves of Kerala: A synthesis on the state of art of knowledge. In: Ramakrishnan, P. S., Saxena, K. G., & Chandrashekara, U. M., (eds.), *Conserving the Sacred for Biodiversity Management* (pp. 193–210). Oxford-IBH Publishing, New Delhi.

Rajalakshmi, S. P., (2016). Determination of nutraceuticals in *Syzygium jambos* L. (Alston) and *Syzygium travancoricum* Gamble. *International Journal of Herbal Medicine, 4*,150–152.

Rajith, N. P., & Ramachandran, V. S., (2010). Ethnomedicines of Kurichyas, Kannur district, Western Ghats, Kerala. *Indian Journal of Natural Products and Resources, 1*, 249–253.

Roy, P., (2002). Working with indigenous communities in the Nilgiris of southern India. In: Bradbear, N., Fisher, E., & Jackson, H., (eds.), *Strengthening Livelihoods: Exploring the Role of Beekeeping in Development* (pp. 99–102). Bees for Development, Monmouth.

Santhosh, K. E. S., Yeragi, S. S., Babu, K. N., & Sreekandan, N. G., (2002). Sacred grove flora of Kerala I: New plant records for Kerala. *Journal of Economic and Taxonomic Botany, 26*,141–143.

Sasi, S., Anjum, N., & Tripathi, Y. C., (2018). Ethnomedicinal, phytochemical and pharmacological aspects of *Flacourtia jangomas:* A review. *International Journal of Pharmacy and Pharmaceutical Sciences, 10*,9–15.

Sharma, A., Thakur, A., Kumari, R., Aarushi, Rolta, R., Sourirajan, A., Dev, K., & Kumar, V., (2020). Comparison of extraction yield, phytochemicals, and *in vitro* antioxidant potential of different extracts of two *ficus* species. *Plant Archives, 20*, 2669–2673.

Shrijani, J. K., Hegde, K., & Shabaraya, A. R., (2018). A review on pharmacological activities of *Vateria indica* Linn. *International Journal of Pharmaceutical and Chemical Research, 4*,5–9.

Singh, V., (2001). *Indian Folk Medicines and Other Plant Based Products.* Scientific Publishers, Jodhpur.

Sreekumar, V. B., Hareesh, V. S., Sreejith, K. A., & Nirmesh, T. K., (2014). Ethnobotanical studies of mudugars in muthikkulam high value biodiversity area with special reference to trees. In: Venkatachalapathy, R., (ed.), *Proc. of 26th Kerala Science Congress* (pp. 2558–2566). Kerala State Council for Science, Technology & Environment, Thiruvananthapuram.

Thakur, G. S., Sharma, R., Sanodiya, B. S., Pandey, M., Prasad, G. B. K. S., & Bisen, P. S., (2012). *Gymnema sylvestre*: An alternative therapeutic agent for management of diabetes. *Journal of Applied Pharmaceutical Science, 2*, 1–6.

Tharakan, G. C., (2007). The Muduga and Kurumba of Kerala, South India and the social organization of hunting and gathering. *Journal of Ecology and Anthropology, 11*,5–24.

Tilburt, J. C., & Kaptchuk, T. J., (2008). Herbal medicine research and global health: An ethical analysis. *Bulletin of the World Health Organization, 86*,594–599.

Vartak, V. D., Kumbhojkar, M. S., & Nipuge, D. S., (1987). Sacred groves in tribal areas of Western Ghats: Treasure trove of medicinal plants. *Bulletin of Medico-Ethno-Botanical Research, 8*,77, 78.

Yadav, S., Yadav, J. P., Arya, V., & Panghal, M., (2010). Sacred groves in conservation of plant biodiversity in Mahendergarh District of Haryana. *Indian Journal of Traditional Knowledge, 9*, 693–700.

CHAPTER 12

Medicinal Agroforestry System: Way Toward Conservation

DHIRAJ KUMAR, RAJENDRA PRASAD, SANGRAM CHAVAN, ASHA RAM, INDER DEV, ASHOK SHUKLA, and PRASHANT SINGH

ICAR–Central Agroforestry Research Institute, Jhansi, Uttar Pradesh, India

ABSTRACT

Since time immemorial, human beings have been using plant-derived products or derivatives for curing their ailments. MAPs are largely collected from forests or fringe areas and scattered plantations. However, we hardly know very few of the traditional MAPs and their uses, and due to population pressure, demand by pharmaceuticals and other industries, forest clearance, MAPs are considered as threatened and, in this context, the idea of their conservation and maintenance is of paramount importance. MAPs not only have the potential to be utilized in homestead or shelter belts but also offer a huge scope for their proliferation in a very well-mannered agroforestry system (AFS). It may be considered a win-win situation, provided the program of their conservation involves stakeholders from growers, industries, forest department, policy makers, researchers, etc. In fact, in the real sense, the conservation and their up-scaling need to be done on a social basis, by creating awareness among growers and prioritizing the steps to be taken as per the demand of users, availability of quality planting materials (QPMs), local climatic conditions, and sharing of benefits with the growers.

Bioprospecting Ethnomedicinal Plant Resources: Sustainable Utilization and Restoration.
Gopal Shukla, Jahangeer A. Bhat, A. P. Das, and Sumit Chakravarty (Eds.)

12.1 INTRODUCTION

In the majority of developing countries, people still depend on the traditional practice of collecting herbs or plant products to cure human and animal diseases. Most of the modern day drugs are prepared from plant-derived chemicals. However, nowadays, due to the dwindling natural sources of these products, a large array of MAPs is under serious threat. The rural population, especially rural and indigenous communities, largely depend on forests for the collection of medicinal plants mainly due to more expensive commercial medicines, unavailability, lack of markets, and knowledge on their uses. Around 90% of MAPs are derived from the wild, and only 10% in active trade are from cultivated lands in developing countries (Karki, 2002). In India, out of 17,000 higher plant species, about 7,500 were reported with medicinal values (Shiva, 1996). In fact, India is home to the largest diversity of medicinal plant species (Table 12.1).

TABLE 12.1 Medicinal Plants and its Distribution

Country/ Region	Medicinal Plant Species in Flora (Total Number)	Reported Number of Medicinal Plant Species	Medicinal Plants (%)	References
World	2,97,000	52,885	10	Schippmann et al. (2002)
India	17,000	7,500	44	Shiva (1996)
India	8,000	1,748	22	Samant et al. (1998)

The mighty Himalayan range is home to the largest resource of medicinal and aromatic plants (MAPs), particularly in Uttarakhand, Sikkim, and the northern region of West Bengal. The great extent of the trans-Himalayan range is home to very rare species of MAPs owing to its congenial climatic conditions and landforms. For smallholders, MAPs not only serve as a good source for better health but also a livelihood source. Herbal-based products have quite a good demand and MAPs are both traded nationally and internationally. As per the WHO estimates, the global market for plant-based medicine will hit the five trillion US$ mark by the year 2050. However, India's share in the global export market of medicinal plants related trade is just 0.5% (Kumar & Janagam, 2011). Recent estimates show that the area under MAPs has increased from 1,31,000 ha in 2004-2005 to 5,57,000 ha in 2013-2014 and decreased to 493 ha in 2014-2015 (Anon., 2015). Cultivation of MAPs may ensure some advantages like enhanced income, management of degraded patches of land, occurrence of fewer pests and diseases, long shelf-life of products, and more foreign exchange (Rao et al., 2000, 2004). Due to the degradation of land resources, forest clearance, increase in mono-cropping of commercial food crops cultivation, cultivation, and preservation of MAPs is quite necessary. Several researchers have highlighted the need for the conservation of MAPs by integrating them into agroforestry systems (AFS) (Lambert et al., 1997; Dery et al., 1999; McNeelay, 2004; Rao et al., 2004; Singh et al., 2019; Jilariya et al., 2019). Under AFS, there is much scope for MAP cultivation by taking advantage of shade, light, nutrients, and space. However, if good marketing facilities, proper management, and QPMs are available, cultivation of MAPs under AFS may be a win-win situation.

12.2 INTEGRATING MAPS IN AGROFORESTRY SYSTEM (AFS)

Many species since time immemorial have been grown in forest areas. These are also found growing naturally on different landforms like farm bunds, homegardens, and scattered plantations. Primarily, the medicinal values of these trees have been exploited at a local scale while their major use on a commercial scale has been secondary. But owing to overdependence on forests for meeting our daily needs, the future of MAP conservation is at stake. Most of the MAPs are grown under wilderness in the shade of trees or scattered conditions. The expansion of MAPs on agricultural lands is meager due to population pressure on food grain requirements. Thus, the utilization of degraded and fallow lands can easily be integrated with trees on boundaries, or farm bunds or AFS may serve as a boon to scale up the idea of MAP cultivation. The process of secondary products and alkaloids formation in MAP occurs due to breakdown and re-synthesis of primary products, and this process is favored under shade conditions (Ram et al., 1999). This may serve as a clue to utilize our trees or AFS for the promotion and conservation of MAPs. Thus, in order to meet the growing global demand of MAPs for the preparation of drugs and other uses, agroforestry offers such a situation where we can easily integrate MAP cultivation along with trees. The medicinal values of some of the prominent tree species are presented in Table 12.2.

TABLE 12.2 Medicinal Values of Tree Species under Tropical Agroforestry Systems

Tree Species	Agroforestry System	Medicinal Use	References
Acacia nilotica(Babul)	Field bunds, scattered trees, woodlots, grazing lands. Middle East, Africa, Indian sub-continent.	Sexually transmitted diseases, diarrhea, dysentery; anti-bacterial, antioxidant, anti-mutagenic, cytotoxic activity, gastrointestinal disorders.	Ali et al. (2012); Donalisio et al. (2018)
Azadirachta indica (Neem)	Woodlots, scattered trees, shelterbelts. Tropical and semi tropical region.	Urinary disorders, leprosy, epitasis, cough, piles, phlegm, analgesic, ulcers, intestinal worms, etc.	Trivedi et al. (2019)
Erythrina spp.	Shade tree in coffee, live fence, forests, hills. Semi-arid, humid, and sub-tropics.	Antipyretic, dysentery, mastitis, asthma, cough, antispasmodic, cytotoxic, anti-tumoral, uterine stimulant, anti-fungal, anti-bacterial, anti-malarial.	Araújo et al. (2012)
Parkia biglobosa(Locust bean tree)	Parklands West Africa	Piles, malaria, stomach disorders, jaundice.	Teklehaimanot (2004)
Prosopis cineraria (Khejri)	Scattered trees. Arid to semi-arid areas of Western Asia and Indian sub-continent.	Leukoderma, leprosy, asthma, dyspepsia, anti-inflammatory, anticonvulsant, antifungal, anticarcinogenic, antidiabetic, hypolipidemic, abortifacient, antioxidant, antimicrobial, wound healing properties.	Girase et al. (2016)
Tamarindus indica (Tamarind)	Scattered trees, field bunds. Indigenous to tropical Africa and exotic to India and Central America.	Jaundice, yellow fever, liver problems, antacid, laxative, haematinic, dermal cleanser, antioxidant, reduce weight, analgesic, antimicrobial, anti-inflammatory, hypolipidemic, anthelmintic, hepatoprotective.	Zohrameena et al. (2017)

Most of the medicinal tree species grow either wildly or as scattered plantations. Thus, they are seldom used for specific medicinal purposes and largely remain unexploited except for their use as a source of food, fuel, fiber, and fodder by nearby forest dwellers. But we need to assess the potentiality of MAPs and then introduce them with a different set of AFS. Many tree species with medicinal values like *bael* (*Aegle marmelos*), *aonla* (*Emblica officinalis*), custard apple (*Annona reticulata*), moringa (*Moringa oleifera*), jamun (*Syzygiumcumini*) and *neem* (*Azadirachta indica*) were grown traditionally particularly in homegardens or as scattered trees and rows by the farmers (Table 12.3).

TABLE 12.3 Important Tree Species and Their Medicinal Uses

Common Name	Scientific Name	Medicinal Uses
Black willow	*Salix nigra*	Quinine alternate, cold-cough, fever, headache.
Teak	*Tectona grandis*	Headache, stomach problems, fever, digestion.
Cork tree	*Quercus suber*	Lung and cough diseases.
Sal	*Shorea robusta*	Dysentery, diarrhea, foot cream, skin diseases.
Tulip tree	*Liriodendron tulipifera*	Swollen joints.
Kumat	*Acacia senegal*	Oral health, throat inflammation, irritation.
Eucalyptus	*Eucalyptus globulus*	Counter cough, cold medications.
Bargad	*Ficus bengalensis*	Gum, tooth disorders.
Peepal	*Ficus religiosa*	Asthma, skin, tooth pains.
Ashok tree	*Saracaasoca*	Bleeding, pain, irritations.
Arjun	*Terminalia arjuna*	Heart disorders.
Amla	*Emblica officinalis*	Digestion, immunity.
Sandalwood	*Santalum album*	Relieves nerves.
Harad	*Terminalia chebula*	Digestion, piles, gastro-intestinal, diarrhea.
Behera	*Terminalia bellerica*	Sore throat, hair, eyes, sedative action.
Audumbar	*Ficus racemosa*	Mosquito bites.
Moringa	*Moringa oleifera*	Blood pressure, hemoglobin, herbal medicines.
Mahagony	*Swietenia mahagoni*	Anemia, fever, dysentery.
Indian rosewood	*Dalbergia sissoo*	Acne treatment, balance oily and dry skins.
Cedar	*Cedrus* spp.	Cold, flu, fever.
Jackfruit	*Artocarpus heterophyllus*	Sedative, induce abortion, fever, diarrhea.
Jamun	*Syzygiumcumini*	Antibiotic, anticarcinogenic, cardio-protective.
Kher	*Acacia catechu*	Diarrhea, nose/throat swelling, dysentery, bleeding.
Apple ring acacia	*Faidherbia albida*	Fever, diarrhea, ocular infections.
Baobab	*Adensonia digitata*	Diarrhea, malaria, microbial infections.

12.3 HOMESTEAD MAPS

A lot of medicinal herbs are used in homestead gardens or in forest fringe areas. Most common among them have been discussed along with the portion of the plant parts used and ailments (Table 12.4).

TABLE 12.4 Uses of Some Common Medicinal Plants

Common Name Maturity Period	Scientific Name	Parts Used	Ailments
Aonla (T) After four years	*Emblica officinalis* Euphorbiaceae	F	Vitamin C, cough, cold, diabetes, laxative, hyper acidity.
Ashok (T) 10 years onward	*Saracaasoca* Caesalpiniaceae	B, Fl	Menstrual pain, uterine disorder, diabetes.
Ashwagandha (H) One year	*Withaniasomnifera* Solanaceae	R, L	Restorative tonic, stress, nerves disorder, aphrodisiac.
Bael/Bilva (T) After 4–5 years	*Aegle marmelos* Rutaceae	F, B	Diarrhea, dysentery, constipation.
Bhumi aonla (H) Within one year	*Phyllanthus amarus* Euphorbiaceae	Wp	Anemic, jaundice, dropsy.
Brahmi (H) One year	*Bacopa monnieri* Scrophulariaceae	Wp	Nervous, memory enhancer, mental disorder.
Chiraita (H) Within one year	*Swertia chirata* Gentianaceae	Wp	Skin disease, burning sensation, fever.
Gudmar (C) After four year	*Gymnemasylvestre* Asclepiadaceae	L	Diabetes, hydrocele, asthama.
Guggul (T) After eight years	*Commiphorawightii* Burseraceae	Gr	Rheumatism, arthritis, paralysis, laxative.
Guluchi/Giloe (C) Within one year	*Tinospora cordifolia* Menispermaceae	St	Gout, pile, general debility, fever, jaundice.
Kalihari (H) Five year	*Gloriosa superba* Liliaceae	Sd, Tu	Skin disease, labor pain, abortion, general debility.
Kalmegh/Bhui neem (H) Within one year	*Andrographis paniculata* Scanthaceae	Wp	Fever, weakness, release of gas.
Long peeper/Pippali (C) After 2–3 years	*Piper longum* Piperaceae	F, R	Appetizer, enlarged spleen, bronchitis, cold.
Makoi (H) Within one year	*Solanum nigrum* Solanaceae	Wp, F	Dropsy, general debility, diuretic, anti-dysenteric.
Patharchur (H) One year	*Coleus barbatus* Lamiaceae	R	Kidney stone, calculi.
Sandal wood (T) 30 years onward	*Santalum album* Santalaceae	Hw, O	Skin disorder, burning sensation, jaundice, cough.

TABLE 12.4 *(Continued)*

Common Name Maturity Period	Scientific Name	Parts Used	Ailments
Sarpagandha (H) After two year	*Rauwolfia serpentina* Apocynaceae	R	Hypertension, insomnia.
Satavari (C) After 2–3 years	*Asparagus racemosus* Liliaceae	Tu, R	Enhance lactation, general weakness, fatigue, cough.
Senna (S) Within one year	*Cassia angustifolia* Liliaceae	TuD	Rheumatism, general debility tonic, aphrodisiac.
Tulsi (perennial) Each three months	*Ocimum sanctum* Lamiaceae	L, Sd	Cough, cold, bronchitis, expectorant.
False black pepper (C) Second years onward	*Embeliaribes* Myrsinaceae	R, F, L	Skin disease, snake bite, helminthiasis.
Pipper mint (H) Perennial	*Mentha piperita* Lamiaceae	L, Fl, O	Digestive, pain killer.
Mehandi (S) 1/25 years	*Lawsoniainermis* Lytharaceae	L, Fl, Sd	Burning, seam, anti-inflammatory.
Ghritkumari (H) Two to five years	*Aloe vera* Liliaceae	L	Laxative, wound healing, skin burns and care, ulcer.
Sadabahar (H)	*Vinca rosea* *Catharanthus roseus* Apocynaceae	Wp	Leukemia, hypotensive, anti-spasmodic, anticarcinogenic.
Bhringraj (H)	*Eclipta alba* Compositae	Sd	Anti-inflammatory, digestive, hair vitalizer.
Swetchitrak (H)	*Plumbago zeylanica* Plumbaginaceae	R	Appetizer, antibacterial, anticarcinogenic.
Raktachitrak (H)	*Plumbago indica* Plumbaginaceae	R	Dyspepsia, colic, inflammation, cough.
Kochila (T) 15 years	*Strychnosnux-vomica* Loganiaceae	Sd	Nervous, paralysis, wound healing.
Harad (T) Eight to 10 years	*Terminalia chebula* Combretaceae	Sd	Triphala, wound ulcer, leprosy, inflammation, cough.
Baheda (T) Seven to eight years	*Terminalia bellerica* Combretaceae	Sd, B	Cough, insomnia, dropsy, vomiting, ulcer, triphala.
Gokhuru (H) One year	*Tribulus terrestris* Zygophyllaceae	Wp	Sweet cooling, aphrodisiac, appetizer, digestive, urinary.

TABLE 12.4 *(Continued)*

Common Name Maturity Period	Scientific Name	Parts Used	Ailments
Neem (T) Four to five years	*Azadirachta indica* Meliaceae	Rh	Sedative, analgesic, epilepsy, hypertensive.
Anantamool (S)	*Hemidesmus indicus* Asclepiadaceae	R, L	Appetizer, carminative, aphrodisiac, astringent.
Bach, Ghorbach (H) One year	*Acorus calamus* Acoraceae	Rh	Sedative, analgesic, epilepsy, hypertensive.
Adusa (S)	*Adhatodavasica* Acanthaceae	Wp	Antispasmodic, respiratory, stimulant.
Nagkesara (T) Four year	*Mesua ferrea* Guttiferae	B, L, Fl	Asthma, skin, burning, vomiting, dysentery, piles.
Khaskhas grass (S)	*Vetiveriazizanioides* Poaceae	R	Hyperdipsia, burning, ulcer, skin, vomiting.
Mandukparni (H)	*Centella asiatica* Umbelliferae	Wp	Anti-inflammatory, jaundice, diuretic, diarrhea.
Kaiwanch (S)	*Mucuna pruriens* Fabaceae	R, Sd, L	Nervous disorder, constipation, nephropathy, strangury, dropsy.
Dalchini (S) Four year	*Cinnamomum zeylanicum* Lauraceae	B, O	Bronchitis, asthma, cardiac disorder, fever.
Kurchi Kutaja (S)	*Holarrhenaantidysenterica* Apocynaceae	B, Sd	Scabies, antipyretic, amoebic dysentery.
Kateli Bhatkataiya (H)	*Solanum xanthocarpum* Solanaceae	F, Sd, Wp	Diuretic, anti-inflammatory, appetizer, stomachic.

Abbreviations: T: Tree; S: Shrub; H: Herb; C: Climber; F: Fruit, B: Bark, Fl: Flower; R: Root, L: Leaves; Wp: Whole plant; Gr: Gum resin; St: Stem; Sd: Seed, Tu: Tuber; TuD: Tubers Hw: Heartwood; O: Oil; and Rh: Rhizome.

Source: Orissa Forest Development Corporation Ltd.

Homestead agroforestry has been mainly practiced in tropical rainfall areas as a part of traditional AFS. Homegardens provide a unique assemblage of a combination of trees, annuals, perennials, MAPs, in multistrata systems with the addition of livestock either temporarily or permanently. It not only serves the purpose of home requirements of food, fuel, fodder, NTFPs but also benefits in terms of carbon sequestration on-site, rich biodiversity, economic sustainability, and ecosystem services. Homegardens serve independence towards the social fabrication of forestry by tree and crop association with the proper utilization of light, moisture, nutrients, and in addition to efficient nutrient cycling, high biodiversity, low use of external inputs, and soil conservation potential (Lakshmi & Joseph, 2017). Homegardens make a substantial contribution to the supply of medicinal plants which may be traded or consumed by the family or locally by the community (Albuquerque & Andrade, 2002). This farming practice is definitely rich in diversity, ecological restoration, socio-cultural, and economic benefits. Thus, local communities, along with the forest department, policy-level stakeholders must come together to conserve this age-old system of eco-restoration, eco-rehabilitation, and ecosystem development.

12.4 CONSTRAINTS IN MAPS CULTIVATION

Most of the growers have to sell their produce to local traders, then to wholesalers followed by processing units. Sometimes they sell to herbal healers, and often this leads to an unorganized marketing structure. There is a lack of technical know-how among the growers regarding processing, preservation, and grading. The price of QPMs is quite high and they are often not available when needed, as most of the growers are not aware of their recognized supplier or source. Irrigation facilities are also not adequate, and the cost of initial installation of drip or sprinkler systems is high. Growers are sometimes not aware of subsidies. There is a lack of processing facilities for the final produce. Also, the demand and supply chain is quite erratic, often leading to loss of the produce. There is also a need for agencies to market the final produce. Finally, the knowledge of MAPs in its overall package of practices is hardly documented and inappropriately regulated. Mostly, the benefits on economics and its large-scale implications are missing. At present, the conservation of MAPs is quite necessary so that they may yield the greatest sustainable benefits for the present generation and fulfill the aspirations of future generations. The conservation must be done through legislative, *in situ*, and *ex situ* methods.

12.4.1 IN SITU CONSERVATION

It includes conservation of species in its natural habitat to have maximum diversity and distinctiveness. It involves conservation through gene banks, botanical parks, biosphere reserves, national parks, sacred groves. These are the sites that will help to maintain maximum genetic, species, and ecosystem level diversity. Other such *in-situ* conservation is like on-farm conservation, home gardens, etc.

12.4.2 *EX SITU CONSERVATION*

It involves the conservation of MAPs outside the natural habitat, may encompass the cultivation and maintenance in botanical parks, gardens, and preservation of plant propagules in repositories, gene banks, cryopreservation, DNA libraries, etc. The institutions like ICAR–National Bureau of Plant Genetic Resources (NBPGR), New Delhi, India under Indian Council of Agricultural Research (ICAR), Ministry of Agriculture and Farmers Welfare, Government of India, Jawaharlal Nehru Tropical Botanic Garden and Research Institute (JNTBGRI), Kerala, and CSIR–Central Institute of MAPs, Lucknow (Uttar Pradesh) under Council of Scientific and Industrial Research (CSIR), Ministry of Science and Technology, Government of India are doing the conservation of these MAPs.

12.5 EXISTING POLICIES AND RECOMMENDATIONS

The medicinal plant sector has been considered an integral part of the indigenous system of medicine and homeopathy in the first five-year plan (Ahuja, 2001). Furthermore, the Planning Commission and the National Medicinal Plant Board (NMPB) have prepared a policy document concerning the commercialization and promotion of the medicinal plant sector. In order to sustainably utilize medicinal plants and mitigate bio-piracy, the Biodiversity Act (2002) framed several rules (Pushpangadan & Kumar, 2005). Several institutions in this regard are also actively working, such as ICMR, ICAR, CSIR, DBT, DST, MoEF, NMPB, DRDO, UGC, NABARD, AICTE, HRDI, and GBPIHED. There is a lot of scope for the promotion and cultivation of MAPs in India. In fact, we need to train people engaged in its cultivation and processing. Regular markets should be provided with insurance and credit facilities, availability of QPMs, and the establishment of proper value chain models. In order to raise awareness among growers about package of practices and weather forecasts, messages should be provided about cultivation practices. Measures should be taken to prioritize MAPs based on location, develop proper agrotechnologies for production and conservation, and meet the demand for MAPs.

12.6 WAY AHEAD

Despite the great benefits associated with MAPs and their direct use for healing human and livestock diseases, their future is at risk, which necessitates their revival. It is high time for researchers, policymakers, farmers, industry people, and medical practitioners to come together and devise a proper plan for moving in the right direction. There is a need to prioritize species of MAPs according to demand so that local growers can benefit. Without proper marketing and demand, local growers may become demotivated if their products do not find a place. Secondly, the drug industry and scientists involved in the process must focus on proper processing, development of drugs with defined uses, exploration of traditional lost species for possible utilization, and documentation. Thirdly, government agencies should provide rights to the growers with a share in the profit, and if possible, the forest department should provide incentives for growing MAPs alongside forest tree species. There is a whole array of activities to be performed at all levels, starting from providing quality planting materials to processing, manufacturing, marketing, and benefit transfer. Everyone must work hand in hand if we want to conserve and maintain MAPs.

12.7 CONCLUSION

MAPs are known to have multifaceted benefits to mankind, apart from its use in culinary, pharmaceutical, and cosmetic industries. However, in recent times there is more threat to these species due to population pressure, increased demand of forest products, overdependence on forests, deforestation, and policy bottlenecks. Thus, conservation and maintenance of MAPs has become quite imperative, and, in this concern, agroforestry plays a pivotal role for its upscaling at all levels. Besides, having benefits in terms of food, fuel, fiber, fodder needs, ecosystem services, and increasing biodiversity, agroforestry may act as a safe haven for conservation of MAPs provided all stakeholders in the chain come together and work in line with priority.

KEYWORDS

- **agroforestry**
- **conservation**
- **diversification**
- **livelihood**
- **medicinal and aromatic plants**

REFERENCES

Ahuja, A. K., (2001). Need for a comprehensive approach to medicinal plants. In: Samant, et al., (ed.), *Himalayan Medicinal Plants: Potential and Prospects* (pp. 1–21) Gyanodaya Prakashan, Nainital.

Albuquerque De, U. P., & Andrade, L. De. H. C., (2002). Conhecimentobotánicotradicional e conservaçãoemumaárea de caatinga no estado de Pernambuco, nordeste do Brasil. *Acta Botanico do Brasil., 16*, 273–285.

Ali, A., Akhtar, N., Khan, B. A., Khan, M. S., Rasul, A., Khalid, N., Waseem, K., Mahmood, T., & Ali, L., (2012). *Acacia nilotica*: A plant of multipurpose medicinal uses. *J. Med. Plant Res., 6*, 1492–1496. doi: 10.5897/JMPR11.1275.

Anonymous, (2015). *Indian Horticulture Database-2014*. Ministry of Agriculture, Government of India.

Araújo, J. X., Oliveira, M. S. G., Aquino, P. G. V., Alexandre-Moreira, M. S., & Sant'Ana, A. E. G. A., (2012). Phytochemical and ethnopharmacological Review of the genus *Erythrina*. In: Rao, V., (ed.), *Phytochemicals – A Global Perspective of Their Role in Nutrition and Health* (pp. 327–352). Intech Publisher Rijeka, Croacia. doi: 10.5772/26997.

Dery, B. B., Otsyina, R., & Ng'tigwa, C., (1999). *Indigenous Knowledge of Medicinal Trees and Setting Priorities for their Domestication in Shinyanga Region, Tanzania*. International Center for Research in Agroforestry, Nairobi, Kenya.

Donalisio, M., Cagno, V., Civra, A., Gibellini, D., Musumeci, G., Ghosh, M., & Lembo, D., (2018). The traditional use of *Vachellianilotica* for sexually transmitted diseases is substantiated by the antiviral activity of its bark extract against sexually transmitted viruses. *J. Ethnopharmacol., 213*, 403–408. doi: 10.1016/j.jep.2017.11.039.

Girase, M. V., Jadhav, M. L., & Jain, A. S., (2016). *Prosopis spicigera*: A nature's gift. *Int. J. Pharm. Chem. Anal., 3*(1), 49–52.

Jilariya, D. J., Thakur, N. S., Singh, N., & Gunaga, R. P., (2019). Economics of cultivation of *Melia dubia*Cav.–*Aloe vera* L. silvi-medicinal model. *Ind. J. Agrofor., 21*(2), 35–40.

Karki, M. B., (2002). *Organic Conservation and Certification: A Strategy for Improved Value Addition and Marketing of Medicinal Plant Products in Himalayas* (pp. 15–20). Paper presented at Regional Workshop at Wise practices and Experimental Learning in the Conservation and Management of Himalayan Medicinal Plants. Kathmandu, Nepal.

Kumar, R. M., & Janagam, D., (2011). Export and import pattern of medicinal plants in India. *Indian J. Sci. Technol., 4*(3), 245–248.

Lakshmi, G., & Joseph, A., (2017). Soil micro-arthropods as indicators of soil quality of tropical home gardens in a village in Kerala, India. *Agrofor. Syst., 91*(3), 439–450.

Lambert, J., Srivastava, J., & Vietmeyer, N., (1997). *Medicinal Plants: Rescuing a Global Heritage*. The World Bank, Washington, DC.

McNeelay, J. A., (2004). Nature vs. nurture: Managing relationships between forests, agroforestry, and wild biodiversity. *Agrofor. Syst., 61*, 155–165.

Pushpangadan, P., & Kumar, B., (2005). Ethnobotany, CBD, WTO, and the Biodiversity Act of India. *Ethnobot., 17*, 2–12.

Ram, M., Ram, D., Singh, S., Naqvi, A. A., & Kumar, S., (1999). Studies on intercropping of patchouli (*Pogostemon patchouli*) with papaya (*Carica papaya*). *Ind. J. Med. Aromat. Plants, 21*, 358–360.

Rao, E. V. S. P., Puttanna, K., Rao, R. S. G., & Kumar, S., (2000). Prospect of commercial mediculture and recent advances in agrotechnologies of aromatic plants in South India. *J. Med. Aromat. Plant Sci., 22*(1B), 207–213.

Rao, M. R., Palada, M. C., & Becker, B. N., (2004). Medicinal and aromatic plants in agroforestry systems. *Agrofor. Syst., 61*, 107–122.

Samant, S. S., Dhar, U., & Palni, L. M. S., (1998). *Medicinal Plants of Indian Himalaya: Diversity Distribution Potential Values*. G.B. Pant Institute of Himalayan Environment and Development, Almora.

Schippmann, U., Leaman, D. J., & Cunningham, A. B., (2002). *Impact of Cultivation and Gathering of Medicinal Plants on Biodiversity: Global Trends and Issues*. Inter-Department Working Group on Biology Diversity for Food and Agriculture, FAO, Rome.

Shiva, M. P., (1996). *Inventory of Forestry Resources for Sustainable Management and Biodiversity Conservation.* Indus Publishing Company, New Delhi.

Singh, M., Singh, V., Tewari, S., Kumar, A., & Singh, R., (2019). Growth and yield of different mint varieties under poplar-based agroforestry system in Terai region of Uttarakhand. *Ind. J. Agrofor., 21*(2), 81–85.

Teklehaimanot, Z., (2004). Exploiting the potential of indigenous agroforestry trees: *Parkia biglobosa* and *Vitellaria paradoxa* in Sub-Saharan Africa. *Agrofor. Syst., 61*, 207–220.

Trivedi, A., Fatima, N., Husain, I., & Misra, A., (2019). An update on the therapeutic potential of neem and its active constituents: A panacea for all diseases. *Era's J. Med. Res., 6*(1), 1–8.

Zohrameena, S., Mujahid, M., Bagga, P., Khalid, M., Noorul, H., Nesar, A., & Saba, P., (2017). Medicinal uses & pharmacological activity of *Tamarindus indica*. *World J. Pharm. Sci., 5*(2), 121–133.

CHAPTER 13

Utilization and Ethnomedicinal Value of Indigenous Green Leafy Vegetables Commonly Consumed in Northern Districts of West Bengal, India

ANJALI SHARMA

Uttar Dinajpur Krishi Vigyan Kendra, Uttar Banga Krishi Viswavidyalaya, Chopra, West Bengal, India

ABSTRACT

The Indian subcontinent offers one of the most diversified genetic resources, and since the beginning of human memory, medicinal plants have been used as a key component of treatment. The northern track of West Bengal is full of genetic resources due to the presence of dense natural forests over a considerable portion of the geological area. Locally available (LA) indigenous vegetables are considered robust, nourishing, and a storehouse of desirable traits. Moreover, these are often grown easily as most of them grow in nature and are resistant to diseases and pests and above all suitable for indigenous local recipes of consumption and tastes. They facilitate the expansion of agricultural production systems, returns, and healthy dishes for year-round consumption. Imparting knowledge of diversification of diets through improved utilization and consumption of LA indigenous green leafy vegetables was the major aim of the present study. The foremost goals of the study were to find the food consumption patterns, practices, and medicinal use of indigenous leafy vegetables by the rural folk of the Northern district of West Bengal with particular reference to Uttar Dinajpur district. Leafy vegetables are either grown or gathered from uncultivated growth from a

Bioprospecting Ethnomedicinal Plant Resources: Sustainable Utilization and Restoration.
Gopal Shukla, Jahangeer A. Bhat, A. P. Das, and Sumit Chakravarty (Eds.)

range of habits and habitats and commonly marketed and used as leaf, tender shoot, or flower vegetables. A sample of 280 households was interviewed. The investigation areas were tired across the socio-economic classes of the upper, the lower upper, middle, lower middle, and lower. There was a considerable difference ($P<0.05$) between the consumption of indigenous green leafy vegetables and livelihood, with more casual laborers and/or non-employed respondents (81%) than those in full-time employment and business (29%) consuming indigenous vegetables. This was suggested based on how much time one has to spend purchasing, preparing, and cooking indigenous vegetables. This indicates the fact that consumption of such local items depends upon buying capacity, preparation, and cooking time. It is evident from the findings that consumption of indigenous vegetables is very meager (8%). Consumption was based on awareness, occupation, income, etc. The most consumed indigenous green leafy vegetables by children were pumpkin and Bottle gourd tender shoots and leaves. The chapter concludes that these vegetable crops have the potential for their utilization in alleviating hunger, malnutrition, and health improvement, thereby making a difference in livelihoods with the realization of the importance and uses of such neglected and underutilized crops.

13.1 INTRODUCTION

India espouses unmatched biodiversity. It is truly a land of life. No wonder it is declared one of the top biodiversity hotspots. This biodiversity has a rich component of medicinal plants also. In fact, our ancestors have taught that every plant possesses medicinal value. Such a blend of philosophy and biodiversity has naturally evolved into powerful medicinal systems. And, therefore, in India medicinal plants have formed an integral part of treatment from the commencement of human memory. West Bengal, being an assorted state, has always something to offer through its glorious traditions and rich cultural heritage. Besides its cultural heritage, the state also occupies an exceptional geographical position and an exceptionally varied climate ranging from extreme cold in the Himalayan regions to the heat of plains in summers. The northern track of West Bengal is full of genetic resources due to the presence of dense natural forests over an extensive portion of the geological area. Vegetables from specific niches that are grown nearby are regarded as strong, nourishing, and full of positive qualities. In addition, they are frequently easier to cultivate because they are typically found growing in nature, are resistant to insects, pests, and illnesses, and are, above all, suitable for use in traditional local cuisine. They support year-round nutrition by diversifying production methods, sources of income, and diets. There is a lot of opportunity to improve nutrition because there are so many different indigenous vegetables to take into consideration (Kalia et al., 2007). The major objective of the survey was to enhance the use and consumption of local leafy vegetables to diversify diets. The primary goal of the survey was to determine how locally grown leafy vegetables were used as medicine by rural residents of West Bengal's northern region, particularly in the Uttar Dinajpur area. These are either cultivated or gathered from the wild and commonly marketed and used as leaf, tender shoot, or flower vegetables.

13.2 STUDY AREA

The Uttar Dinajpur district came into existence after the division of the former West Dinajpur district (Figure 13.1). The district occupies an area of 3,142 sq km. The district borders Bangladesh on the east, Bihar on the west, Darjeeling and Jalpaiguri districts in the north, and Malda and South Dinajpur districts in the south. The district forms part of the basin lying between the Rajmahal hills on the east. Main rivers include the Kulik, Nagar, and Mahananda. Uttar Dinajpur district has two sub-divisions, Raiganj and Islampur including four municipalities and nine blocks. Around 27% of the population belongs to scheduled castes and some scheduled tribes (ethnic groups) like Kisku, Mardi, Soren, Murmu, Tudu, Hasda, Chaune, Besara, and Urao. Muslims constitute around 47% of the population and the rest are Hindus.

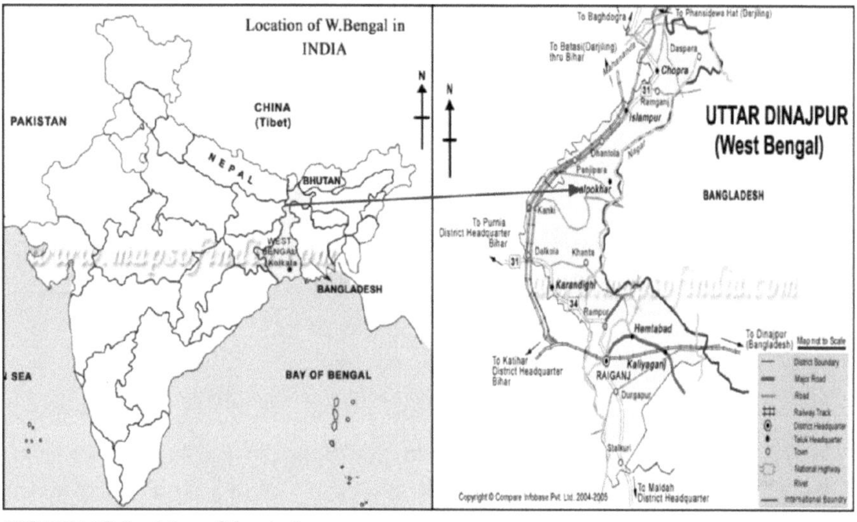

FIGURE 13.1 Map of the study area.

13.3 MATERIALS AND METHOD

An exploratory cross-sectional research design was adopted for this survey to cover the populations of both subdivisions of Raiganj and Islampur of Uttar Dinajpur of West Bengal. A total of eight villages from two subdivisions of the district were visited and a sample of 280 households (including 25% households of indigenous communities) was interviewed. For the selection of households, random sampling technique was used. The upper, lower upper, middle, lower, and lower socioeconomic levels were represented by different enumeration areas. This was done to track variations in the intake of local green vegetables. Focus group discussions, picture book forms, interview schedules, and an observation method were all used in the data collection process. Both quantitative and qualitative analyses were done on the data that had been gathered. Furthermore, descriptive summary statistics were used to determine the most popular native green leafy vegetables, how frequently they were consumed, and their known therapeutic benefits. Chi-square tests were used to obtain p-values. Comparison of means in consumption among the residents of both subdivisions was seen through analysis of variance (ANOVA) and a p-value of < 0.05 was considered significant.

13.4 RESULTS AND DISCUSSION

13.4.1 HOUSEHOLD CONSUMPTION PATTERNS OF INDIGENOUS GREEN LEAFY VEGETABLES

It was shown that households consumed a variety of local green vegetables. Table 13.1 provides an overview of the native leafy vegetable consumption patterns at the time of the study and other commonly consumed foods.

TABLE 13.1 Availability of Indigenous Leafy Vegetables Round the Year

Scientific Name	English Name	Local Name	Availability in the Year
Ipomoea batatas	Sweet potato	*Misthi Aloo*	Sept.–Nov.
Corchorus olitorius	Jute	*Paat*	April–May, Oct.–Nov.
Cucurbita maxima	Pumpkin	*Misthi Kumra*	June–Sept.
Amaranthus spp.	Amaranthus	*Lal sag*	Oct.–Feb.
Colocasia antiquorum	Colocasia	*Kochu*	May–Oct.
Lagenaria siceraria	Bottle gourd	*Lau*	Oct.–June
Diplazium esculentum	Fern	*Dehki/ Dehkia*	July–Oct.
Chenopodium album	Chenopodium	*Bathua*	Dec.–Jan.
Solanum tuberosum	Potato	*Aloo*	Dec.–Jan.
Lathyrus sativus	Lathyrus pea	*Khesari*	Jan.–Feb.

No significant group differences showed in ANOVA ($p > 0.05$) in mean monthly consumption between residents of both subdivisions (Table 13.2). Only a small number of dishes made from native vegetables were consumed at the time of the survey, and when they were, they were typically consumed as a relish with staples like rice, puffed rice, or flakes, and chapatti. This was revealed by the results of the 24-hour recall and the 7-day food frequency. The following indigenous leafy vegetables were consumed round the year during lunch and dinner in the form of preparation called saak or sag, i.e., leaves/foliage or tender shoots of sweet potato (Misthi aloo; 2.3%), Jute (Pat; 9.2%), pumpkin (Misthi Kumro; 4.6%), Amaranthus (Laal sag; 8.4%), Colocasia (Kochu; 5.6%), Bottle gourd (Lau sag; 5.8%), edible fern (Dehki sag; 3.2%), Chenopodium (Bathua; 2.8%), and Lathyrus pea (Khesari sag; 1.8%; Table 13.2). However, as shown in Table 13.2 a variety of indigenous vegetables were eaten at other times of the year, once a month or rarely. It was also reported by near about half of those who consumed these vegetables that the vegetables were bought as not enough produced or collected in case of both subdivisions.

TABLE 13.2 Distribution of Indigenous Leafy Vegetables Consumed by Frequency and Source

Leafy Vegetable	C	F				S		
		1	2	3	4	OP	B	SE
Ipomoea batatas	2.3	0	2.0	0.3	0	2.0	0.1	0.2
Corchorus olitorius	9.2	0.8	6.8	1.6	0	0.6	8.2	0.4
Cucurbita maxima	4.6	0.6	2.8	1.0	0.2	1.4	3.0	0.2
Amaranthus spp.	8.4	4.6	3.2	0.4	0.2	1.8	6.2	0.4
Colocasia antiquorum	5.6	3.0	2.3	0.3	0	1.2	4.2	0.2
Lagenaria siceraria	5.8	0.6	4.0	1.1	0.1	2.8	1.8	1.2
Diplazium esculentum	3.2	0.0	2.0	1.0	0.2	0.0	2.9	0.3
Chenopodium album	2.8	0.3	1.2	1.7	0	1.8	0.2	0.8
Solanum tuberosum	0.6	0	0.0	0.4	0.2	0.4	0.0	0.2
Lathyrus sativus	1.8	0	1.1	0.4	0.3	0.9	0.1	0.8

Abbreviations: C: Consumption in percent; F: Frequency in percent (frequency codes: 1–once a week; 2–twice a week; 3–once a month; and 4–rarely); S: Source in percent (OP: own produced/collected; B: bought from market or others; and SE: social exchange).

13.4.2 FOOD CONSUMPTION PATTERNS AMONG CHILDREN

Only a few children (21.2%), according to the 24-hour recall's findings, ate conventional plant vegetables. Consumption of leafy vegetables mostly for lunch was as follows: pumpkin leaves and tender shoots, Bottle gourd leaves and tender shoots and jute with 7.2%, 8.6% and 2.3%, respectively while, *Colocasia* were the most consumed for dinner at 12.3% followed by *Bathua* and *Amaranth* at 2.5% and 0.6%, respectively. Most of the children shared their indigenous leafy vegetables with their parents or other family elders.

Indigenous green leafy vegetables are increasingly being produced and consumed because research has shown that they are both nutritious and medicinal. In both urban and rural contexts, indigenous leafy greens have a substantial impact on the diet and health of the underprivileged (Gackowski et al., 2003).

13.4.3 FACTORS INFLUENCING CONSUMPTION OF INDIGENOUS LEAFY VEGETABLES

75% of respondents reported that the cost of the food was a concern, but 43% said that they had no issues with food selection, preparation, or consumption. Additionally, 21.1% of the respondents mentioned they didn't have enough time to shop for and cook the local leafy vegetables, and 1.1% said they didn't actually know how to prepare it. Results from the Focus Group Discussions and observations indicated that these vegetables were readily available in local markets, though the availability of such vegetables varied from place to place. Most of these vegetables were sold in local *haats* fresh but mostly of poor quality in terms of hygiene. When handled poorly, these vegetables' attributes were occasionally seen as being subpar, which turned off many potential consumers from consuming them. The most commonly sold traditional leafy vegetables were Jute, Pumpkin, Amaranthus, Colocasia, and Bottle gourd. The availability round the year of these leafy vegetables is given in Table 13.1. Deforestation, excessive grazing, desertification, excessive agricultural intensification, excessive fishing, and agriculture on marginal lands all contribute to the depletion of the natural resource base, which lowers agricultural production potential and lowers agriculture's carrying capacity. Improved management is required to stop the deterioration of this resource base and create agricultural production systems that support our ecosystem because natural resources (soil, water, and nutrients) have limits. The survey also sought to find out the culinary methods for preparing recipes with these leafy vegetables. Most of these traditional green leafy vegetables are nutritious if properly prepared. In the present study, it was found that respondents used preparation and cooking procedures that might lessen the nutritive value of prepared items. After observation, some of the procedures used in the preparation of leafy vegetables which need to be discouraged are described in subsections.

13.4.3.1 CHOPPING OF VEGETABLES BEFORE WASHING

Nearly 85% of the respondents were found chopping their vegetables prior to washing them. Respondents washed their chopped vegetables many times. Water-soluble vitamins, namely vitamin B complex and C, were generally lost during the washing stage if the vegetables were chopped prior to washing, as indicated by the green pigmentation of the water after washing. The green pigmentation of the water after washing vegetables relates to nutrient loss from the washed vegetables.

13.4.3.2 OVER COOKING OR BOILING

Many respondents (75.6%) overcooked or boiled their vegetables and thereafter discarded the used water, thereby losing nutrients. Boiling or cooking vegetables for more than 40 minutes results in loss of vitamin C, which is the most unstable vitamin.

13.4.3.3 USING SODIUM BICARBONATE

A remarkable 76% of respondents reported adding sodium bicarbonate when boiling or cooking some veggies because of their delicate flavor. Although the vitamin B complex, specifically vitamins B1, B2, and niacin, are lost, sodium bicarbonate aids in the tenderization and preservation of the veggies' vibrant green color, making it the preferred method. Therefore, there was a severe lack of various vegetables, pulses, and legumes in the rural community's diets, which contributed to the high prevalence of undernutrition and micronutrient deficiency (Arlappa et al., 2007).

13.4.3.4 OTHER FACTORS

Occupation, gender, wealth, and education levels also played a role in the selection and consumption of indigenous green leafy vegetables. It was observed that occupation affected the choice and consumption of local leafy vegetables. There was a significant association between consumption of these indigenous green leafy vegetables and occupation ($P < 0.05$), with respondents in casual labor and/or unemployment (81%) having higher consumption rates than those in full-time work and business (29%). This was shown to rely on how much time is available for purchasing, preparing, and cooking these local vegetables. In all instances, there was no statistically significant difference between men and women in terms of their impressions of consuming local veggies ($p > 0.05$). The preference for eating vegetables was found to be lower in men than in women. The survey also revealed no association between household income and consumption of vegetables, with the exception of Amaranthus, jute, and bottle gourd, which differed significantly ($P < 0.05$) and were consumed at higher rates by households with lower total monthly incomes than those with higher incomes (Table 13.3). Except for Chenopodium and edible fern, there were no associations between education and intake of indigenous vegetables. The survey's results revealed that occupation, sex, income, and educational attainment are some of the key variables influencing the consumption and use of these veggies. Jana (2003) discovered accessibility to be a limiting factor in a related study. In general, just 38% of indigenous leafy vegetables were consumed. The most frequent times to consume and use indigenous leafy vegetables were when they were in season and cheaper. Familiarity of adults with these vegetables contributed to their widespread intake and use.

TABLE 13.3 Household Indigenous Leafy Vegetables as Influenced by Total Monthly Household Income

Leafy Vegetable	Consumption (%)							
	MI1	MI2	MI3	MI4	MI5	MI6	MI7	X2
Ipomoea batatas	1.6	1.2	1.4	0.8	0.7	0.2	0.1	–
Corchorus olitorius	4.5	6.6	4.2	3.2	2.0	0.4	1.6	0.03
Cucurbita maxima	4.5	3.0	3.5	2.2	1.0	0.6	0.1	–
Amaranthus spp.	5.2	4.1	3.2	1.3	1.0	0.3	1.3	0.02
Colocasia antiquorum	3.8	1.2	2.0	0.8	0.6	0.2	0.1	–
Lagenaria siceraria	6.6	3.2	2.3	2.0	0.4	1.0	1.2	0.03
Diplazium esculentum	2.0	2.2	1.4	0.4	0.2	0.1	0.1	–
Chenopodium album	2.5	1.5	0.6	0.6	0.3	0.2	0.1	–
Lathyrus sativus	1.2	1.0	0.4	0.5	0.2	0.1	0.1	–

Abbreviations: N: Total household (i.e., 280); MI: Monthly income in Indian National Rupee; MI1: < 3,000 (n = 32.3% of N); MI2: 3,000–5,000 (n = 23.0%); MI3: 5,000–10,000 (n = 20.7%); MI4: 10,000–20,000 (n = 11.2%); MI5: 20,000–30,000 (n = 6.3%); MI6: 30,000–40,000 (n = 2.3%); MI7: > 40,000 (n = 4.2%); and X2: Chi-square tests with significance at $p < 0.05$.

13.5 ETHNOMEDICINAL VALUE OF INDIGENOUS LEAFY VEGETABLES

These traditional leafy vegetables are highly recommended due to their comparatively high nutritional value and are also important in ensuring food security during times of drought and poor harvest and are also very important for income generation, as indigenous vegetables provide both nutritional and economic security. These vegetables bring variety to daily food consumption and are high in vitamins, minerals, trace elements, dietary fiber, and proteins. Vegetables are valuable for their ecological, agronomic, and cultural aspects in addition to their importance as food and a source of medicine (Arunachalam et al., 2007). The respondents reported that these indigenous vegetables also have ethnomedicinal importance to them ensuring them good general health and also curing some day-to-day common health problems or ailments and diseases (Table 13.4). Similar ethnomedicinal importance of indigenous vegetables was also reported by Kimiywe et al. (2007).

TABLE 13.4 Ethnomedicinal Value of Indigenous Leafy Vegetables

Indigenous Vegetable	Medicinal Value	F
Ipomoea batatas	High blood pressure, general health.	8.6
Corchorus olitorius	Constipation, diabetes, high blood pressure, general health, appetizer.	9.2
Cucurbita maxima	Diabetes, high blood pressure, backache, anemia, appetizer.	10.0
Amaranthus spp.	Anemia, improves eyesight, general health.	10.0
Colocasia antiquorum	Anemia, general health.	9.5
Lagenaria siceraria	Oedema, constipation, high blood pressure, cleanses blood, general health.	10.0
Diplaziumesculentum	High blood pressure, diabetes, anticarcinogenic, general health.	7.5
Chenopodium album	Diabetes, jaundice, cough, cold, high blood pressure, general health.	8.5
Lathyrus sativus	General health.	6.3

*Abbreviations:*F: Frequency (per cent of N, i.e., 280).

To determine the rationale for these views, additional field trial-based scientific investigations are required. Malaria, diarrhea, anemia, colds, and coughs, diabetes, cancer, and high blood pressure were among the most frequently reported ailments. The study looked for additional qualities associated with these foods in addition to establishing the perceived therapeutic value of indigenous green vegetables among respondents. The most prevalent views about vegetables were their ancestors' dietary practices, their exquisiteness, their ability to increase appetite, and their worth as a means of extending life. Other benefits included raising hemoglobin levels, cleaning the blood, and many more (Table 13.4).

In addition to providing vital nutrients, the variety of indigenous plants and wild plants found in most tropical areas is thought to have numerous health benefits (Johns & Sthapit, 2004; Onyango et al., 2005). According to studies, nations that maintain their native vegetable diets and consume large amounts of these vegetables are far less likely to experience cardiovascular disease, diabetes, and other negative effects of the nutrition in transition (Maxwell & Frankenberger, 1992). In the Sub-Himalayan Terai region of West Bengal's Jalpaiguri District, Arunachalam et al. (2007) also noted a possible connection between local vegetables and the capacity to treat diabetes, high blood pressure, gastrointestinal tract diseases, and protozoan parasites. Consumption of indigenous vegetables was reported to cure diseases like high blood pressure, diabetes, urinary, and stomach disorders, bronchitis, and diarrhea (Sureja et al., 2007). These vegetables were also traded in the local market ensuring income to the local residents as was also reported from Arunachal Pradesh by Sureja et al. (2007). Consumption and utilization of underutilized indigenous green leafy vegetables was encouraged among respondents through this intervention.

13.6 CONCLUSION

Traditional leafy vegetables are those plants whose leaves, flowers, or aerial parts have been incorporated into a community's traditional cuisine culture and food habits. But lack of awareness of medicinal and nutritive values of a particular crop and lack of transfer of this knowledge from one generation to another replaced many indigenous crops traditionally cultivated and consumed by local farmers. It is evident from these findings that consumption of indigenous vegetables is very low. Consumption was based on awareness; occupation, income, etc., are other influencing factors. Frequently consumed indigenous green leafy vegetables by children were Pumpkin and Bottle gourd tender shoots and leaves. It was observed that price, lack of time, and knowledge in culinary preparation are the major constraints to the consumption and utilization of indigenous leafy vegetables. Generally, these vegetables were preferred because of their ethno-medicinal and nutritional potential. The respondents indicated enthusiasm towards the consumption of indigenous leafy vegetables if they were sensitized about their value and utilization. Promotion of more and more cash crops among farmers, merely any awareness program on indigenous leafy vegetables as well as a gradual increase in population load over land is posing a challenge to relevant stakeholders for development in this area.

ACKNOWLEDGMENTS

The author acknowledges the support and help of Mrs. Sarathi Singha, Mrs. Rita Singha, Mrs. Chameli Mandal, Shri Bhupati Roy, Shri Kailash Sarkar, Shri Harimohan Barman, and other knowledgeable persons for providing information and their support in documentation of required information during the study. The author is also thankful to Senior Scientist and other staff of Uttar Dinajpur KVK for their support during the course of work.

KEYWORDS

- **consumption**
- **indigenous green leafy vegetable**
- **nutrition**
- **sodium bicarbonate**
- **stakeholders**
- **utilization**

REFERENCES

Arlappa, N., Balakrishna, N., Laxmaiah, A., & Brahmam, G. N. V., (2007). Pattern of indigenous vegetables and legume consumption in India. In: *Proc. of International Conference on Indigenous Vegetables and Legumes. Prospectus for Fighting Poverty, Hunger and Malnutrition.* Hyderabad, India.

Arunachalam, V., Sit, A. K., & Bhattacharya, M., (2007). Edible leaves of Jalpaiguri district of West Bengal, India. In: *Proc. International Conference on Indigenous Vegetables and Legumes. Prospectus for Fighting Poverty, Hunger and Malnutrition.* Hyderabad, India.

Gackowski, J., Mbazo, J., Mbah, G., & Moulende, T., (2003). African traditional leafy vegetables and the urban and peri-urban poor. *Food Policy*, *28*, 221–235.

Jana, J. C., (2007). Use of traditional and underutilized leafy vegetables of sub-Himalayan Terai region of West Bengal. In: *Proc. of International Conference on Indigenous Vegetables and Legumes. Prospectus for Fighting Poverty, Hunger and Malnutrition*. Hyderabad, India.

Johns, T., & Sthapit, B., (2004). Bio-cultural diversity in the sustainability of developing country food systems. *Food and Nutrition Bulletin*, *25*(2), 143–155.

Kalia, P., Sharma, A., Singh, S., & Singh, Y., (2007). Locally adapted indigenous vegetables of Himachal Pradesh and their role in alleviating poverty, hunger, and malnutrition. In: Proc. *International Conference on Indigenous Vegetables and Legumes. Prospectus for Fighting Poverty, Hunger and Malnutrition*. Hyderabad, India.

Kimiywe, J., Waudo, J., Mbithe, D., & Maundu, P., (2007). Utilization and medicinal value of indigenous leafy vegetables consumed in urban and peri-urban Nairobi. *African Journal of Food, Agriculture, Nutrition and Development*, *7*(4), 123–135.

Maxwell, S., & Frankenberger, T. R., (1992). *Household Food Security: Concepts, Indicators, Measurement: A Technical Review*. UNICEF and IFAD.

Onyango, M. A., Tushaboomwe-Kazoobz, C., Onyango, J. M., & Macha, S. E., (2005). *Improved Community Land Use for Sustainable Production and Utilization of African Indigenous Vegetables in the Lake Victoria Region*. VICRES Report.

Sureja, A. K., Singh, R. K., Bhardwaj, R., Rai, A. K., & Singh, D., (2007). Underexploited vegetables of the Adi tribe of Arunachal Pradesh. In: *Proc. International Conference on Indigenous Vegetables and Legumes. Prospectus for Fighting Poverty, Hunger and Malnutrition*. Hyderabad, India.

CHAPTER 14

Stevia rebaudiana (Bert).: A Natural Sweetener

KASIREDDY SIVASANKARREDDY,[1] BABLU PAUL,[1] GOPAL SHUKLA,[2,3] and HOSSAIN ALI MONDAL[4]

[1]*Department of Genetics and Plant Breeding, Uttar Banga Krishi Viswavidyalaya, Pundibari, West Bengal, India*

[2]*Department of Forestry, Uttar Banga Krishi Viswavidyalaya, Pundibari, West Bengal, India*

[3]*Department of Forestry, North Eastern Hill University, Tura Campus, Meghalaya, India*

[4]*School of Crop Improvement, College of Postgraduate Studies in Agricultural Sciences, Umiam, Meghalaya, India*

ABSTRACT

Studies uncovered that stevia has been utilized throughout the world since ancient times for different purposes, for instance, as sugar and a medication. In this chapter, the history, taxonomy, botanical description, chemical properties, traditional and medicinal uses, cultivation including propagation, commercialization, and genetic improvement of the stevia crop have been summarized systematically. *Stevia rebaudiana* (Bertoni) is a highly valued perennial medicinal plant (2n = 22). The genus stevia has 230 species of stevia plants. As we realize that the leaves of stevia have diterpene glycosides (stevioside and rebaudiosides), non-nutritive, non-harmful, high-intensity sugars. Its white crystalline compound (stevioside) is the regular natural sugar without any calories and is more than 100–300 times better than table sugar. The development of new varieties of *Stevia rebaudiana* with a higher substance of stevioside and rebaudiosides is the essential point

Bioprospecting Ethnomedicinal Plant Resources: Sustainable Utilization and Restoration.
Gopal Shukla, Jahangeer A. Bhat, A. P. Das, and Sumit Chakravarty (Eds.)

of plant breeders. Regular crop improvement approaches like composites and synthetics development can also be utilized to capture some portion of the available heterosis in the genotypes. Mutation breeding also helps to develop new variants. Polyploidy breeding also helps to develop polyploids like tetraploids that have bigger leaf size, thickness, and have possible use in expanding biomass furthermore, yield in correlation with diploid strains. Biotechnological approaches are also helpful for crop improvement, for example, tissue culture for the mass propagation of first-class genotypes, and another culture for improvement of pure homozygous lines.

14.1 INTRODUCTION

Stevia rebaudiana (Bert) Bertoni is a small, semi-bushy, perennial shrub of the *Asteraceae* family. It is indigenous to Paraguay and Brazil of South America in the Rio Monday valley highlands, where stevia grows in sandy soils close to streams (Katayama et al., 1976; Sivaram & Mukundan, 2003; Jain et al., 2009). It is commonly known as 'Honey Leaf,' 'Sweet Leaf,' 'Sweet Herbs,' and 'Sweet Weed.' Initially, it was called *Eupatorium rebaudiana*, but in the year 1905, its name changed to *Stevia rebaudiana* (Bertoni). It was renamed in tribute to Paraguayan Chemist Dr. Rebaudi in 1905 by Dr. Moises Santiago Bertoni (Sakaguchi & Kan, 1982; Lemus Mondaka et al., 2012). It has been used to sweeten tea for centuries by Guarani Indians (Soejarto et al., 1982; Lemus Mondaka et al., 2012). Stevia is found in the tribe *Eupatorium* among the large family of *Asteraceae* containing 950 genera (Soejarto et al., 1983; Lester, 1999). The taxonomic classification of *S. rebaudiana* (Bartoli) is given in Table 14.1.

TABLE 14.1 Taxonomic Classification of *S. rebaudiana*

Domain:	Eukaryota
Kingdom:	Plantae
Sub-kingdom:	*Tracheobionta*
Super division:	*Spermatophyta*
Division:	*Magnoliophyta*
Class:	Magnoliopsida
Sub-class:	*Asteridae*
Group:	Monochlamydae
Order:	Asterales
Family:	Asteraceae (Compositae)
Subfamily:	Asteroideae
Tribe:	*Eupatorieae*
Genus:	*Stevia*
Species:	*Rebaudiana*

The genus Stevia consists of 230 species of herbaceous plants, but *S. rebaudiana* (Bertoni), a native of Paraguay, offers the most sweetening properties of the genus, whereas other species are also noted (Frederico et al., 1996). Some other species are *S. micrantha* (annual stevia), *S. eupatoria* or *S. lemmonii* (Lemmon's stevia), *S. plummerae* (Plummer's stevia), *S. salicifolia* (Willow leaf stevia), *S. serrata* (sawtooth stevia), *S. viscida* (viscid stevia), *S. plummerae* var. alba, *S. rhombifolia* (Kunth), *S. ovata var. texana* (round leaf candy leaf), *S. commixta*, *S. eptophylla*, *S. myriadenia*, *S. nepetifolia*, *S. oligophylla*, *S. ophryphylla*, *S. origanoides*, *S. satureiaefilia*, *S. selloi*, and *S. triflora*. It can grow up to 100 cm tallness; small, elliptical sessile leaves 2–3 cm in length, a granulated upper surface and slightly pubescent oppositely arranged, elongate-lanceolate or spatulate shaped, oblongata, serrate margin from middle to tip and below with blunt-tipped lamina; the stem is woody, fragile, and weak at the bottom with the extensive root system (Yadav et al., 2011; Lemus Mondaka et al., 2012). Trichomes of varying sizes (from small, i.e., 80–150 μm to large, i.e., 300–500 μm) are present on the leaf surface (Jain et al., 2014).

14.2 FLORAL BIOLOGY

Flower initiation (Plate 14.1) will take place after the formation of four true leaves in stevia (Carneiro, 1990). Stevia is a short-day plant. The flowers of stevia (Plate 14.2) are small (7–15 mm), white with a pale purple throat, and pentamerous (Robinson, 1930). The compound inflorescence is a cyme of corymbs differentiating in the apex capitula with five white tubular flowers (Carneiro, 1990). The anthers of the stevia flower are small and five in number, while the stigma is bifurcated or bilobed from the middle and the style is enclosed by anthers (Yadav et al., 2011). The species is sensitive to day length with a critical day length of 13 hours as it is an obligate short-day plant (Brandle et al., 1998). Plant architecture, morphogenetic responses, synthesis of important bioactive compounds, and productivity are greatly influenced by light quality and quantity, i.e., photoperiodism and irradiation levels (Metivier & Viana, 1979; Ermakov & Kochetov, 1994, 1996; Tariq et al., 2014). If the plant flowers early, its vegetative growth attenuates and leads to a reduction of leaf yield and steviol glycosides in stevia. Under long-day conditions, the vegetative stage persists for a long time leading to an increase in leaf biomass as well as in the stevioside content (Ceunen et al., 2012; Ceunen & Geuns, 2013).

PLATE 14.1 Developmental stages of stevia flower.

PLATE 14.2 Stevia flower in different angles.

14.3 REPRODUCTION AND POLLINATION BEHAVIOR

Stevia is a highly cross-pollinated plant and the amount of selfing is in the range of 0–0.5% and crossing 0.7–98.7% while, the species shows self-incompatibility (Brandle et al., 1998). Stevia seeds are enclosed in achenes of three millimeter with persistent pappus bristles which are helpful for its dispersal (Goettemoeller & Ching, 1999). The seed viability varies with a maximum of 50% (Miyazaki & Watanabe, 1974; Duke, 1993; Carneiro et al., 1997; Lester, 1999). Fertile seeds of stevia are dark and infertile seeds are pale or clear in color (Plate 14.3; Oddone, 1997, 1999; Goettemoeller & Ching, 1999). Seed test weight (1,000 seeds) of stevia is 0.15–0.30 gm and seed yield are at the most 8.1 kg ha^{-1} depending on plant density (Carneiro, 1990; Madan et al., 2010).

a) Fertile seeds: Dark coloured | b) Infertile seeds: Pale/clear colour | c) Infertile seeds: Pale/clear colour

PLATE 14.3 Steviaseed color and viability.

14.4 PROPAGATION

14.4.1 SEED PROPAGATION

Storing the seeds at low temperature or in containers cannot prevent germination loss, which is 50% or more in three years (Shuping & Shizhen, 1995; Brandle et al., 1998). Seed germination is poor due to infertile seeds (Goettemoeller & Ching, 1999; Kumar, 2013) and undersized endosperm (Yadav et al., 2011). Self-incompatibility is also a major problem in stevia raised by seed (Brandle & Telmer, 2007; Kovylyaeva et al., 2007).

14.4.2 VEGETATIVE PROPAGATION

Stevia is propagated through stem cuttings easily (Smitha & Umesha, 2011); however, the major limitations of this method are poor rooting and a low number of new plants produced. Growth regulators were reported to stimulate root initiation (Ibrahim et al., 2008; Gantait et al., 2015). Stem cutting is, however, a destructive method for QPM production.

14.4.3 MICROPROPAGATION

Plant tissue culture is the only method helpful for producing high-quality planting materials (QPMs) avoiding the problems of seed and vegetative propagation (Khalil et al., 2014). Plant regeneration from in-vitro culture can be attained by either organogenesis or embryogenesis. The leaf explant was useful for callus induction (Babu et al., 2011; Ahmad et al., 2016; Golkar et al., 2018). Moreover, multiple shoot production from nodal and internodes by using various plant growth regulators (Debnath, 2008; Hossain et al., 2008; Fakhrul et al., 2014; Bayraktar et al., 2016).

14.5 CULTIVATION

14.5.1 CLIMATE

Stevia is a semi-humid subtropical plant that may be grown in a wide range of climates, from semi-humid to temperate (Ramesh et al., 2006). However, the roots of the plant get damaged in mild frost. Ideal growing areas are with long photoperiods with a diurnal temperature above 25°C but below 35°C with uninterrupted light or partial shade (Midmore & Rank, 2002; Ramesh et al., 2006; Lankes & Pude, 2008; Angelini et al., 2016). Stevia grows well in light-textured, well-drained soil with 6.5–7.5 pH and enriched organic matter free from waterlogging conditions (Pal et al., 2015). The species is also moderately tolerant to salt stress, but saline soils must be avoided to ensure optimum yield and steviol glycosides in leaves (Angelini et al., 2016).

14.5.2 IRRIGATION AND FERTILIZER

Unless the growing area has reliable rainfall throughout the year, supplementary irrigation is essential to avoid any water stress on plants (Midmore & Rank, 2002). It can't grow in dry conditions. It is particularly susceptible to water stress and requires regular light irrigation to reduce the effect of drought. The addition of mulches around the plant minimizes the effects of dryness and hot temperatures (Hossain et al., 2017). It requires a moderate amount of fertilizer due to its adaptation to poor-quality soils. High rates of fertilizer trials reported yield reductions (Midmore & Rank, 2002). It needs low nitrogen and high phosphorus and potassium. The nitrogenous fertilizer (urea) must be applied in three split applications, i.e., (i) basal dose; (ii) after first cutting of leaves; and (iii) after second cutting of leaves (Hossain et al., 2017).

14.5.3 HARVESTING

The crop is harvested just before flowering to ensure the highest steviol glycoside substance in the leaves (Sumida, 1980). The leaf yield increases with increasing plant density from 83,000 to 111,000 plants per hectare in the first-year production.

14.5.4 POSTHARVEST

After harvesting, the whole crop is allowed to drying facilities, i.e., either sun drying or (artificial) drying. If harvested material is not dried rapidly, the quality of the leaf can deteriorate via oxidation, which leads to the loss of one-third of the stevioside substance within three days (Oddone, 1999). Immediately after harvest, the herb can be dried on a net or glass sheet. In the case of large-scale production, the stem and soft leaf material will be dried completely by using a drying wagon and kiln. High temperature and a long time of drying minimize the stevioside substance in the final product (Hossain et al., 2017).

14.6 CHEMICAL CONSTITUENTS OF STEVIALEAF

Stevia consists of many chemical constituents like Steviol glycosides, Polyphenols, Flavonoids, Polyhydroxy indolizidine alkaloid, strains A to H, triterpenes, Volatile oil components, pigments, gums, and inorganic constituents but steviol glycosides are present in abundance (Kinghorn, 1987; Geuns, 2003; Taleie et al., 2012; Kumari & Chandra, 2015). Soejarto et al. (1982) tested 110 species for sweetness, out of which only 18 were found to have sweet features. The sweet principle was first isolated in 1909 but stevioside was produced from the extract in 1931 and its chemical structure was established as a diterpene glycoside in 1952. Other compounds were isolated including Rebaudioside A during the 1970s (Barriocanal et al., 2008).

Stevia contains more than 30 steviol glycosides with different concentrations which constitute up to 20% of the dry weight of the leaf. Some of these are Steviolmonoside, Rubusoside, Steviolbioside, Dulcoside A, Stevioside, Dulcoside B, Rebaudioside A to O, Rebaudioside VIII and Rebaudioside IX (Kim & Dubois, 1991; Brandle et al., 1998; Prakash et al., 2008; Ohta et al., 2010; Chaturvedula et al., 2011a–d; Kim et al., 2011; Chaturvedula & Prakash, 2011; Zimmermann, 2011; Ma et al., 2018). Stevioside is the major constituent (4–13% w/w) followed by Rebaudioside A (2–4% w/w), Rebaudioside C (1–2% w/w), and nucleoside A (0.4–0.7% w/w) and it is about 300 times sweeter than sucrose (Lemus Mondaka et al., 2012). The diterpene glycosides isolated from leaves of stevia have the same backbone, i.e., Steviol and mainly differ in the number of carbohydrate residues, mono-, di-, and tri-saccharides containing glucose or rhamnose at positions C13 and C19 (Kochikyan et al., 2006).

The phenolic compounds and chlorogenic acids discovered in *Stevi-arebaudiana* are pyrogallol, 4-methoxybenzoic acid, p-coumaric acid, 4-methylcatechol, Sinapic acid, cinnamic acid, 3-caffeoylquinic acid (3-CQA), 4-caffeoylquinic acid (4-CQA), 5-caffeoylquinic acid (5-CQA), 3,4-dicaffeoylquinic acid (3,4-diCQA), 3,5-dicaffeoylquinic acid (3,5-diCQA), 4,5-dicaffeoylquinic acid (4,5-diCQA), cis-4,5-dicaffeoylquinic acid (cis-4,5-diCQA), A cis-3,5-dicaffeoylquinic acid (a cis-3,5-diCQA), A cis-4,5-dicaffeoylquinic acid (a cis-4,5-diCQA), 5-p-coumaroylquinic acid (5-p-CoQA), caffeoyl-feruloylquinic acid (CFQA), 4-caffeoyl-5-feruloylquinic acid (4-C,5FQA), 3-caffeoylshikimic acid (3-CSA), 4-caffeoylshikimic acid (4-CSA), 5-caffeoylshikimic acid (5-CSA), feruloylquinic acid (FQA), 5-feruloylquinic acid (5-FQA), tricaffeoylquinic acid (triCQA), 1,3,5-tricaffeoylquinic acid (1,3,5-triCQA) and 3,4,5-tricaffeoylquinic acid, i.e., 3,4,5-triCQA (Kim et al., 2011; Karaköse et al., 2011). The leaves of stevia also contain flavonoids which were divided into two subgroups. The first subgroup is flavonols viz., quercetin, quercetin-3-O-glucoside, quercetin-3-O-rutinoside, quercetin-3-O-β-D-arabinoside, quercetin-3-O-β-D-rhamnoside, quercetin-3-O-(4-O-trans-caffeoyl)-α-L-rhamno-pyranosyl-(1-6)-β-D-galactopyranoside, Kaempferol-3-O-rhamnoside, and the second sub-group was flavones viz., Apigenin, Apigenin-4'-O-β-D-glycoside, Apigenin-7-O-β-D-glycoside, luteolin, and luteolin-7-O-β-D-glycoside (Ghanta et al., 2007; Tadhani et al., 2007; Li et al., 2009; Cacciola et al., 2011).

14.7 UTILIZATION OF STEVIA

14.7.1 TRADITIONAL

Guarani tribes of Paraguay and Brazil used stevia species especially *S. rebaudiana*, as a sweetener in medicinal teas for treating heartburn and other ailments (Brandle & Telmer, 2007; Goyal et al., 2010). The leaves of stevia are used as a sweetener in teas, salads, and coffee, among others.

14.7.2 MEDICINAL USES

Stevia has a good energy value of approximately 2.7 kcal/g dry matter which is comparable to other low-caloric sweeteners available on the market. Due to its low caloric value, it may be advisable for consumption by replacing the commonly used sucrose compounds for control of obesity, type 2 diabetes plus other socially-related diseases (Anton et al., 2010; Marcinek & Krejpcio, 2016). Studies have recommended a daily intake of 5 mg/kg body weight for each day of steviol glycosides is secure and has no carcinogenic effects. The higher doses lead to toxic action and have negative side effects on hematological and biochemical parameters of the blood as well as enzymatic activity (Chatsudthipong & Muanprasat, 2009; Awney et al., 2011).

Stevia has a high antioxidant activity due to its phenols and flavonoids (Taylor, 2005; Shukla et al., 2009; Marcinek & Krejpcio, 2016). The phenols and secondary metabolites help to suppress cardiac and cancer diseases (Dragović-Uzelac et al., 2010). Stevia prevents ulceration in the gastro-intestinal tract (Kochikyan et al., 2006) including antihypertensive (Chan et al., 2000; Lee et al., 2001; Jeppesen et al., 2002), antihyperglycemic (Suanarunsawat & Chaiyabutr, 1997; Jeppesen et al., 2002; Chen et al., 2006) activities. It also helps to clear up skin troubles like dermatitis, skin blemishes, eczema, acne outbreaks, wrinkles, scarring, rashes, and itchiness (Ranjan et al., 2011). Additionally, stevia also offers other health benefits like antibacterial, antimicrobial, antiseptic, anti-inflammatory, vasodilator, and immunomodulatory activities (Taylor, 2005; Marcinek & Krejpcio, 2016). Stevioside does not have any mutagenic, allergic, carcinogenic effects or teratogenic effects (Pól et al., 2007). Stevioside was also reported to reduce dental caries (Basu, 2014).

14.8 COMMERCIALIZATION OF STEVIA

Commercial cultivation of stevia was first reported in Paraguay (1964; Lewis, 1992). Stevioside as a sweetener in the food and drug industry was first used in Japan (Chatsudthipong & Muanprasat, 2009). In Japan, Sumida (1968) began a major attempt aimed at cultivating stevia as a crop. Now it is the largest consumer of stevioside extracted from leaves of stevia (Midmore & Rank, 2002). Japan, China, Taiwan, Thailand, Korea, Brazil, Malaysia, and Paraguay are the world's leading stevia producers. At present, it is consumed in several countries including Argentina, China, Brazil, Korea, Paraguay, Mexico,

United States, Indonesia, Israel, and Canada (Yadav et al., 2011). During the late 1990s, stevia was introduced in India at the University of Agricultural Sciences, Bangalore. Later it was introduced by many institutes and research focused on cultivation and crop improvement (Yadav et al., 2011).

14.9 GENETIC IMPROVEMENT OF STEVIA

14.9.1 CONVENTIONAL APPROACHES

Plant introduction is the process of introducing a genotype or group of genotypes of plants to new environments where they have not previously been produced. The introduced material may cultivate without any change in the original genotype or subject to selection. The isolation of plant types from the population is called selection and it is essential for any crop improvement. Stevia genetic improvement through selection was reported by Abdelsalam et al. (2016) by selecting 19 different accessions. The main aim of the breeder is to increase the frequency of desirable alleles by selecting superior recombinants in the population by the population improvement method. Recurrent selection is the most appropriate population improvement method for stevia yield development (Yadav et al., 2011). A synthetic variety is created by crossing multiple lines in all possible combinations that can combine well with each other. A composite variety is created by mixing the seeds of numerous phenotypically superior lines and encouraging them to open-pollination to generate crosses in all combinations within the mixed lines.

The mutation is the sudden heritable change other than the Mendelian segregation and gene recombination in an organism. In this method, radiation or chemical agents are used randomly to modify the DNA in seeds of plants to generate mutants. In some cases, these mutants may express some desirable traits. In the physical method of mutation, the X-ray, gamma-ray, and neutron beams are used regularly. In a chemical method of mutation, Ethyl Methane Sulfonate (EMS) is commonly used for producing mutants. With the help of EMS and Gamma-ray, Khan et al. (2016) successfully generated stevia mutants with higher Reb-A and Stevioside content (200% increase).

The organisms that contain more than two sets of chromosomes (normal diploids) are called polyploids. In the case of stevia, the somatic chromosome number in a normal diploid variety is $2n = 22$ [$n = x = 11$]. It is possible to develop progeny plants with a greater number of chromosomes using chemical treatment in seeds or meristem (growing tissues) of plants.

Colchicine is generally used to produce polyploids in plants. Tetraploids (4n) in stevia were produced by Yadav et al. (2013) who had shown significantly bigger leaf size, thickness in leaf, more chlorophyll content, and reduced internode length. Similarly, Hegde et al. (2015) also reported polyploids with elevated steviol glycoside content in the leaf, a high number of secondary branches, extra leaf thickness, and area and flowering delay.

14.9.2 BIOTECHNOLOGICAL APPROACHES

14.9.2.1 ANTHER CULTURE

The anther present in a flower produces pollen. Anther culture is a way of producing haploids from pollen grains. The pollen of a plant contains half of the chromosome number as that in somatic cells. The haploid progenies developed by anther culture can be used for selecting plants with different traits and interests. The desirable haploid plants can be converted to diploid by colchicine treatment. Flachsland et al. (1996) tried to regenerate whole plants from anthers of stevia but in the end, they ended up with diploid plants, which indicates that the diploid somatic cells of anthers responded.

14.9.2.2 TISSUE CULTURE

Plant tissue culture can play a significant role in stevia propagation because of low germination potential and micro-sized seeds. Tissue culture also produces disease-free materials rapidly.

14.9.2.3 MARKER-ASSISTED SELECTION (MAS)

Marker-assisted selection (MAS) is simply referred to as MAS. It involves the selection of plants containing genomic regions of the traits of interest. The individual gene(s) with significant influence on a definite target trait can be recognized based on their linkage to molecular markers. The effectiveness of integrating the desirable traits into the selected germplasm could be greatly enhanced. Through MAS, the desirable plants can be selected early (Yadav et al., 2011).

14.9.2.4 TRANSGENIC PLANT

Transgenic technology is a biotechnological tool by which scientists can artificially mobilize desirable coding sequences into a plant and the resultant plant is called a transgenic plant. The inserted gene may arrive from another dissimilar plant or entirely different organism like bacteria or animals. The wanted genes from broadly different organisms can be combined into a single plant. The transgene can be transferred by two types of methods viz. vector-mediated transfer (Agrobacterium-mediated gene transfer) and direct gene transfer (gene gun method). Agrobacterium-mediated gene transfer is one of the biotechnological tools used in stevia to improve the steviol glycosides. Agrobacterium mediated transformation in stevia was done by using *Agrobacterium tumefaciens* (Khan et al., 2016) and *A. rhizogenes* (de Jesús Sánchez-Cordova et al., 2019). Gene gun method is also known as micro-projectile bombardment or biolistic method. Mubarak et al. (2015) transferred barge into *in vitro* propagated plant using this method producing the transgenic plant resistance to the herbicide.

KEYWORDS

- **biolistic method**
- **clonal propagation**
- **genetic improvement**
- **herbicide**
- **medicinal uses**
- **Stevia rebaudiana**
- **transgenic plant resistance**

REFERENCES

Abdelsalam, N. R., Haraz, A. S., Haraz, A. E., Saleh, M. S., & Elsheikh, A. E., (2016). Genetic improvement through selection of different *Stevia rebaudiana* genotypes. *Alexandria Science Exchange Journal*, *37*, 10–25.

Ahmad, N., Rab, A., & Ahmad, N., (2016). Light-induced biochemical variations in secondary metabolite production and antioxidant activity in callus cultures of *Stevia rebaudiana* (Bert). *Journal of Photochemistry and Photobiology B: Biology*, 51–56.

Angelini, L. G., Martini, A., Passera, B., & Tavarini, S., (2016). *Cultivation of Stevia Rebaudiana Bertoni and Associated Challenges.* https://doi.org/10.1007/978-3-319-26478-3_8-1.

Anton, S. D., Martin, C. K., Han, H., Coulon, S., Cefalu, W. T., Geiselman, P., & Williamson, D. A., (2010). Effects of *Stevia*, aspartame, and sucrose on food intake, satiety, and postprandial glucose and insulin levels. *Appetite*, *55*(1), 37–43.

Awney, H. A., Massoud, M. I., & El-Maghrabi, S., (2011). Long-term feeding effects of stevioside sweetener on some toxicological parameters of growing male rats. *J. Appl. Toxicol.*, *31*(5), 431–438.

Babu, P., Chikkasubbanna, V., Prasad, T. G., & Radhakrishna, D., (2011). *In vitro* studies on the bearing ability of *Stevia* for stevioside biosynthesis. *Bioscience Biotechnology Research Communications*, *4*, 19–22.

Barriocanal, L., Palacios, M., Benitez, S., Jimenez, J., Jimenez, N., et al., (2008). Apparent lack of pharmalogical effect of steviol glycosides used as sweeteners in humans, a pilot study of repeated exposures in some normatensive and hypotensive individuals and type1 and type 2 diabetics. *Regulatory Toxicology and Pharmacology*, *51*, 37–41.

Basu, P., (2014). *Stevia: All About a Sweet Miracle Plant*. http://bdnews24.com/health/2014/10/02/Stevia-all-about-a-sweet-miracle-plant (accessed on 18 December 2023).

Bayraktar, M., Naziri, E., Akgun, I. H., Karabey, F., Iihan, E., Akyol, B., Bedir, E., & Gurel, A., (2016). Elicitor induced stevioside production, *in vitro* shoot growth, and biomass accumulation in micropropagated *Stevia rebaudiana*. *Plant Cell, Tissue Organ Culture*, *127*(2), 289–300.

Brandle, J. E., & Telmer, P. G., (2007). Steviol glycoside biosynthesis. *Phytochemistry*, *68*(14), 1855–1863.

Brandle, J. E., Starratt, A. N., & Gijzen, M., (1998). *Stevia rebaudiana*: Its agricultural, biological, and chemical properties. *Can. J. Plant Sci.*, *78*, 527–536.

Cacciola, F., Delmonte, P., Jaworska, K., Dugo, P., Mondello, L., Rader, J. I., et al., (2011). Employing ultra-high pressure liquid chromatography as the second dimension in a comprehensive two-dimensional system for analysis of *Stevia rebaudiana* extracts. *J. Chromatogr. A*, *1218*, 2012–2018.

Carneiro, J. W. P., (1990). *Stevia rebaudiana (Bert.) Bertoni, Production of Seed*. M. Sc. Thesis., State University of Maringa, Brazil.

Carneiro, J. W. P., Muniz, A. S., & Guedes, T. A., (1997). Green house bedding plant production of *Stevia rebaudiana* (Bert) Bertoni. *Can. J. Plant Sci.*, *77*, 473–474.

Ceunen, S., &Geuns, J. M. C., (2013). Influence of photoperiodism on the spatio-temporal accumulation of steviol glycosides in *Stevia rebaudiana* (Bertoni). *Plant Sci.*, *198*, 72–82.

Ceunen, S., Werbrouck, S., &Geuns, J. M. C., (2012). Stimulation of steviol glycoside accumulation in *Stevia rebaudiana*by red LED light. *J. Plant Physiol.*, *169*, 749–752.

Chan, P., Tomlinson, B., Chen, Y. J., Liu, J. C., Hsieh, M. H., & Cheng, J. T., (2000). A double-blind placebo-controlled study of the effectiveness and tolerability of oral stevioside in human hypertension. *British Journal of Clinical Pharmacology*, *50*(3), 215–220.

Chatsudthipong, V., & Muanprasat, C. H., (2009). Stevioside and related compounds: Therapeutic benefits beyond sweetness. *Pharmacol. Ther.*, *121*, 41–54.

Chaturvedula, V. S., & Prakash, I., (2011). Structures of the novel diterpene glycosides from *Stevia rebaudiana. Carbohydr. Res.*, *346*, 1057–1060.

Chaturvedula, V. S., Clos, J. F., Rhea, J., Milanowski, D., Mocek, U., Du Bois, G. E., et al., (2011a). Minor diterpenoid glycosides from the leaves of *Stevia rebaudiana*. *Phytochem. Lett.*, *6*, 175–178.

Chaturvedula, V. S., Rhea, J., Milanowski, D., Mocek, U., & Prakash, I., (2011d). Two minor diterpene glycosides from the leaves of *Stevia rebaudiana*. *Nat. Prod. Commun.*, *6*, 175–178.

Chaturvedula, V. S., Upreti, M., & Prakash, I., (2011b). Diterpene glycosides from *Stevia rebaudiana*. *Molecules*, *16*, 3552–3562.

Chaturvedula, V. S., Upreti, M., & Prakash, I., (2011c). Structures of the novel a-glycosyl linked diterpene glycoside from *Stevia rebaudiana*. *Carbohydr. Res.*, *346*, 2034–2038.

Chen, J., Jeppesen, P. B., Abudula, R., Dyrskog, S. E., Colombo, M., & Hermansen, K., (2006). Stevioside does not cause increased basal insulin secretion or β-cell desensitization as does the sulphonylurea, glibenclamide: Studies *in vitro*. *Life Sciences*, *78*(15), 1748–1753.

De Jesús Sanchéz-Cordova, Á., Capataz-Tafur, J., Barrera-Figueroa, B. E., López-Torres, A., Sanchez-Ocampo, P. M., García-López, E., & Huerta-Heredia, A. A., (2019). Agrobacterium rhizogenes-mediated transformation enhances steviol glycosides production and growth in *Stevia rebaudiana* plantlets. *Sugar Tech.*, *21*(3), 398–406.

Debnath, M., (2008). Clonal propagation and antimicrobial activity of an endemic medicinal plant *Stevia rebaudiana*. *J. Med. Plants Res.*, *2*, 45–51.

Dragovi-Uzelac, V., Savi, Z., Brala, A., Levaj, B., Kovaevi, B. D., & Bisko, A., (2010). Evaluation of phenolic content and antioxidant capacity of blueberry cultivars (*Vaccinium corymbosum* L.) grown in the northwest Croatia. *Food Technol. Biotechnol.*, *48*(2), 214–221.

Duke, J., (1993). *Stevia rebaudiana*. In: Duke, J., (ed.), *CRC Handbook of Alternative Cash Crops* (pp. 422–424). CRC Press, Boca Raton, FL.

Ermakov, E. I., & Kochetov, A. A., (1994). Growth and productivity of *Stevia* under regulated conditions, depending on the photoperiod and light intensity. *Russ. Agric. Sci.*, *11*, 11–14.

Ermakov, E. I., & Kochetov, A. A., (1996). Specificities of the growth and development of *Stevia* plants under different controllable light conditions. *Russ. Agric. Sci.*, *1*, 9–11.

Fakhrul, R. H., Norrizah, J. S., Jaapar, S. S., & Noor, A. S., (2014). The effect of nitrogen concentrations on the growth and development of *Stevia rebaudiana* (Bertoni) and production of stevioside and rebaudioside. *Australian Journal of Basic and Applied Sciences*, *8*(10), 500–509.

Flachsland, E., Mroginski, L., & Davina, J., (1996). Regeneration of plants from anthers of *Stevia rebaudiana* Bertoni (Compositae) cultivated *in vitro*. *Biocell*, *20*, 87–90.

Frederico, A. P., Ruas, P. M., Marin-Morales, M. A., Fuas, C. F., & Nakajima, J. N., (1996). Chromosome studies in some *Stevia* Cav. (Compositae) species from Southern Brazil. *Braz. J. Genet.*, *19*, 605–609.

Gantait, S., Das, A., & Mandal, N., (2015). *Stevia*: A comprehensive review on ethnopharmacological properties and *in vitro* regeneration. *Sugar Tech.*, *17*, 95–106.

Geuns, J. M., (2003). Stevioside. *Phytochemistry*, *64*(5), 913–921.

Ghanta, S., Banerjee, A., Poddar, A., & Chattopadhyay, S., (2007). Oxidative DNA damage preventive activity and antioxidant potential of *Stevia rebaudiana* (Bertoni) Bertoni, a natural sweetener. *J. Agric. Food Chem.*, *55*, 10962–10967.

Goettemoeller, J., & Ching, A., (1999). Seed germination in *Stevia rebaudiana*. In: Janick, J., (ed.), *Perspectives on New Crops and New Uses* (pp. 510–511). ASHS Press, Alexandria VA.

Golkar, P., Moradi, M., & Garousi, G. A., (2018). Elicitation of *Stevia* glycosides using salicylic acid and silver nanoparticles under callus culture. *Sugar Tech.*

Goyal, S. K., Samsher, & Goyal, R. K., (2010). *Stevia* (*Stevia rebaudiana*) a bio-sweetener: A review. *International Journal of Food Sciences and Nutrition*, *61*(1), 1–10.

Hegde, S. N., Rameshsing, C. N., & Vasundhara, M., (2015). Characterization of *Stevia rebaudiana* Bertoni polyploids for growth and quality. *Medicinal Plants - International Journal of Phytomedicines and Related Industries*, *7*(3), 188–195.

Hossain, M. A., Shamim, Kabir, A. H. M., Jahan, T. A., & Hasan, M. N., (2008). Micropropagation of *Stevia*. *Int. J. Sustain. Crop. Prod.*, *3*, 1–9.

Hossain, M. F., Islam, M. T., Islam, M. A., & Akhtar, S., (2017). Cultivation and uses of *Stevia* (*Stevia rebaudiana* Bertoni): A review. *African Journal of Food, Agriculture, Nutrition and Development*, *17*(4), 12745–12757.

Ibrahim, I. A., Nasr, M. I., Mohammed, B. R., & El-Zefzafi, M. M., (2008). Plant growth regulators affecting *in vitro* cultivation of *Stevia rebaudiana*. *Sugar Tech.*, *10*, 254–259.

Jain, P., Kachhwaha, S., & Kothari, S. L., (2009). Improved micro-propagation protocol and enhancement in biomass and chlorophyll content in *Stevia* [*Stevia rebaudiana* Bert.] by using high copper levels in the culture medium. *Scientia Horticulturae, 119*, 315–319.

Jain, P., Kachhwaha, S., & Kothari, S. L., (2014). Biotechnology and metabolic engineering of *Stevia rebaudiana* (Bert.) Bertoni: Perspective and possibilities. *International Journal of Life Sciences Biotechnology and Pharma Research*, *3*(3), 15.

Jeppesen, P. B., Gregersen, S., Alstrup, K. K., & Hermansen, K., (2002). Stevioside induces anti-hyperglycaemic, insulinotropic and glucagonostatic effects *in vivo*: Studies in the diabetic Goto-Kakizaki (GK) rats. *Phytomedicine*, *9*(1), 9–14.

Karaköse, H., Jaiswal, R., & Kuhnert, N., (2011). Characterization and quantification of hydroxycinnamate derivatives in *Stevia rebaudiana* leaves by LC-MSn. *J. Agric. Food Chem.*, *59*, 10143–10150.

Katayama, O., Sumida, T., Hayashi, H., & Mitsuhashi, H., (1976). *The Practical Application of Stevia and Research and Development Data* (p. 747). I.S.U Company, Japan.

Khalil, S. A., Zamir, R., & Ahmad, N., (2014). Selection of suitable propagation method for consistent plantlets production in *Stevia rebaudiana* (Bertoni). *Saudi Journal of Biological Sciences*, *21*, 566–573.

Khan, S. A., Rahman, L., Verma, R., & Shanker, K., (2016). Physical and chemical mutagenesis in *Stevia rebaudiana*: Variant generation with higher UGT expression and glycosidic profile but with low photosynthetic capabilities. *Acta Physiologiae Plantarum*, *38*(1), 4.

Kim, I. S., Yang, M., Lee, O. H., & Kang, S. N., (2011). The antioxidant activity and the bioactive compound content of *Stevia rebaudiana* water extracts. *LWT Food Sci. Technol.*, *44*, 1328–1332.

Kim, S. H., & Dubois, G. F., (1991). Natural high potency sweeteners. In: Marie, S., & Piggott, J. R., (eds.), *Handbook of Sweeteners*. Blacki Avi, Glasgow, New York.

Kinghorn, A. D., (1987). Biologically active compounds from plants with reputed medicinal and sweetening properties. *Journal of Natural Products*, *50*(6), 1009–1024.

Kochikyan, V. T., Markosyan, A. A., Abelyan, L. A., Balayan, A. M., & Abelyan, V. A., (2006). Combined enzymatic modification of stevioside and rebaudioside A. *App. Biochem. Microbiol.*, *42*(1), 31–37.

Kovylyaeva, G. I., Bakaleinik, G. A., Strobykina, I. Y., Gubskaya, V. I., Sharipova, R. R., Al'Fonsov, V. A., Kataev, V. E., & Tolstikov, A. G., (2007). Glycosides from *Stevia rebaudiana*. *Chemistry of Natural Compounds*, *43*, 81–85.

Kumar, R., (2013). Seed germination of *Stevia rebaudiana* influenced by various potting media. *Octa Journal Biosciences*, *1*, 143–146.

Kumari, M., & Chandra, S., (2015). Phytochemical studies and estimation of major steviol glycosides in varied parts of *Stevia rebaudiana*. *Int. J. Pharm. Pharm. Sci.*, *7*, 62–65.

Lankes, C., & Pude, R., (2008). Possibilities for growth of *Stevia* in Europe temperate zones. In: Geuns, J. M. C., (ed.), *Steviol Glycosides: Technical and Pharmacological Aspects;*

Proc. 2nd Stevia Symposium 2008 of European Society of Stevia (EUSTAS) (pp. 103–116). KU Leven, Leuven.

Lee, C. N., Wong, K. L., Liu, J. C., Chen, Y. J., Cheng, J. T., & Chan, P., (2001). Inhibitory effect of stevioside on calcium influx to produce antihypertension. *Planta Medica, 67*(9), 796–799.

Lemus, M. R., Vega, G. A., Zura, B. L., & Ah-Hen, K., (2012). *Stevia rebaudiana* Bertoni, source of a high potency natural sweetener: A comprehensive review on the biochemical, nutritional, and functional aspects. *Food Chemistry, 132*(3), 1121–1132.

Lester, T., (1999). *Stevia rebaudiana(Sweet Honey Leaf)* (Vol. 2, p. 120). Aust. new crops news Lett. 11. Nat. Prod. Radiance.

Lewis, W. H., (1992). Early uses of *Stevia rebaudiana* leaves as sweetener in Paraguay. *Econ. Bot., 46*, 336, 337.

Li, J., Jiang, H., & Shi, R., (2009). A new acylated quercetin glycoside from the leaves of *Stevia rebaudiana* Bertoni. *Nat. Prod. Res., 23*, 1378–1383.

Ma, G., Bechman, A., Bunders, C., Devkota, K. P., Charan, R. D., Ramirez, C., Snyder, T. M., et al., (2018). New diterpene glycosides from *Stevia rebaudiana* Bertoni: Rebaudioside VIII and rebaudioside IXd. *Natural Product Communications, 13*(10), 1243–1250.

Madan, S., Ahmad, S., Singh, G. N., Kohli, K., Kumar, Y., Singh, R., & Garg, M., (2010). *Stevia rebaudiana* (Bert.) Bertoni: A review. *Indian Journal of Natural Products and Resources, 1*(3), 267–286.

Marcinek, K., & Krejpcio, Z., (2016). *Stevia rebaudiana* Bertoni: Health promoting properties and therapeutic applications. *Journal fur Verbraucherschutz und Lebensmittelsicherheit, 11*(1), 3–8.

Metivier, J., & Viana, A. M., (1979). The effect of long and short-day length upon the growth of whole plants and the level of soluble proteins, sugars, and stevioside in leaves of *Stevia rebaudiana* Bert. *J. Exp. Bot., 30*, 1211–1222.

Midmore, D. J., & Rank, A. H., (2002). *A New Rural Industry- Stevia- to Replace Imported Chemical Sweeteners. A Report for the Rural Industries Research and Development Corporation.* Rural Industries Research and Development Corporation. Publication No. W02/022.

Miyazaki, Y., & Watanabe, H., (1974). Studies on the cultivation of *Stevia rebaudiana* Bertoni: On the propagation of the plant. *Jap. J. Trop. Agric., 17*, 154–157.

Mubarak, M., El Halmouch, Y., Belal, A., Elfadeel, M., EL-Din, T., Mahmoud, S. F., et al., (2015). Improving sweet leaf (*Stevia rebaudiana* var. Bertoni) resistance to bialaphos herbicide via 'bar' gene transfer. *Plant Omics, 8*(3), 232.

Oddone, B., (1997). How to grow *Stevia*? *Technical Manual.* Edn. 1, Guarani Botanicals INC., Pawtucket, CT.

Oddone, B., (1999). How to grow *Stevia*? *Technical Manual.* Edn. 2, Guarani Botanicals, Inc., Pawcatuck, CT.

Ohta, M., Sasa, S., Inoue, A., Tamai, T., Fujita, I., Morita, K., et al., (2010). Characterization of novel steviol glycosides from leaves of *Stevia rebaudiana* Morita. *J. Appl. Glycosci., 57*, 199–209.

Pal, P. K., Kumar, R., Guleria, V., Mahajan, M., Prasad, R., Pathania, V., Gill, B. S., et al., (2015). Crop-ecology and nutritional variability influence growth and secondary metabolites of *Stevia rebaudiana* Bertoni. *BMC Plant Biol., 15*, 67.

Pól, J., Hohnov, Á., Hyotyl, B., & Äinen, T., (2007). Characterization of *Stevia rebaudiana* by comprehensive two-dimensional liquid chromatography time-of-flight mass spectrometry. *Journal of Chromatography A, 1150*(1, 2), 85–92.

Prakash, I., Dubois, G. E., Clos, J. F., Wilkens, K. L., & Fosdick, L. E., (2008). Development of rebiana, a natural, non-caloric sweetener. *Food Chem. Toxicol., 46*(7), S75–82.

Ramesh, K., Singh, V., & Megeji, N. W., (2006). Cultivation of *Stevia* [*Stevia rebaudiana* (Bert.) Bertoni]: A comprehensive review. *Adv. Agron., 89*, 137–177.

Ranjan, R., Jaiswal, J., & Jena, J., (2011). *Stevia* as a natural sweetener. *International Journal of Research in Pharmacy and Chemistry, 1*(4), 1199–1202.

Robinson, B. L., (1930). *Contributions from the Gray Herbarium of Harvard University*. The Gray Herbarium of Harvard University, Cambridge, MA.

Sakaguchi, M., & Kan, T., (1982). Japanese *research* on *Stevia rebaudiana. Cl. Cult., 34*, 235–248.

Shukla, S., Mehta, A., Bajpai, V. K., & Shukla, S., (2009). *In vitro* antioxidant activity and total phenolic content of ethanolic leaf extract of *Stevia rebaudiana* Bert. *Food and Chemical Toxicology, 47*, 2338–2343.

Shuping, C., & Shizhen, S., (1995). Study on storage technique of *Stevia rebaudiana* seed. *Acta Agron. Sin., 21*, 102–105.

Sivaram, L., & Mukundan, U., (2003). *In vitro* culture studies on *Stevia rebaudiana. In Vitro Cellular and Developmental Biology-Plant, 39*(5), 520–523.

Smitha, G. R., & Umesha, K., (2011). Vegetative propagation of *Stevia* [*Stevia rebaudiana* (Bertoni) Hemsl.] through stem cuttings. *Journal of Tropical Agriculture, 50*, 72–75.

Soejarto, D. D., Compadre, C. M., Medon, P. J., Kamath, S. K., & Kinghorn, A. D., (1983). Potential sweetening agents of plant origin. II. Field search for sweet-tasting *Stevia* species. *Econ. Bot., 37*, 71–79.

Soejarto, D. D., Kinghorn, A. D., & Fansworth, N. R., (1982). Potential sweetening agents of plant origin III, organoleptic evaluation of *Stevia* leaf herbarium samples for sweetness. *Journal of Natural Products, 45*, 590–599.

Suanarunsawat, T., &Chaiyabutr, N., (1997). The effect of stevioside on glucose metabolism in rat. *Canadian Journal of Physiology and Pharmacology, 75*(8), 976–982.

Sumida, T., (1968). Reports on *Stevia* introduced from Brazil as a new sweetness resource in Japan. *J. Cent. Agric. Exp. Stn., 31*, 1–71.

Sumida, T., (1980). Studies on *Stevia rebaudiana* (Bertoni): Introduced from Brazil as a new sweetness resource in Japan. *J. Cent. Agric. Exp. Stn., 31*, 1–71.

Tadhani, M. B., Patel, V. H., & Subhash, R., (2007). *In vitro* antioxidant activities of *Stevia rebaudiana* leaves and callus. *J. Food Compos. Anal., 20*, 323–329.

Taleie, N., Hamidoghli, Y., Rabiei, B., & Hamidoghli, S., (2012). Effects of plant density and transplanting date on herbage, stevioside, phenol and flavonoid yield of *Stevia rebaudiana* Bertoni. *Intl. J. Agri. Crop. Sci., 4*(6), 298–302.

Tariq, U., Ali, M., & Abbasi, B. H., (2014). Morphogenic and biochemical variations under different spectral lights in callus cultures of *Artemisia absinthium* L. J. *Photochem. Photobiol. B Biol., 130*, 264–271.

Taylor, L., (2005). *The Healing Power of Natural Herbs*. Square One Publishers, Inc., Garden City Park, New York.

Yadav, A. K., Singh, S., Dhyani, D., & Ahuja, P. S., (2011). A review on the improvement of *Stevia* [*Stevia rebaudiana* (Bertoni)]. *Canadian Journal of Plant Science, 91*(1), 1–27.

Yadav, A. K., Singh, S., Yadav, S. C., Dhyani, D., Bhardwaj, G., Sharma, A., & Singh, B., (2013). Induction and morpho-chemical characterization of *Stevia rebaudiana* colchiploids. *Ind. J. Agric. Sci.*, *83*(2), 159–165.

Zimmermann, B. F., (2011). Tandem mass spectrometric fragmentation patterns of known and new steviol glycosides with structure proposals. *Rapid Commun. Mass Spectrom.*, *25*, 1575–1582.

CHAPTER 15

Traditional and Pharmacological Values of *Oroxylum indicum*

T. KALAIVANI, N. KANAGAM, and C. RAJASEKARAN

Department of Biotechnology, School of Bio-Sciences and Technology, Vellore Institute of Technology, Vellore, Tamil Nadu, India

ABSTRACT

Oroxylum indicum (L.) Vent. belongs to the family Bignoniaceae, is one of the widely utilized folklore medicinal species in India and Southeast Asia. The phytochemical compounds isolated from *O. indicum* are found to have several pharmacological activities like antifungal, antibacterial, anticarcinogenic, anti-inflammatory, antioxidant, and anti-hyperglycemic. It has a high impact on traditional medicinal formulations such as *Chyawanprash* and *Dashmula*. *O. indicum* is traditionally exploited for treating disorders like abdominal pain, ulcer, throat infection, piles, and bronchitis. Globally, despite the discovery of many drugs, cancer cure is still a challenge. Though many novel synthetic chemotherapeutic agents were identified in the last decade, most of them were not economically affordable. Therefore, it is the need of the hour to develop a potent and affordable anticarcinogenic drug. This study gives an account of the phytochemical and pharmacological activities of the plant extract and its derived compounds extracted from *O. indicum*.

15.1 INTRODUCTION

Oroxylum indicum belongs to the family Bignoniaceae synonym names are *Bignonia indica* L., *Calosanthes indica* (L.) Blume, *Spathodeaindica* (L.)

Bioprospecting Ethnomedicinal Plant Resources: Sustainable Utilization and Restoration.
Gopal Shukla, Jahangeer A. Bhat, A. P. Das, and Sumit Chakravarty (Eds.)

Pers. It has different vernacular names in different regional languages. Few known English and vernacular names are Broken Bones Plant, Midnight horror, Indian Trumpet Flower, Tree of Damocles, *Phari*, and *Andurlu*. It is named as 'midnight horror' because of its lengthy pods hanging down the branches making it appear like swinging sickles in the night (Dev et al., 2010; Plate 15.1(a)). Inside these pods, seeds look like butterflies and are used in Chinese herbal medicine. Thus, the white and paper-thin "wings" of the seeds are known to the Chinese as *muhudie* (*mu* = tree; *hudie* = butterfly) or the tree butterfly (Plate 15.1(b)).

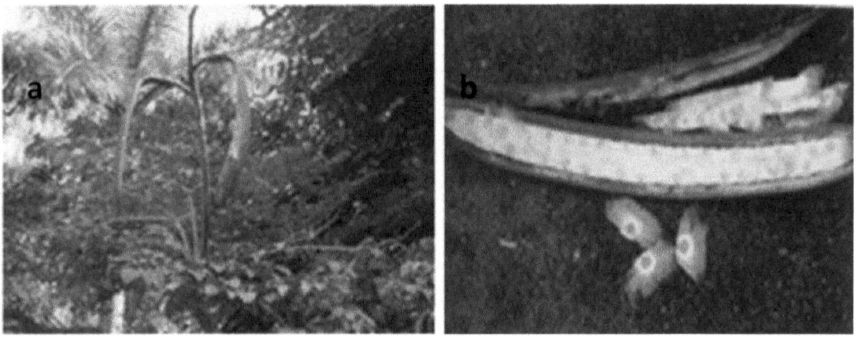

PLATE 15.1 *Oroxylum indicum* – whole tree with pods (a); and open pods showing seeds (b).

After an episode portrayed in an ancient story by Cicero, this tree is called 'tree of Damocles,' where a sword was hanging above the head of Damocles. The tree blooms at night and undergoes natural pollination by bats. Apart from that, the bulky leaf stem of the plant weakens and falls down, and accumulates at the bottom of the tree trunk, making it appear like a stack of fractured limb bones, giving another name "broken bones plant." This tree is habitually grown as ornamental due to its strange look. Though it's a tropical tree, it can survive even at low temperature (32°F) for a short term. This tree is native to India and Southern China but also distributed throughout China, South, and Southeast Asia. It is distributed all over India, mainly covering ravine and humid forest ranges up to an altitude of 1,200 m. From ancient times, many terrestrial plants are used as medicines in Egypt, China, India, and Greece, some of which were also developed into drugs. Cancer is a globally widespread significant disorder increasing day by day due to a lack of early detection methods, poor prognosis, and unavailability of cost-effective drugs. For cancer treatment, a paradigm on ethnobotanical and ethno pharmacological data is very beneficial to identify potential anticarcinogenic molecules (Chanda & Nagani, 2013). *Oroxylum indicum* is enriched with flavonoids and phenols, possesses free radical scavenging activity thereby has many biological and pharmacological activities, including antioxidant activity (Kalaivani & Mathew, 2009). Previous reports revealed that *O. indicum* bark acts as a powerful anticarcinogenic remedy for nasopharyngeal malignancy (Mao, 2002) and abdominal tumors due to the antioxidant potential of the plant. India, the most leading region in drug discovery from natural flora has to put in more efforts to explore efficient anticarcinogenic drugs from plants to save humans from cancer (Chanda & Nagani, 2013).

15.2 ECONOMIC IMPORTANCE

The economic value of the plant is listed as ornamental due to its peculiar look. This tree is often grown as an ornamental plant. Young shoots, unripe fruits, and flowers are used as vegetables. The leaves are chopped as fodder. The stem, trunk bark, and fruits are used as a mordant because the stem bark synthesizes a Khaki color pigment. The timber of the plant is cast-off to make match cartons. Seeds produce non-drying oils suitable for perfumery, and the entire plant is utilized for medicinal purposes (Dey et al., 1978; Mao, 2002; Kalaivani & Mathew, 2009; Dev et al., 2010; Raghu et al., 2013).

15.3 MEDICINAL VALUE

Since ancient times, different parts of *Oroxylum indicum* have been used as traditional Ayurvedic medicines. It is used in traditional or indigenous medicinal systems to cure various diseases/disorders such as leukoderma, cancer, snakebite, rheumatism, diabetes, and asthma which could not be cured by allopathy (Kalaivani & Mathew, 2009). In this chapter, we discuss some information related to such diseases and their traditional treatments followed since olden days (Plate 15.2).

15.3.1 NASOPHARYNGEAL CANCER

In India, *Oroxylum indicum* is a widely used tree for the healing of numerous disorders. The decoction of *O. indicum* bark acts as a potent anticarcinogenic drug especially against nasopharyngeal cancer based on the traditional

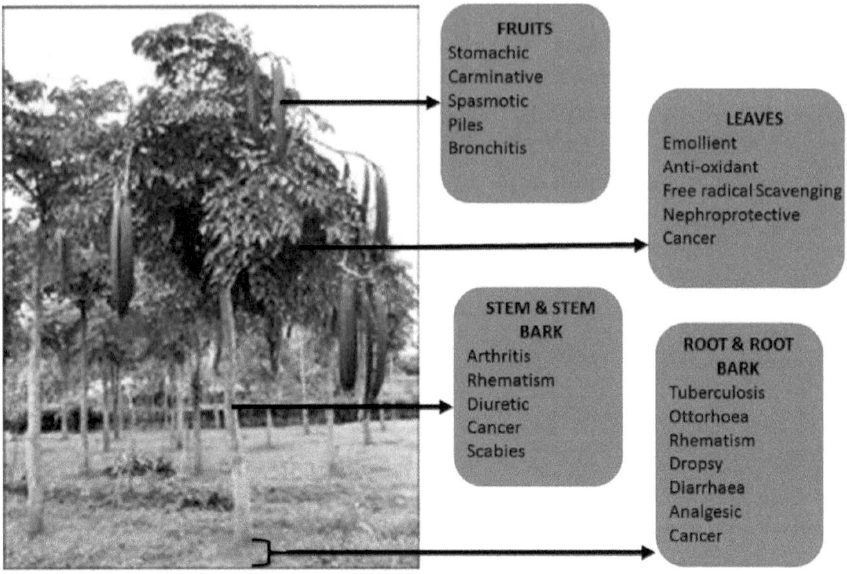

PLATE 15.2 Traditional medicinal uses of *Oroxylum indicum.*

knowledge (TK) of the Maram Naga village population (Mao, 2002). About 1 kg of fresh bark boiled in five liters of water for 40 minutes, the decoction is filtered and given with honey three times a day to treat nasopharyngeal cancer (Mao, 2002).

15.3.2 SNAKE BITE

In the central Himalayan region, the traditional system of medicine has many antidotes for treating snake bite. Customarily, naturopathy has many remedies and medications to heal the late poisoning of snake poison. At an interval of 30 minutes, the solution of *O. indicum* is administered to the patient till the patient regains consciousness (Singh & Ali, 1998).

15.3.3 DIARRHEA

Root bark is administered to cure blood dysentery by indigenous communities of Keonjhar and the Mayurbhanj region of Odisha, India (Kambaska, 2006). The fermented powdered root bark is sieved, filtered, and the liquid filtrate is administered orally along with sugar candy or honey thrice a day for three days to control blood dysentery.

15.3.4 JAUNDICE

O. indicum is used as a traditional medicine against jaundice in some parts of Nepal (Kunwar et al., 2006). The root powder mixture of *Ficus auriculata* and bark of *O. indicum* is consumed to treat jaundice.

15.3.5 FEVER AND PNEUMONIA

The seeds of the tree are used from ancient times for treating fever and pneumonia in Nepal. It is called Nepali *Totola* or *Baldane*. In every Thangmi ritual house, there must be at least one *Baldane* tree (Kunwar et al., 2006). The seeds are finely beaten, mixed with water, and strained to increase its medicinal value. This concoction is fed to patients suffering from high fever and pneumonia (Turin, 2003).

15.3.6 TONSILLITIS AND DIURETIC

Chinese have this species for treating tonsillitis and diuretic problems in humans. Other medicinal features of different tree parts are mentioned in Ayurvedic medicines, and the list has been summarized in Table 15.1.

TABLE 15.1 Ayurvedic Applications of *Oroxylum indicum*

Part	Active Ingredient	Medicinal Uses	References
Stem and root bark	Ellagic acid, flavones such as oroxylin A, scutellarin-7-rutin-oside, chrysin, and traces of alkaloids like tannic acid, sitosterol, and galactose; chrysin.	Tuberculosis, dropsy, alleviate anxiety, astringent, rheumatism, tonsillitis, scabies, deworming, mouth cancer, gastric ulcer, diaphoretic, anodyne, bitter tonic, sudorific, stomachic, otor-rhea, as *Dashmula* in ayurvedic preparations and to heal animal wounds.	Subramanian & Nair (1972); Dey et al. (1978); Yoganarasimhan (1996); Bhattacharjee (2000); Suresh Babu et al. (2006);Raghu et al. (2013)
Leaves	Aloe-emodin, anthraquinone.	Emollient	Dey et al. (1978)
Seeds	Terpenes, saponins, alkaloids, baicalein.	Purgative, lungs diseases, anti-microbial activity.	Bhattacharje& Das (1969); Rasadah et al. (1998); Roy et al. (2007)
Tender fruits	–	Carminative and stomachic.	Dey et al. (1978)

15.3.7 POST-PARTUM TREATMENT

Postpartum treatment is essential to cleanse or help complete evacuation to get relief from pain. This method of treatment is differing from the treatment given to control abortion. The natural herbs given here are generally *ushna* (temperature), *teekshna* (incisive), and *paachana*, i.e., digestion (Bhuvad & Nishteswar, 2015).

15.4 AYURVEDIC PRODUCTS

Some of the commercial Ayurveda products using *Oroxylum indicum* as one of the ingredients are described in subsections.

15.4.1 DASAMOOLARISHTA (AYUR ELIXIR)

Dasamool is a combination of 10 different herbal plant roots called "TEN ROOT" elixir is the mainstay of traditional lore for the ancient Ayurvedic herbalists. It is used as a restorative agent by the Ayurvedic system of medicine for treating diseases like nervous disorders, cardiac problems, and asthma (Kambaska, 2006; Gupta et al., 2008).

15.4.2 CHYAWANPRASH

Chyawanprash is a popular traditional Ayurvedic dietary supplement. This *rasayana* is used as an anti-aging to nourish the blood, lymph, muscles, nerves, heart, brain, and lungs. Since time immemorial, Chyawanprash was used to gain energy and as an immune-booster. It is a rejuvenating support system to maintain the human body system (Parle & Bansal, 2006).

15.4.3 MAHA NARAYAN VATTA MASSAGE OIL

According to the Ayurvedic concept of *Tridosha*, Maha Narayan *vatta* massage oil is highly recommended to treat/heal *vata* (gas) conditions. "MUSCLEGARD" is a massage oil that instantly gives relief for mild abrasions, contusions or bruises, general muscular aches, strains, sprains, stiffness of joints, lumbago, neuralgic pains, and backaches. It could also be used for all kinds of common muscular weaknesses, atrophies, muscular wasting, and facial paralysis. The oil intensifies muscular activities by giving strength and nourishment to muscle fibers.

15.4.4 DEEP LOVE HERBAL EXTRACT

Deep love herbal drug provides emotional strength, clearness, and compassion in day-to-day relationship problems.

15.4.5 ASHWAGANDHA GHEE

Traditionally, Ashwagandha ghee can be used for both internal and external applications to promote *prana*, *ojas* (life, force), vigor, male virility, and vitality. It is helpful to calm down *vata* and it is also considered tridoshic in moderation. It also enhances the tone of the *dhatus* (metal), slows down the aging process, and supports cellular longevity.

15.5 PHARMACOLOGICAL ACTIVITIES

Apart from the common applications, many studies reported the biological and pharmacological activities of *Oroxylum indicum* (Table 15.2).

15.5.1 ANTI-INFLAMMATORY

Aqueous and alcoholic extracts of *O. indicum* significantly controlled the release of myeloperoxidase which causes inflammation. Both root and stem bark decoction were found to suppress edema formation against carrageenan and leaves were reported to have anti-inflammatory activity (Habtemariam, 1997). In rats, long-lasting inflammation was controlled by using the root bark. Similarly, inflammation was also controlled significantly when pretreatment with n-butanol (Sowjanya et al., 2019). The species acts as a major component to control chronic inflammatory conditions like arthritis (Dev et al., 2010).

15.5.2 ANTICARCINOGENIC

Based on traditional ethnobotanical uses of *O. indicum,* it is very clear that the entire plant is having significant anticarcinogenic activity against human cancer cell lines. The plant extracts of *Artemesia monosperma* (Asteraceae),

TABLE 15.2 Pharmacological Activities of *Oroxylum indicum*

Plant Part	Activity	Target	Name of the Extract	References
Root and stem bark.	Antiulcer, immuno-stimulant activity; anticarcinogenic to t breast cancer, antiallergic, antioxidant, antiproliferative, estrogen receptor inhibitor, genotoxicity, antimutagenic, anti-inflammatory, antiarthritic.	Gastric ulcer in animals, pyloric mucosa presents in the stomach of rat, breast cancer cell lines – MCF7 and MDA-MB-231; *in vitro* system.	Aqueous and alcoholic extract, n-butanol fraction, nitrosated fraction.	Tepsuwan et al. (1992); Khandhar et al. (2000); Nakahara et al. (2001, 2002); Jiwajinda et al. (2002); Khan et al. (2002); Laupattarakasem et al. (2003); Lambertini et al. (2004); Suresh Babu et al. (2006); Zaveri et al. (2006)
Twigs, leaves, flowers	Anticarcinogenic, cytotoxic, antioxidant, antimutagenic, antibacterial, antitumor.	Wister rats, against 2-amino anthracene, blood (HL-60) cancer cells; foodborne pathogens, superoxide generation using a xanthine–xanthine oxidase assay system (HL-60), against heterocyclic amine Trp-P-1.	Ethanolic extracts, ethanol-CH_2Cl_2-soluble fraction, methanol extract, buffered methanol (80%) and 20% phosphate-buffered saline, acetone extracts.	Wall et al. (1988); Na Thalang et al. (2000, 2001); Jiwajinda et al. (2002); Nakahara et al. (2002); Alzoreky& Nakahara (2003); Shalini et al. (2016)
Fruits (baicalein)	Antibacterial, antimutagenic, cell proliferation assay, antiproliferative, cytotoxic, anticarcinogenic, apoptosis, and anti-tumor.	Human Hep2 cell line, against heterocyclic amine Trp-P-1, human breast cancer cell lines, leukemia cell line – HL-60, Murine melanoma cell line – B-16 and human colon carcinoma cell line – HCT-8.	Methanolic extract	Habtemariam (1997); Nakahara et al. (2001, 2002); Lambertini et al. (2004); Costa-Lotufo et al. (2005); Kamkaen et al. (2006); Roy et al. (2007)
Seed	Antioxidant, anticarcinogenic activity.	Heinz body induction	Aqueous extract	Palasuwan et al. (2005)

Origanumdayi post (Labiatae), and *Urtica membranacea* (Urticaceae) exhibited time-dose dependent response in human cancer cell lines. Because of resistance against cancer chemotherapy, researchers' ultimate choice is anticarcinogenic molecules/compounds or formulations from plant sources (Solowey et al., 2014). Since ancient times, the traditional Chinese system of medicine treated cancer by fruit methanol extracts which also possess antipyretic, anti-inflammatory, and anti-hypersensitivity properties. Baicalein is reported recently to inhibit *in vitro* proliferation of HL-60 cancer cells. Extracts can act as an effective supplement for cancer treatment, due to the anti-tumor effects of baicalein on human cancer cells (Roy et al., 2007). The plant is used in several herbal formulations to treat different types of cancer in the Indian system of medicine also. Methanolic extract possesses Trp-P-1 immunogenicity, genotoxic, and cell proliferative activity. The extracts were found to have cell-specific toxic effects on tumor cell lines of CEM, HL-60, B-16, and HCT-8 (Prakash et al., 2013).

15.5.3 ANTI-MICROBIAL

Different solvent extracts of *O. indicum* exhibit antimicrobial activity against various pathogens and it explores the possibility of phytochemicals for therapeutics. The crude extracts as well as purified compounds such as 1 (2,5-dihydroxy-6,7-dimethoxy flavone) and 2 (3,7,3,'5'-tetramethoxy-2-hydroxy flavone) inhibited the growth of both gram-positive and -negative pathogenic bacteria including pathogenic fungi (Dossou-Yovo et al., 2014). Major flavonoids namely 2,5-dihydroxy-6,7-dimethoxy flavone and 3,7,3,'5'-tetramethoxy-4'-hydroxy flavone were isolated from the species which exhibited prominent antimicrobial activity against *Bacillus subtilis*, *Escherichia coli*, *Staphylococcus aureus*, and *Shigella dysenteriae* (Rasadah et al., 1998; Suresh Babu et al., 2006). Lapachol and b-lapachone isolated from dichloromethane extracts were reported to possess antifungal activity against dermatophytes and wood rot fungi. Microscopical studies confirmed that its extracts inhibited mycelia and conidia development in filamentous fungi (Chaturvedi & Nikhil, 2016).

15.5.4 CYTOTOXICITY ASSAY

Major flavonoids 2,5-dihydroxy-6,7-dimethoxy flavone and 3,7,3,'5'-tetramethoxy-4-hydroxy flavone were isolated from root ethyl acetate extract of *O. indicum*. LC_{50} values of these two flavonoids with petroleum ether, ethyl acetate, and methanol extracts were found to be 14.12, 3.80, 3.97, 6.02, and 8.70 μg/ml, respectively (Habtemariam, 1997).

15.6 PHYTOCHEMICAL COMPOUNDS

Chrysin, baicalein, baicalein-7-O-glucoside, baicalein-7-O-diglucoside (Oroxylin B) were isolated from the seeds (Zhang et al., 2013). Various phytochemical constituents present in different parts of the tree are presented in Table 15.3.

TABLE 15.3 Prominent Phytochemical Constituents of *Oroxylum indicum*

Plant Part	Group	Compound	References
Leaves	Flavonoids	Baicalein, Scutellarein, Baicalein-7-glucuonide, Baicalein-6-glucuronide, apigenin, Scutellarein-7-glucuronide, cyclohexane-1,4-diol, quercetin-3-o-alpha-arabino pyranoside, 1-(2-hydroxyethyl).	Kumar et al. (2011); Sowjanya et al. (2019)
Seeds	Polyphenolic compound, flavonoids.	Ellagic acid, terpenes, alkaloids, and saponins, baicalein-7-odiglucoside, baicalein-7-o-glucoside, baicalein, chrysin, apigenin.	Habtemariam (1997); Dossou-Yovo et al. (2014)
Root bark	Flavonoids	Ellagic acid, oroxylin-A, chrysin, scutellarin, baicalein, stigma-5-en-3-ol, 5-hydroxy 8-methoxy 7-o-beta-D-glucopyranuronosyl flavone, pratensol, 3-(4-hydroxy phenyl) 2-propenoic acid and flavonoid 3,4,'5,7-tetrahydroxy-flavonol, 5-hdroxy 4,'7-dimethoxy flavone, 5-hydroxyl-7-methoxy-2-2-methoxy-6-(3,4,5-trihydroxy-6-(hydroxymethyl), tetrahydro-2H-pyran2-yoloxy, phenyl-4H-chromen-4-one, 7-o-methyl chrysin, dihydrooroxylin A, methyl-3,4,5-trihydroxy-6-(5-hydroxy-6-methoxy-4-oxo-2-phenylchroman-7-yloxy)-tetrahydro-2H-pyran-2-carboxylate.	Sowjanya et al. (2019)
Wood, leaves, fruit, stem bark.	Sterol, glycosides, flavonoids.	Prunetin, β-sitosterol, anthraquinone glycosides, aloe-emodin, chrysin, oroxylin-A, ursolic acid, baicalein, scutellarein, chrysin, baicalien-7-glucuronide, p-coumaric acid.	Habtemariam (1997); Raghu et al. (2013)

15.7 CONCLUSION

Oroxylum indicum is a multipurpose medicinal plant with a unique source of multiple compounds serving diverse pharmacological actions. Since meager work has been carried out on the evaluation of biological activity and the toxicity of the compound isolated, a broad investigation is required to unlock other therapeutic opportunities to battle against diseases. Initiation of the drug development program is essential to discover new modern drugs using components isolated from the species. Though crude extracts of its several parts have curative properties, modern drug development with the wide-ranging investigation of its mode of action, bioactivity, pharmacotherapeutics, proper standardization of clinical trials, and toxicity is needed. Recently, there has been a growing trend and awareness in its R&D. Similar therapeutic and revenue-generating herbal products available in the market give inspiration among the scientists to explore more information about this medicinal plant. The species is profusely used in the Homeopathic, Ayurveda, and Unani medicinal systems to cure both diseases and disorders. Previous reports on their biologically active molecules specify that their physiological and biochemical importance modifies during developmental stages. These aspects are very positive for the commercial exploration of this plant. Understanding the mode of action of different active chemical constituents may provide a special opportunity to develop its various formulations for health benefits. Aerial parts of the plant like bark, stalks, and also roots possess rich flavonoids that might be used against different disorders in the future. Whether it is also endowed with an inimitable property, this natural scientific curiosity was also enthusiastically attended by undertaking a detailed exploration of the active part and chemical constitution. Modern biotechnological approaches are required for the enhancement of active ingredients of the plant. The mechanism of action of the plant ingredients in treating various ailments is to be explored. Further studies are needed to evaluate anticarcinogenic properties.

ACKNOWLEDGMENTS

The authors are thankful to the Vellore Institute of Technology Management and the Dean School of Biosciences and Technology for providing proper infrastructure, constant support, and encouragement.

KEYWORDS

- **anticarcinogenic drug**
- **cytotoxicity assay**
- **medicinal plants**
- **Oroxylum indicum**
- **pharmacological activities**
- **phytochemical compounds**
- **secondary metabolites**
- **traditional medicine**

REFERENCES

Alzoreky, N. S., & Nakahara, K., (2003). Antibacterial activity of extracts from some edible plants commonly consumed in Asia. *International Journal of Food Microbiology, 80*(3), 223–230. doi: 10.1016/s0168-1605(02)00169-1.

Bhattacharje, A. K., & Das, A. K., (1969). Phytochemical screening of some Indian plants. *Q. J. Crude Drug Res., 9*, 1408.

Bhattacharjee, S. K., (2000). *Handbook of Aromatic Plants*. Pointer Publishers, Jaipur.

Bhuvad, S., & Nishteswar, K., (2015). Antioxidant activity perspectives. Rasayana Karma. *Joinsysmed, 3*(2), 69–81.

Chanda, S., & Nagani, K., (2013). *In vitro* and *in vivo* methods for anticancer activity evaluation and some Indian medicinal plants possessing anticancer properties: An overview. *Journal of Pharmacognosy and Phytochemistry, 2*(2), 140–152.

Chaturvedi, V., & Nikhil, K., (2016). Effects of algal bio-fertilizer on the growth of *Vigna radiata. International Journal of Engineering and Technical Research, 4*(1), 111–115.

Costa-Lotufo, L. V., Khan, M. T., & Ather, A., (2005). Studies of the anticancer potential of plants used in Bangladeshi folk medicine. *J. Ethnopharmacol., 99*(1), 21–30.

Dev, L. R., Anurag, M., & Rajiv, G., (2010). *Oroxylum indicum*: A review. *Pharmacognosy Journal, 2*(9), 304–310. https://doi.org/10.1016/S0975-3575(10)80121-X.

Dey, A. K., Mukherjee, A., Das, P. C., & Chatterjee, A., (1978). Occurrence of aloe-emodin in the leaves of *Oroxylum indicum* Vent. *Ind. J. Chem., 16B*, 1042.

Dossou-Yovo, H. O., et al., (2014). Medicinal plant research. *Journal of Medicinal Plant Research, 8*(1), 368–377.

Gupta, R. C., Sharma, V., Sharma, N., Kumar, N., & Singh, B., (2008). *In vitro* antioxidant activity from leaves of *Oroxylum indicum* (L.) vent. – a north Indian highly threatened and vulnerable medicinal plant. *Journal of Pharmacy Research, 1*(1),65–72.

Habtemariam, S., (1997). Flavonoids as inhibitors or enhancers of the cytotoxicity of tumor necrosis factor- in L-929 tumor cells. *J. Nat. Prod., 60*(8), 775–778.

Jiwajinda, S., Santisopasri, V., Murakami, A., Kim, O. K., Kim, H. W., & Ohigashi, H., (2002). Suppressive effects of edible Thai plants on superoxide and nitric oxide generation. *Asian Pacific Journal of Cancer Prevention, 3*, 215–223.

Kalaivani, T., & Mathew, L., (2009). Phytochemistry and free radical scavenging activities of *Oroxylum indicum*. *International Journal of Science & Technology*, *4*, 45–52.
Kambaska, K. B., (2006). Ethnomedicinal plants used by the tribals of Similipal Bioreserve, Orissa, India: A pilot study. *Ethnobotanical Leaflets*, *10*, 149–173.
Kamkaen, N., Wilkinson, J. M., & Cavanagh, H. M. A., (2006). Cytotoxic effect of four Thai edible plants on mammalian cell proliferation. *Thai Pharmaceutical and Health Science Journal*, *1*(3), 189–195.
Khan, M. T. H., Lampronti, I., Martello, D., Bianchi, N., Jabbar, S., Choudhuri, M. S., Datta, B. K., & Gambari, R., (2002). Identification of pyrogallol as an antiproliferative compound present in extracts from the medicinal plant *Emblica officinalis*: Effects on *in vitro* cell growth of human tumor cell lines. *International Journal of Oncology*, *21*, 187–192.
Khandhar, M., Shah, M., Santani, D., & Jain, S., (2000). Antiulcer activity of the root bark of *Oroxylum indicum* against experimental gastric ulcers. *Pharmaceutical Biology, 44*(5), 363.
Kumar, V., Chaurasia, A. K., Naglot, A., Gopalakrishnan, R., Gogoi, B. J., Srivastava, R. B., & Deka, D. C., (2011). Antioxidant and antimicrobial activities of stem bark extracts of *Oroxylum indicum* vent. (*Bignoniaceae*)- a medicinal plant of northeastern India. *South Asian Journal of Experimental Biology, 1*(3), 152–157.
Kunwar, R. M., Nepal, B. K., Kshhetri, H. B., Rai, S. K., & Bussmann, R. W., (2006). Ethnomedicine in Himalaya: A case study from Dolpa, Humla, Jumla and Mustang districts of Nepal. *Journal of Ethnobiology and Ethnomedicine*, *2*, 1–6. https://doi.org/10.1186/1746-4269-2-27.
Lambertini, E., Piva, R., Khan, M. T. H., Lampronti, I., Bianchi, N., Borgatti, M., & Gambari, R., (2004). Effects of extracts from Bangladeshi medicinal plants on *in vitro* proliferation of human breast cancer cell lines and expression of estrogen receptor alpha gene. *International Journal of Oncology, 24*, 419–423.
Laupattarakasem, P., Houghton, P. J., Hoult, J. R., & Itharat, A., (2003). An evaluation of the activity related to inflammation of four plants used in Thailand to treat arthritis. *Journal of Ethnopharmacology, 85*, 207–215.
Mao, A. A., (2002). *Oroxylum indicum* vent. – a potential anticancer medicinal plant. *Indian Journal of Traditional Knowledge*, *1*(1), 17–21.
Na Thalang, V., Nakahara, K., & Trakoontivakorn, G., (2000). Functional evaluation of edible plants in Thailand by using several animal cell lines. *Japan International Research Center for Agricultural Sciences (ITRCAS) Fellowship Program Report.*
Na Thalang, V., Trakoontivakorn, G., & Nakahara, K., (2001). Determination of antioxidant activity of some commonly consumed leafy vegetables in Thailand. *Japan International Research Center for Agricultural Sciences (ITRCAS) Fellowship Program Report*, *9*, 39–46.
Nakahara, K., Onishi-Kameyama, M., Ono, H., Yoshida, M., & Trakoontivakorn, G., (2001). Antimutagenic activity against trp-P-1 of the edible Thai plant, *Oroxylum indicum* vent. *Bioscience, Biotechnology and Biochemistry, 65*, 2358–2360.
Nakahara, K., Trakoontivakorn, G., Alzoreky, N. S., Ono, H., Onishi, Kameyama, M., & Yoshida, M., (2002). Antimutagenicity of some edible Thai plants, and a bioactive carbazole alkaloid, mahanine, isolated from *Micromelum minutum*. *Journal of Agricultural and Food Chemistry, 50*, 4796–4802.
Palasuwan, A., Soogarun, S., Lertlum, T., Pradniwat, P., & Wiwanitkit, V., (2005). Inhibition of Heinz body induction in an invitro model and total antioxidant activity of medicinal Thai plants. *Asian Pacific Journal of Cancer Prevention*, *6*(4), 458–463.

Parle, M., & Bansal, N., (2006). Traditional medicinal formulation, Chyawanprash: A review. *Indian Journal of Traditional Knowledge, 5*(4), 484–488.

Prakash, O., Kumar, A., Kumar, P., & Ajeet, A., (2013). Anticancer potential of plants and natural products: A review. *American Journal of Pharmacological Sciences*, *1*(6), 104–115. https://doi.org/10.12691/ajps-1-6-1.

Raghu, A., George, S., Krishna, R. V., & Sindhu, K., (2013). Bioactive properties of phenolics present in *Oroxylum indicum*: A review. *Journal of Pharmacognosy and Phytochemistry, 2*(3), 23–27. https://doi.org/10.4149/endo.

Rasadah, M. A., Houghton, P. J., Amala, R., & Hoult, J. R. S., (1998). Antimicrobial and anti–inflammatory activity of extracts and constituents of *Oroxylum indicum* vent. *Phytomedica, 5*, 375–381.

Roy, M. K., Nakahara, K., Na, T. V., Trakoontivakorn, G., Takenaka, M., Isobe, S., & Tsushida, T., (2007). Baicalein, a flavonoid extracted from a methanolic extract of *Oroxylum indicum* inhibits proliferation of a cancer cell line *in vitro* via induction of apoptosis. *Pharmazie, 62*(2), 149–153.

Shalini, M., Kalaivani, T., Balaji, S., Gurung, V., Barpande, M., Agarwal, S., & Rajasekaran, C., (2016). Alleviation of 4-nitroquinoline 1-oxide induced oxidative stress by *Oroxylum indicum* (L.) leaf extract in albino Wistar rats. *BMC Complementary and Alternative Medicine*, *16*(1),229. doi: 10.1186/s12906-016-1186-x.

Singh, V. K., & Ali, Z. A., (1998). *Herbal Drugs of Himalaya*. Today & Tomorrow's Printers and Publishers.

Solowey, E., Lichtenstein, M., Sallon, S., Paavilainen, H., Solowey, E., & Lorberboum-Galski, H., (2014). Evaluating medicinal plants for anticancer activity. *Scientific World Journal*, 721402. doi: 10.1155/2014/721402.

Sowjanya, K. S., Swati, M., & Manasa, S. S., (2019). Review on *Oroxylum indicum*. *Journal of Pharmaceutical Sciences and Research, 11*(8), 2905–2909.

Subramanian, S. S., & Nair, A. G. R., (1972). Flavanoids of the leaves of *Oroxylum indicum* and *Pajanelia longifolia*. *Phytochemistry*, *11*, 439.

Suresh, B. K., Hari, B. T., Srinivas, P. V., Hara, K. K., Murthy, U. S. N., & Rao, J. M., (2006). Synthesis and biological evaluation of novel C (7) modified chrysin analogues as antibacterial agents. *Bioorganic and Medicinal Chemistry Letters*, *16*(1), 221–224. https://doi.org/10.1016/j.bmcl.2005.09.009.

Tepsuwan, A., Furihata, C., Rojanapo, W., & Matsushima, T., (1992). Genotoxicity Indicum, and cell proliferative activity of a nitrosated *Oroxylum* mutation, vent fraction in the pyloric mucosa of rat stomach. *Mutation Research, 281*, 55–61.

Turin, M., (2003). Ethnobotanical notes on Thangmi plant names and their medicinal and ritual uses. *Contributions to Nepalese Studies*, *30*(1), 19–52.

Wall, M. E., Wani, M. C., Hughes, T. J., & Taylor, H., (1988). Isolation, plant antimutagenic agents, 1. General bioassay and procedures. *Journal of Natural Products, 51*, 866–873.

Yoganarasimhan, S. N., (1996). *Medicinal Plants of India*. Karnataka Interline.

Zaveri, M., Gohil, P., & Jain, S., (2006). Immunostimulant activity of n-butanol fraction of root bark of *Oroxylum indicum*, vent. *Journal of Immunotoxicology, 3*(2), 83–99. https://doi.org/10.1080/15476910600725942.

Zhang, L., Wang, Y., Liu, H., Yu, L., Liu, X., Chen, L., & Zhang, N., (2013). Developing hydroxypropyl methylcellulose/hydroxypropyl starch blends for use as capsule materials. *Carbohydrate Polymers, 98*(1), 73–79. https://doi.org/10.1016/j.carbpol.2013.05.070.

CHAPTER 16

Livelihood Improvement Through Turmeric in Eastern Himalayan Foothills of West Bengal, India

SAYAN CHOWDHURY, KUMARESH PAL, ROSALIN LAISHRAM, SRINJOY GHOSH, DIPA KUNDU, TATA RAGHU, LITAN DAS, PRITHUSAYAK MONDAL, GOUTAM KUMAR PANDIT, NANDITA SAHANA, and SOMNATH MANDAL

Uttar Banga Krishi Viswavidyalaya, Pundibari, Cooch Behar, West Bengal, India

ABSTRACT

Turmeric is a multipurpose spice crop of the Indian subcontinent from the *Zingiberaceae* family. The golden-colored rhizome of the crop is loaded with a number of functional bioactive compounds. Apart from curcuminoids, the major bioactive compound, several other medicinally important compounds have been reported in *Curcuma longa* rhizome powder and in essential oil (EO) extracted from the rhizome. Beside the domesticated *Curcuma longa*, there are around 28 well-characterized wild turmeric species which are also believed to have several medicinal properties. Some of these wild species are rich in nutritional quality, whereas some possess a natural aroma which has important applications in food and condiment industries. The essential oils (EOs) of these species with proven insect repellent activity have been used as bioinsecticides against different storage pests as well as insect vectors for field application. In recent studies, the turmeric rhizome extracts have shown antifungal and antibacterial properties against both animal and plant pathogenic organisms. This multi-benefit crop is grown widely in the Eastern Himalayan (Terai) of West Bengal, India. The crop has a very low cost of

Bioprospecting Ethnomedicinal Plant Resources: Sustainable Utilization and Restoration.
Gopal Shukla, Jahangeer A. Bhat, A. P. Das, and Sumit Chakravarty (Eds.)

cultivation and is suitable for intercropping. The crop has immense potential in organic farming and can be introduced in indigenous community belts, forest-surrounded uplands, large plantation crops, and orchards as an intercrop. The availability of good cultivars and true-to-type planting material remains a challenge for turmeric cultivation. Parallelly, the turmeric-based industry is relatively scanty and unorganized in this part of the state. Altogether, turmeric is an ideal choice as a low-input cost and high-impact industrial crop for socio-economically poor farming communities of this region.

16.1 INTRODUCTION

Spices are the plant products with EOs and other aromatic compounds which are widely used as food additives to supplement aroma and flavor in the foods. There are different types of spices, some of which are used only as food additives while, some are used as medicines. India blessed with the diverse agro-climatic conditions grows 63 different spice crops at various parts of the country. *Spices have very high demand* both in the domestic and international market and therefore have important contribution in the agricultural market of the country. One of the major spices grown in various parts of India is turmeric, *Curcuma longa* L., representing the spice family *Zingiberaceae.* Turmeric, a native rhizomatous perennial herb of Southeast Asia was first described by Carolus Linnaeus in the year 1753 in his book "*Species Plantarum*" (Ravindran et al., 2007). The term *Curcuma* is derived from the Arabic word for yellow, "*kurkum*," while *longa* indicates the elongated shape of its rhizome. It is designated as the "Golden spice" due to the color of its rhizome (Ravindran et al., 2007). The *Curcuma* species are widely distributed in the tropical countries of Asia like India, the southern part of China, Papua New Guinea, including northern Australia. Globally, India is the leading producer of turmeric and contributes almost 80% of the total production (Yanai et al., 1998) while fulfilling around 90% of the global need (Olojede et al., 2009). About 92% of the total turmeric production in India is consumed domestically, and the rest is exported (Anon., 2018).

Turmeric is one of the ancient and prominent important spices in the world. *Curcuma longa* L. (syn. *C. domestica* Val.), commonly known as turmeric or *Haldi* (Hindi), *Haridra* (in Sanskrit) or *Halud* (in Bengali) is a predominant cash crop of India, Pakistan, Bangladesh, China, and Sri Lanka. The rhizomes of turmeric are cleaned, boiled, dried, polished, and ground to powder as a processed product. Turmeric has wide applications in everyday culinary practices in household kitchens to the dyeing industry since time immemorial. This multi-use primeval and revered spice of India, popular as *'Indian Saffron,'* offers sacred values and can be used as food additives, making turmeric an ideal choice as a food colorant. Turmeric has been used as an Ayurvedic remedy for many diseases. Recently, turmeric was reported with anticarcinogenic, antibacterial, antifungal, and anti-aging activities and extensively used in conventional Ayurveda as well as the modern drug and cosmetic industry. The health-promoting benefits are associated with the presence of bioactive constituents due to their known positive impact on cell components (Mandal et al., 2018). In the case of turmeric, the medicinal properties are mainly attributed to polyphenolic pigments curcuminoids in its rhizome.

16.2 BIOCHEMICAL COMPOSITION OF CULTIVATED TURMERIC AND ASSOCIATED THERAPEUTIC PROPERTIES

The principal coloring constituent of *Curcuma longa* is curcumin (diferuloyl methane) and two of its closely related compounds namely demethoxycurcumin (*p*-coumaryl feruloyl methane) and bisdemethoxycurcumin (di-*p*-coumaryl methane). Together they are called curcuminoid, present in approximately in a ratio of 3.5:2:1 (Ammon & Wahl, 1991) within the rhizome. The chemical structures of the three components of curcuminoids are enumerated in Figure 16.1.

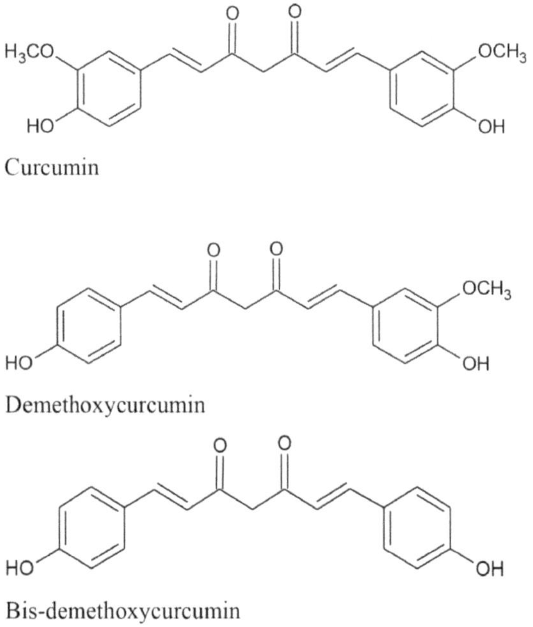

FIGURE 16.1 Structure of individual isomers of curcuminoids.

The most important bioactive component of turmeric is curcuminoids, which make up 2–6% of the spice (Sandur et al., 2007). The commercially available curcuminoids with therapeutic potential contain approximately 70% curcumin, 17% demethoxycurcumin, and 3% bisdemethoxycurcumin (Aggarwal et al., 2007). In addition, the rhizomes of turmeric contain a wide array of secondary metabolites that impart its quality as well as medicinal properties. Earlier research by different groups of scientists indicates the presence of flavonoids (Kim et al., 2011), tannins, and anthocyanins (Niranjan et al., 2003), vitamins such as thiamine, pyridoxine, niacin, and riboflavin (Kubmarawa et al., 2007), EOs that are rich in terpenoids, short-chain alcohols, ketones, and fatty acids (Braga et al., 2003), and considerable amounts of minerals like iron, calcium, zinc, potassium, magnesium, phosphorus, and sodium (Mane et al., 2018) in the turmeric rhizomes. The major bioactive compounds present in *Curcuma longa* rhizome are summarized in Figure 16.2.

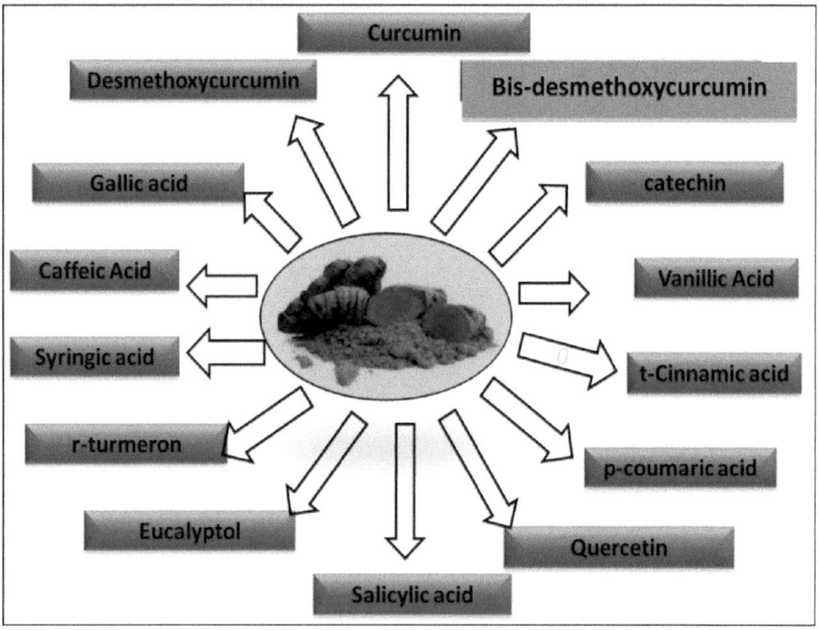

FIGURE 16.2 Important bioactive compounds present in turmeric rhizome.

The volatile constituents of turmeric contain bioactive compounds like turmerone, zingiberene, ar-turmerone, curcumene, cineole, p-cymene, and curlone (Chainani-Wu, 2003). A recent study with 45 cultivated turmeric genotypes of Terai Bengal has proved the color intensity of *Curcuma longa* rhizomes is directly correlated with the medicinal property of the rhizome (Pal et al., 2020). Dark, orange-colored rhizomes with high TRI (turmeric redness index) have been found to have significantly higher curcuminoids content, total phenol content (TPC), total flavonoids content (TFC), high iron content, and antioxidant activity. A popular genotype named TCP 2 (Turmeric Collection Pundibari 2) with dark orange color rhizomes was found excellent with its high antioxidant potential, curcuminoid, total phenol, flavonoid, and iron content.

Turmeric was reported with profound medicinal importance due to its anti-inflammatory, anti-mutagenic, anti-carcinogenic, anti-bacterial, anti-fungal, anti-parasitic, and detox properties possibly due to the presence of curcuminoids and volatile oils in its rhizomes. Curcumin has a global market value as a potential component for therapeutic uses. These constituents of turmeric rhizomes have been reported to promote liver and kidney functions; alleviate biliary disorders and cure diabetic and hepatic disorders (Hermann & Martin, 1991; Osawa et al., 1995; Sugiyama et al., 1996; Nakamura et al., 1998). The ethanolic extract of turmeric has the potential to reduce the ulcer index and acidity of stomach contents. Pre-treatment with turmeric extract lowered the intensity of ulceration. Hypothermic restraint stress of the gastric wall mucus was prevented by treatment of the ethanolic extract of turmeric and decreased the negative impact of lesions induced by necrotizing agents (Rafatullah et al., 1990). Curcumin has a positive effect in maintaining the level of cholesterol and lipid peroxides. Oral administration of 500 mg/day curcumin for a week significantly increased HDL (high-density lipoprotein) cholesterol (29%) while a 33% decrease in the level of serum lipid peroxides and 12% in total serum cholesterol (Soni et al., 1992).

Curcumin is recognized as an epigenetic modulator, which functions as a promising chemo-preventive agent by inhibiting the enzyme cytochrome P450 and thereby preventing chemically induced toxicity (Commandeur & Vermeulen, 1996; Kaypee et al., 2015). The anticarcinogenic property exhibited by curcuminoids is due to their oxygen radical scavenging activity (Soudamini & Kuttan, 1989). The regulation of programmed cell death pathway by the bioactive compounds of turmeric plays an important role in its anticarcinogenic effect (Chen & Huang, 1998). Turmeric influences apoptosis positively and inhibits cell cycle progression, which stops the spread of cancerous cells and induces death of the affected cells. In previous studies, turmeric extracts have been shown effective in arresting cancerous cell growth in rat aortic smooth muscle cells (Gautam et al., 1998). *Curcuma longa* extract and its EO prevent the growth of different bacteria, parasites, and pathogenic fungi. A study of chicks affected with the fecal parasite *Eimeria maxima* revealed that diets supplemented with turmeric resulted in reduction of small intestinal lesion scores and increased weight gain (Allen et al., 1998). Turmeric may function in medication of boils, bruises, blistering, ulcers, eczema, parasitic infections, hemorrhages, and skin diseases (Anon., 1999). The protective characteristics of turmeric shown on the cardiovascular system include decreasing cholesterol and triglyceride levels, lowering susceptibility of LDL (low density lipoprotein) to lipid peroxidation (LPO) and preventing platelet aggregation in incorporation to lower plasma cholesterol and triglyceride levels (Ramirez et al., 1999). The multipurpose role of curcumin in the human body is summarized in Figure 16.3.

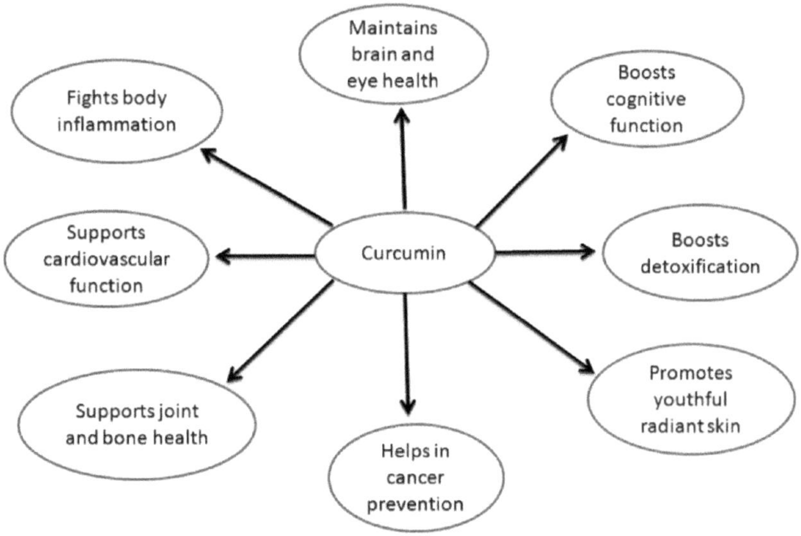

FIGURE 16.3 Outline of major functions of curcumin in human body.

16.3 TURMERIC AS PLANT PROTECTANTS

Combined research conducted in different parts of the world has presented conclusive evidence proving that turmeric can be utilized as a potential natural pesticide for possible use in crop protection and management (Chahal et al., 2005; Murthy et al., 2015; Chowdhury et al., 2016). Pesticide compositions based on turmeric and its derivatives would not only help in safeguarding plants and plant products from different harmful diseases and pests but also aid organic food production systems, where only a handful of alternative pesticides are available. Rhizome extracts and essential oils (EOs) from cultivated and wild turmeric show a broad-spectrum activity against insect pests and plant pathogenic fungi and have extensive accounts of application in the protection strategies of the crop from stored grains to many agricultural products. Recent progress in this field indicates that specific chemical compounds of EOs interfere with the nervous system of the insects (Capinera et al., 2008; Raje et al., 2015). The binding sites of these volatile compounds in the insects are quite unique and have high dissimilarity with their mammalian counterparts. Thus, most of these EO volatiles are relatively non-toxic to plants, fish, and mammals. By virtue of this, the pesticides based on the EOs qualify as "reduced risk" pesticides. Many EOs are exempted from pesticide registration and already popular as flavoring agents and condiments in foods and beverages. The "special status" of these oils makes them hassle-free and promising raw materials for fast-tracked pesticide production in commercial ventures. The wide availability of the turmeric rhizome for the production of these EOs substantiates these business ventures as more lucrative and sustainable. Consumers use these "green pesticides" against home and garden pests, but they are also effective in large-scale agricultural ventures, particularly for organic food production. Additionally, while resistance development in the insect pests and pathogens against many synthetic pesticides continues to be an issue, it is anticipated that resistance will be difficult to achieve against essential oil-based pesticides due to the complex mixtures of a number of compounds in these EOs (Avanço et al., 2017). Developing countries like India, which have wide biodiversity in turmeric species, may benefit and will have a wider impact on its future IPM programs provided that emphasis is given toward research development of non-toxic, environmentally friendly turmeric-based pesticides.

16.3.1 ANTIBACTERIAL ACTIVITY

Although turmeric has been known to show strong antibacterial activity against many animal diseases causing bacteria, but much work has not been conducted against plant pathogenic bacteria. Very few studies of turmeric rhizome extracts have been reported to enhance the average zone of inhibition of bacterial wilt causing strains of *Ralstonia solanacearum* (Murthy et al., 2015).

16.3.2 STORAGE GRAIN PEST

Turmeric oil has been reported to provide protection against red flour beetle, i.e., *Tribolium castaneum* in case of wheat (Chahal et al., 2005). Mortality of adult *Tribolium castaneum* was found directly proportional to increasing concentration of turmeric leaf extracts with organic solvents which reduced the larval, pupal, and adult emergence along with weight loss of the insect (Chander et al., 1994). Compounds extracted from turmeric powder like ar-turmerone (dehydroturmerone, a major constituent of turmeric rhizome powder oil) exhibited a strong repulsive action against *Tribolium castaneum*(Su et al., 1982). In some south Asian countries, food grains were traditionally stored by mixing with 2% of turmeric powder (Golob &Webley, 1980; Saxena et al., 1988).

16.3.3 INSECTICIDE

Turmeric plant oil has been proven to be very useful as an insecticide. Leaves and upper parts of the turmeric plant, on hydro distillation, yield oil rich in 2-phellandrene (70%) that has proven potential to inhibit the growth and development of the diamond back moth (*Plutella xylostella*) even at a 1% concentration (Govindaraddi, 2005). Turmeric extract also exhibited feeding inhibition and mortality of Bihar hairy caterpillar (*Spilosoma oblique*; Chowdhury et al., 2016) in third instar stage of larvae. Effectiveness of extracts prepared from rhizomes against rice weevil (*Sitophilus oryzae*) has shown potential to control the pest in considerably low concentration (Rani, 2017). The turmeric leaf oil is also having ovicidal and nymphicidal activity against *Dysdercus koenigii* (red cotton strainer). Turmeric oil was tested against *Rhyzopertha dominica* F., lesser grain borer of rice. The adult members of the insect were found to be most susceptible to fumigation of *C. longa* leaf oil at 0.1% and 0.05% concentration (Gangwar & Tiwari, 2017) was found to be the most effective against adult insects. Fly ash in combination with turmeric dust (10%) indicated promising activity against rice pest – *Cnaphalocrosis medinalis*, some pests of eggplant namely – *Epilachna vigintioctopunctata*, *Aphis gossypii*, *Urentius hystricellus*, and *Coccidohystrix insolitus*, and against several pests of okra namely *Spodoptera litura*, *Tetranychus neocaledonicus*, *Amrascade vastans*, *Dysdercus cingulatus*, *Anomis flava*, *Oxycarenus hyalinipennis*, and *Earias vittella* showing up to 80% mortality depending on species (Sankari & Narayanasamy, 2007).

16.3.4 FUNGICIDE

Curcuma longa is blessed with extraordinary antifungal activity, supported by research findings. *In vitro* study was conducted to test mycelial growth inhibition effect of Plant extracts on *Macrophomina phaseolina* causing stem rot of jute. At 5,000 ppm, turmeric oil recorded 70% mycelial growth inhibition of the fungus (Gotyal et al., 2016). Antifungal property of Curcuma longa was determined against pathogens like *Fusarium graminearum, Fusarium chlamydosporum, Alternaria alternata, Fusarium tricinctum, Sclerotinia sclerotiorum, Botrytis cinerea, Fusarium culmorum, Rhizopus oryzae, Cladosporium cladosporioides, Fusarium oxysporum*, and *Colletotrichum higginsianum*, and EC50 values were determined (Chen et al., 2018). Differential protein expression of *F. graminearum* was also compared after treatment with turmeric extracts and pure compounds of the constituents of turmeric extracts, and at least 46 candidate proteins were identified. These proteins were involved in basic metabolic pathways like respiration and RNA biogenesis. The antifungal effect was manifested by cell membrane disruption and inhibition of key enzymes of ergosterol synthesis (Chen et al., 2018).

16.4 POTENTIAL OF WILD CURCUMA

Apart from the cultivated Curcuma longa, several wild species of turmeric have been identified, characterized, and analyzed for their biochemical composition. The matured plant of some of these wild turmeric species along with their rhizome is presented in Plate 16.1.

PLATE 16.1 Some unique and important wild species of *Curcuma* and their rhizome.

Among the wild species, *Curcuma amada* has been found to have very good medicinal properties. Amada (*Curcuma amada* Roxb.) has a mango-like smell and flavor and is commonly known as mango ginger (Policegoudra et al., 2011). Its use in chutney and pickle preparations is well known. Ayurveda and Unani systems of treatment use mango ginger as an appetizer, diuretic, emollient, expectorant, antipyretic, aphrodisiac, and laxative, and to cure asthma, hiccough, biliousness, skin diseases, restlessness, and injuries. The major chemical components in mango ginger include starch, phenols, curcuminoids, although in very small quantities in comparison to the cultivated turmeric. Mango ginger is abundant with terpenoids, a class of compounds like difurocumenonol, amadannulen, and amadaldehyde (Ravindran et al., 2004).

Curcuma caesia Roxb. (Black turmeric) is a perennial herb with bluish black colored rhizomes and is well known as '*kali haldi*' in different parts of India. This species is found in forests of Madhya Pradesh and in north-eastern India. *Kali haldi* is considered very pious due to its unusual color of the rhizome and used in religious occasions. Apart from its religious use by different indigenous communities in northeastern India, it is also used as an anti-aging tonic. It is a recognized medicinal herb with a range of medicinal properties such as anti-fungal, anti-asthmatic, and anti-depressant (Devi et al., 2015). Black turmeric extracts also possess anti-convulsant and anti-helminthic activity along with anti-fungal and anti-bacterial activities (https://vikaspedia.in/agriculture/crop-production/package-of-practices/medicinal-and-aromatic-plants/curcuma-caesia). *C. Caesia* has crucial volatiles in its EOs such as camphor, ar-turmerone, (Z) 1,8-cineole, elemene, borneol, bornyl acetate, curcumene, ocemene, and ar-curcumene. Effective *in vitro* micro-propagation method for *C. caesia* using rhizome bud as explants was also characterized for this wild turmeric species (Chowdhury et al., 2020).

Curcuma zedoaria on the other hand is curcuma species with starch rich rhizome and commonly known as '*Shoti.*' The dried and powdered rhizomes of this species are indigenously used as baby food (Rahman, 2012). The amylose contents of the *Shoti* starches are similar to potato starch (Jyothi et al., 2003). These rhizomes with enriched composition of carbohydrates have good potential as a diet for infants, chicks, and domestic animals. The *Shoti* starch can be further enriched by mixing with cow milk, vitamin A and D as a low-cost (LC) dietary supplement for infants. Furthermore, the rhizomes are known for their antioxidant activities and are already used in food coloring and pharmaceutical industries (Mishra et al., 2018).

16.5 PROBLEMS AND CHALLENGES IN TURMERIC CULTIVATION

Turmeric is propagated exclusively through rhizomes. About 2,000–2,500 kg seed rhizomes each weighing around 20–25 grams ('setts') are required for planting one hectare of land. 'Mother rhizomes' or 'primary tubers' (the swollen underground part at the starting point of aerial shoots) as well as larger side rhizomes (3–6 cm in length with two to three number of eyes) are used as planting materials. Conventionally, it is believed that bigger-sized rhizomes are better planting material for giving higher yields. For sowing, both the mother rhizomes and fingers can be used. The finger rhizomes are split into 4-5 cm long pieces, after which the mother rhizomes are planted unchanged or cut into two pieces; each having at least one sound bud. Around 5–10 folds multiplication by the traditional propagation method can be achieved annually, which is not profitable. In addition, the conventional method of propagation suffers from three major disadvantages:

- 30–40% of the yield has to be kept for next year's crop raising, which reduces the income of the farmers;
- Transmission of soil borne diseases imposes serious restriction on the export quality of the harvest;
- The seed rhizomes need proper preservation to prevent rhizome rot disease caused by bacteria and fungi, and storage insect attacks.

Turmeric crop improvement by breeding approaches is difficult due to rare flowering and difficulty of seed set formation. The only method available for the improvement of the crop is through selection of desirable genotypes from natural population or generation of new genotype by somaclonal variation through tissue culture approaches. The desirable criteria for the improvement of turmeric are betterment on yield, curcumin content, disease and insect-pest resistance, and better adaptability. Nowadays, *in vitro* cell and tissue culture methodology is widely used to ensure the survival of endangered plant species and rapid mass propagation for large-scale cultivation. Various protocols are available for micropropagation of turmeric varieties. These micropropagation techniques for the rapid clonal multiplication of turmeric species are widely used for conservation of genotypes and production of microrhizomes *in vitro*. Tissue culture-generated microrhizomes have many advantages as compared to traditional cultivation systems. It helps to save a large number of rhizomes from the previous harvest which are used to serve as seed material for the next cultivation. The microrhizomes generated through tissue culture are disease-free, so the risk of crop loss due to soil and seed-borne pathogens is less. The unavailability of elite variety seed material is an obstruction for quality cultivation and high economic output. Tissue culture is a perfect solution to this problem as many plantlets and microrhizomes can be produced from a small number of explants. Hence, the establishment of micropropagation techniques for elite and region-specific genotypes can solve the problem of scarcity of quality planting materials (QPMs) in the case of turmeric (Pal et al., 2020).

Another major technological intervention for promotion of scientific turmeric cultivation can be the introduction of an improvised single bud technology (SBT). In SBT, only 4–5 grams of rhizome with a single bud is recommended as planting material. Initially, the turmeric plantlets are grown in pots for 45 days and then fully grown plantlets with 3 to 4 leaves are transferred to the fields. The seed rate is much lower with this method in contrast to the higher seed cost of primary rhizome. In addition, it can reduce the field time of the crop, as the initial emergence of the plants is done off the field, thereby shortening the crop cycle and offering an opportunity to introduce another short duration crop to the cropping pattern and generating additional income for the farmers. Originally adopted from TNAU, Coimbatore, this technology is being modified for suitability to the northern Bengal region by UBKV, Pundibari. A stepwise diagram of the SBT method and its comparative advantage over the traditional method is presented in Figure 16.4.

Parameters	Traditional	SBT
Seed Requirement	2.5 ton per ha	700 kg per ha
Crop establishment	80%	98-100%
Rhizome maturity	8-9 month	6-7 months

FIGURE 16.4 Improvised SBT method (stepwise) and its comparative advantage.

16.6 TURMERIC CULTIVATION IN SUB-HIMALAYAN REGION

The Eastern sub-Himalayan plains of northern Bengal have the unique positional advantage of being in the juncture of three major phyto-geographical regions, namely the northeast, the eastern peninsula, and the eastern

Himalaya. This makes the region very rich in diversified crop species. The plain is surrounded by forests and tea gardens from all sides and consists of three districts namely Jalpaiguri, Alipurduar, and Cooch Behar. With the disadvantage of being the most distant districts from the state capital Kolkata, the region is fighting against low agricultural productivity and industrial backwardness and consequently affected with high unemployment and rural poverty. The major livelihood activity of the region is traditional farming but subsistence. In particular, the forest adjoining villages deserve special attention as infiltration of wild animals imposes a serious challenge for the cultivation of rice, wheat, potato, or any other traditional crops. Also, these crops demand heavy inputs like fertilizer and pesticides which the farmers can't afford. Turmeric is an ideal crop to revitalize the economy of this region as the terai agro-climate and soil conditions favor its cultivation.

Turmeric has wider adaptability and provides an alternative choice to farmers of this region. Turmeric can be grown even in marginal and degraded soils, barren lands, homestead gardens, or even any unused upland. Turmeric is a shade-loving crop, thus it can be grown as an intercrop with areca nut and other woody perennials. Participation of women farmers can be encouraged for turmeric cultivation and processing; eventually enhancing the total family income. Careful selection of genotypes with optimum allocation of resources available in the farming system improves the dynamic nature of competition with cost-effectiveness and efficiency of farm operations. As a low-input crop, it can be an attractive model crop for the tribal-dominant villages of these areas for earning hard cash as well as effective utilization of available cultivable land. Organic farming can be another viable alternative for turmeric due to high demand and befitting agro-climatic suitability and socio-economic characteristics. On average, organic turmeric powder fetches 100–200 Indian rupees higher price per kg than conventional turmeric powder. The nearby presence of environmentally sustainable countries like Bhutan and Nepal, as well as the 'Organic state' Sikkim, improves the chances of export potential of organic turmeric or *'Herbal Haldi'* grown in this region. The sole agricultural university of northern Bengal, UBKV, Pundibari, has a devoted research wing for the improvement of turmeric. The local and introduced genotypes of turmeric collected from the eastern and the northeastern geographical locations are maintained under the aegis of AICRP on Spices at UBKV. With the growing market for organically grown agri-products in states like Sikkim as well as neighboring countries like Bhutan and Nepal, it is high time to motivate farmers of terrain plains towards an ecosystem approach to turmeric production as well as crop protection. In this direction,

UBKV can act as a nodal center to undertake the research activities and as well as transmission of improved technologies to the stakeholders.

One of the genotypes of UBKV, Pundibari Center, TCP 2 (Turmeric Pundibari Collection 2) locally known as Suranjana, was found to provide good yield (29 tons per ha). TCP 2 has been recommended for intercropping with areca nut and other woody perennials (Plate 16.2).

TCP2 (Turmeric collection Pundibari)

TCP2 Intercropping in mango orchard

TCP2 Intercropping in coconut orchard

PLATE 16.2 Elite genotype (TCP 2) of terrain plain of West Bengal can be grown with intercropping.

The curcuminoid content of TCP 2/Suranjana is quite high (5.2%). Apart from good curcuminoid content, TCP 2 has good iron content. It is also rich in other bioactive phenols and flavonoids like *p*-coumaric acid, catechin, sinapic acid, and quercetin. The EO of TCP 2 possesses 33 volatile compounds like high ar-turmeron, curlone, eucalyptol, and chemigran. A rapid and improved tissue culture protocol has been developed for superior TCP 2 genotype using a single bud from rhizome sprout as explants (Pal et al., 2020). Preservation and reproduction by multiplication of the true-to-the-type plantlets of TCP 2 and popularization of the genotype among local farmers for large-scale farming of the genotype may illustrate different approaches to guarantee economic security of the otherwise resource-deprived farming communities of the Terai region of West Bengal, India. Moreover, this university has also developed Uttar Rangini (TCP 129) variety having leaf spot and leaf blotch disease resistance property. In addition, UBKV has recently released high-yielding (32 tons per ha) Uttar Rupanjana (TCP 64) variety suitable for cultivation in rainfed and irrigated conditions of the region. All three varieties developed by UBKV are suited for fallow lands, intercropping, multi-storyed cropping in arecanut orchard, and newly established mango orchards, thereby providing a solution to the land shortage problem for turmeric cultivation.

16.7 VALUE ADDITION

The value-added products of turmeric are dehydrated turmeric powder, EO, dye for cosmetic industries, and oleoresin. EOs are very frequently used for the preparation of detergents, soaps, cosmetics, pharmaceuticals, perfumes, confectionery food products, soft drinks, and insecticides. With the increasing affinity toward bio-products, the importance of EOs is increasing at a very fast pace throughout the globe. EOs are environment-friendly. For example, EO from turmeric is a powerful antiseptic and anti-inflammatory agent that revitalizes the skin. Along with the cultivated *Curcuma longa*, different wild species of turmeric like *curcuma amada*, *curcuma zedoaria*, and *curcuma caesia* produce a good yield for EOs. These wild curcuma species fetch a good market price and can be used as a potential source of both EO and natural dye. Oleoresins are the concentrated product obtained from dry spices. The extraction is assisted using an organic solvent or solvent mixture like acetone, alcohol, and hexane. The flavor of the concentrated fresh spice is completely retained here. It also retains the complete volatile as well as non-volatile compound profile present in the spice. The content of residual solvent in the oleoresin should not exceed 30 ppm. Oleoresin can be promoted for possible use in lipid-based food products like processed meat, fish cheese, baked food, and vegetables. It has greater heat stability than normal EO.

16.8 TURMERIC-BASED SKILL AND ENTREPRENEURSHIP DEVELOPMENT

Entrepreneurship development acts as a catalyst for accelerating the economic development of any region. Change in the socio-economic status of rural youth of the sub-Himalayan region is the need of the hour, and it is possible through skill development by various focused training programs. Three types of business models can be developed based on turmeric cultivation, namely, turmeric powder production, EO production, and bioactive pigment, i.e., curcumin production. The comparative financial modalities of the three turmeric-based business ventures are presented in Table 16.1.

TABLE 16.1 Comparative Financial Modalities of the Three Turmeric-based Business Venture

Product	Recommended for	Fixed Assets	B:C Ratio	Recovery of Fixed Cost
Powder	SHG/FPO	10.0 lakhs INR	1.49	6.5 months
Essential oil	Woman farmer	1.0 lakh INR	1.90	10 months
Curcumin	FPC/entrepreneur	50.0 lakhs INR	1.90	27 months

Since turmeric is an industrial crop, turmeric cultivation could be integrated with entrepreneurship development by a clustered approach. Farmers club, SHGs (self-help groups), FPOs (farmer producer company), cooperatives, entrepreneurs have a pivotal role to play in this business model. Central and State governments, financial institutions (NABARD/Bank), academic institutions (universities/KVKs) will act as facilitating agencies by promoting backward and forward linkage with the stakeholders. Group farming, cooperative farming, contract farming, marketing, branding, certification (GACP (good agriculture collection practices)/FSSAI), online trading (eNAM) can be explored for socio-economic improvement of the region. It is possible when farmers and rural youths of this region are empowered with the required knowledge, skill, and entrepreneurship and are linked to value addition of economic activities in the agri-business domain. The government, semi-government or NGOs can create awareness among the farmers regarding the scope of entrepreneurship development. A proposed model on the development of value chain dependent turmeric-based economic activities is summarized in Figure 16.5.

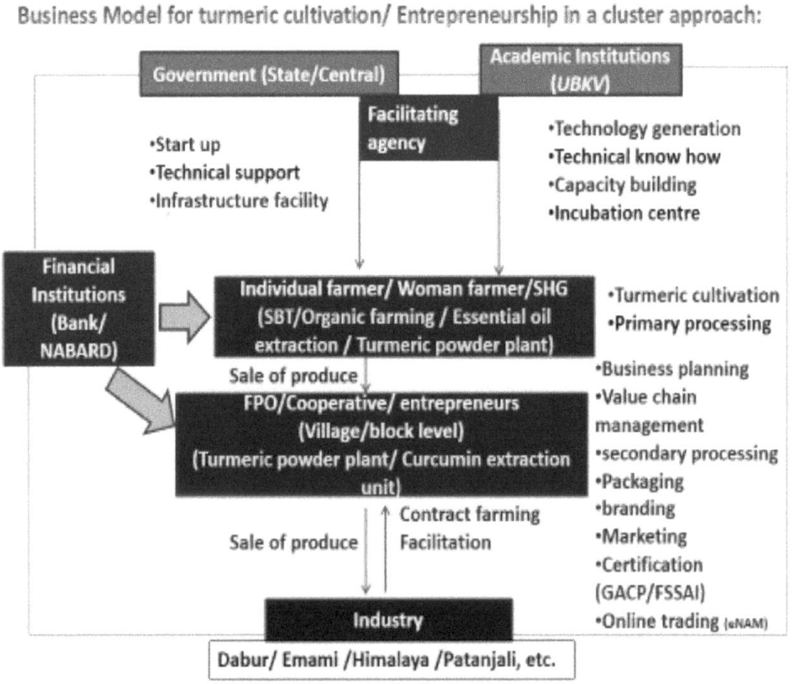

FIGURE 16.5 A proposed business model integrating turmeric cultivation with entrepreneurship development.

16.9 CONCLUSION

People of the sub-Himalayan or Terai region of West Bengal are simple but hardworking. The livelihood of the local people is predominantly agrarian and subsistence. Farmers are adaptive to new technologies in agriculture, and acceptance of TCP 2 or Suranjana among the farmers replacing of age-old local turmeric cultivars that are grown in these regions. Availability of good quality disease-free, true to the type seed is the key bottleneck for wide cultivation and popularization of TCP 2 turmeric. Technological intervention by introducing elite genotype with the development of a new package of practices for cultivation, processing, packaging, and marketing of turmeric will boost the farm income in rural areas. The presence of areca nut, jackfruit plantations in every household indicates the possibility of multi-tier cropping using turmeric. Export potential turmeric cultivation to nearby countries like Bhutan can be achieved by introducing organic cultivation of turmeric in this region. Proper training and demonstration of technologies may encourage relatively young groups to pursue turmeric-based entrepreneurship development. Terai plain is very rich in its natural biodiversity. Wild species like *Curcuma amada*, *Curcuma caesia*, and *Curcuma zedoaria* are present bountifully in the region. Conservation of the species as well as wide cultivation can open a new avenue towards the herbal industries for making condiments, cosmetics, and skincare products. Farmers with very low land holding may improve their quality of life if proper training can be given to them by the generation of extra income from these activities. Simultaneous implementation of all these components will surely improve the socio-economic status of the rural people of the area and help to achieve a better quality of life.

ACKNOWLEDGMENT

The authors sincerely acknowledge financial support received under the project "STI Hub as a Platform for Socioeconomic Development of targeted Schedule Caste Communities of Terai Dooars Region of West Bengal" sanctioned vide Ref. no. DST/SEED/SCSP/STI/2019/116 by SEED Division, Department of Science and Technology (DST), Government of India.

KEYWORDS

- **ayurveda**
- **Curcuma spp.**
- **curcumin**
- **entrepreneurship**
- **livelihood**
- **micropropagation**
- **turmeric cultivation**

REFERENCES

Aggarwal, B. B., Sundaram, C., Malani, N., & Ichikawa, H., (2007). Curcumin: The Indian solid gold. *Adv. Exp. Med. Biol., 595*, 1–75.

Allen, P. C., Danforth, H. D., & Augustine, P. C., (1998). Dietary modulation of avian coccidiosis. *Int. J. Parasitol., 28*, 1131–1140.

Ammon, H. P., & Wahl, M. A., (1991). Pharmacology of *Curcuma longa. Planta Med., 57*, 1–7.

Anonymous, (1999). *Rhizoma Curcumae Longae* (Vol 1). WHO monographs on selected medicinal plants. World Health Organization.

Anonymous., (2018). *AgriXchange Profile of Turmeric*. Agricultural and Processed Food Products Export Development Authority (APEDA), New Delhi, India. Available at http://agriexchange.apeda.gov.in (accessed on 18 December 2023).

Avanço, G. B., Ferreira, F. D., Bomfim, N. S., De Souza, P. A., Santos, R., Peralta, R. M., Brugnari, T., et al., (2017). *Curcuma longa* L. essential oil composition, antioxidant effect, and effect on *Fusarium verticillioides* and *fumonis* in production, *Food Control*, *73*(B), 806–813. https://doi.org/10.1016/j.foodcont.2016.09.032.

Braga, M. E. M., Leal, P. F., Carvalho, J. E., & Meireles, M. A. A., (2003). Comparison of yield, composition, and antioxidant activity of turmeric (*Curcuma longa* L.) extracts obtained using various techniques. *J. Agric. Food. Chem., 51*, 6604–6611.

Capinera, J. L., Hoy, M. A., Paré, P. W., Farag, M. A., Trumble, J. T., Isman, M. B., & Stocks, I., (2008). Neurological effects of insecticides and the insect nervous system. *Encyclopedia of Entomology*, 2596–2605. doi: 10.1007/978-1-4020-6359-6_2202.

Chahal, K. K., Arora, M., Joia, B. S., & Chhabra, B. R., (2005). Bio-efficacy of turmeric oil against *Tribolium castaneum* (Herbst) under laboratory conditions. In: *Proc. 1st Congress on Insect Science*. PAU Ludhiana.

Chainani-Wu, N., (2003). Safety and anti-inflammatory activity of curcumin: A component of turmeric (*Curcuma longa*). *J. Altern. Comple. Med.*, *9*, 161–168.

Chander, H., Nagender, A., Ahuja, D. K., & Berry, S. K., (1994). Laboratory evaluation of plant extracts as repellents to the rust red flour beetle, *Tribolium castaneum* (Herbst) on jute fabric. *Int. Pest Control, 41*, 18–20.

Chen, C., Long, L., Zhang, F., Chen, Q., Chen, C., Yu, X., Liu, Q., et al., (2018). Antifungal activity, main active components, and mechanism of *Curcuma longa* extract against *Fusarium graminearum*. *PLoS One, 13*(3), e0194284.

Chen, H. W., & Huang, H. C., (1998). Effect of curcumin on cell cycle progression and apoptosis in vascular smooth cells. *Br J. Pharmacol., 124*, 1029–1240.

Chowdhury, H., Gotyal, B. S., Selvaraj, K., & Sarkar, S. K., (2016). Bio-efficacy of plant extracts on stem rot, *Macrophomina phaseolina* (Tassi) Goid and Bihar hairy caterpillar, *Spilosoma oblique* Walker in jute crop. *Journal of Applied and Natural Science, 8*(1), 191–195.

Chowdhury, S., Pal, K., Chakraborty, M., Chakraborty, S., Mandal, S., Pandit, G. K., Maitra, S., & Sahana, N., (2020). Conservation and *in vitro* propagation of an endangered wild turmeric (*Curcuma CaesiaRoxb.*) species from sub-Himalayan Terai region of West Bengal. *Int. J. Curr. Microbiol. App. Sci., 9*(2), 2132–2140.

Commandeur, J. N., & Vermeulen, N. P., (1996). Cytotoxicity and cytoprotectivc activities of natural compounds. The case of curcumin. *Xenobiotica, 26*, 667–680.

Devi, H. P., Mazumder, P. B., & Devi, L. P., (2015). Antioxidant and antimutagenic activity of *Curcuma caesia*Roxb. rhizome extracts. *Toxicol. Rep., 2*, 423–428. doi: 10.1016/j.toxrep.2014.12.018.

Gangwar, P., & Tiwari, S. N., (2017). Insecticidal activity of *Curcumalonga* essential oil and its fractions against *Sitophilus oryzae L.* and *Rhyzopertha dominica F.* (*Coleoptera*). *Int. J. Pure App. Biosci., 5*(3), 912–921.

Gautam, S. C., Xu, Y. X., Pindolia, K. R., Janakiraman, N., & Chapman, R. A., (1998). Non-selective inhibition of proliferation of transformed and nontransformed cells by the anticancer agent curcumin (diferuloylmethane). *Biochem. Pharmacol., 55*, 1333–1337.

Golob, P., & Webley, D. J., (1980). *The Use of Plants and Minerals as Traditional Protectants of Stored Products*. Tropical Products Institute, London.

Gotyal, B. S., Chowdhury, H., Selvaraj, K., & Sarkar, S. K., (2016). Bio-efficacy of plant extracts on stem rot, *Macrophomina phaseolina* (Tassi) Goid and Bihar hairy caterpillar, *Spilosom aobliqua Walker* in jute crop. *J. Appl. Nat. Sci., 8*(1), 191–195.

Govindaraddi, K., (2005). *Antifeedant and Insecticidal Properties of Essential Oils of Turmeric (Curcuma Longa L.) and Garlic (Allium Sativum L.) Against Diamond Back Moth, PlutellaXylostella (L).* M. Sc. Thesis., CCS Haryana Agricultural University, Hisar.

Hermann, P. T. A., & Martin, A. W., (1991). Pharmacology of *Curcuma longa*. *Planta Med., 57*, 1–7.

Jyothi, A. N., Mooorthy, S. N., & Vimala, B., (2003). Physicochemical and functional properties of starch from two species of *Curcuma*. *International Journal of Food Properties, 6*(1),135–145.

Kaypee, S., Mandal, S., Chatterjee, S., & Kundu, T. K., (2015). Emerging epigenetic therapies: Lysine acetyltransferase inhibitors. In: Gray, S. G., (ed.), *Epigenetic Cancer Therapy* (pp. 471–493). Academic Press, London UK. https://doi.org/10.1016/B978-0-12-800206-3.00021-5.

Kim, S., Yang, M., Lee, O., & Kang, S., (2011). Antioxidant activities of hot water extracts from various spices. *Int. J. Mol. Sci., 12*, 4120–4131.

Kubmarawa, D., Ajoku, G. A., Enwerem, N. M., & Okorie, D. A., (2007). Preliminary phytochemical and anti-microbial screening of 50 medicinal plants from Nigeria. *Afr. J. Biotechnol., 6*, 1690–1696.

Mandal, S., Paul, P. K., & Sahana, N., (2018). Extraction and concentration methods for bio-actives components in fruits and vegetables. In: *Technological Interventions in the Processing of Fruits and Vegetables* (pp. 301–321). https://doi.org/10.1201/9781315205762. Apple Academic Press, Palm Bay, FL, USA.

Mane, R. P., Kshirsagar, R. B., Sawate, A. R., Patil, B. M., & Kale, R. G., (2018). Studies on evaluation of physicochemical and nutritional properties of fresh turmeric rhizome. *J. Pharma. Phytochem., 7*, 2895–2897.

Mishra, J., Bhardwaj, A., & Mishra, K., (2018). *Curcuma sp.: The Nature's Souvenir for High-Altitude Illness: Management of High-Altitude Pathophysiology* (pp. 153–169). Academic Press.

Murthy, N., Soumya, K., & Srinivas, K. C., (2015). Antibacterial activity of *Curcuma longa* (Turmeric) plant extracts against bacterial wilt of tomato caused by *Ralstonia solanacearum*. *International Journal of Science and Research, 4*(1), 2136–2141.

Nakamura, Y., Ohto, Y., Murakami, A., Osawa, T., & Ohigashi, H., (1998). Inhibitory effects of curcumin and tetrahydrocurcuminoids on tumor promoter induced reactive oxygen species generation in leukocytes *in vitro* and *in vivo*. *Jpn. J. Cancer Res., 89*, 361–370.

Niranjan, A., Dhan, P., Tewari, S. K., Pande, A., Pushpangadan, P., & Prakash, D., (2003). Chemistry of *Curcuma* species, cultivated on sodic soil. *J. Med. Aromat. Pl. Sci., 25*, 69–75.

Olojede, A. D., Nwokocha, C. C., Akinpelu, A. O., & Dalyop, T., (2009). Effect of variety, rhizome, and seed bed types on yield of turmeric (*Curcuma longa* L.) under a humid tropical agro-ecology. *Adv. Biol. Res., 3*, 40–42.

Osawa, T., Sugiyama, Y., Inayoshi, M., & Kawakishi, S., (1995). Antioxidative activity of tetrahydrocurcuminoids. *Biosci. Biotech. Biochem.,59*, 1609–1612.

Pal, K., Chowdhury, S., Dutta, S. K., Chakraborty, S., Chakraborty, M., Pandit, G. K., Dutta, S., et al., (2020). Analysis of rhizome color content, bioactive compound profiling and ex-situ conservation of turmeric genotypes (*Curcuma longa L.*) from Sub-Himalayan Terai region of India. *Ind. Crops &Products, 150*, 112401. https://doi.org/10.1016/j.indcrop.2020.112401.

Policegoudra, R. S., Aradhya, S. M., & Singh, L., (2011). Mango ginger (*Curcuma amada* Roxb.)- a promising spice for phytochemicals and biological activities. *J. Biosci., 36*, 739–748.

Rafatullah, S., Tariq, M., Al, Y. M. A., Mossa, J. S., & Ageel, A. M., (1990). Evaluation of turmeric (*Curcuma longa*) for gastric and duodenal antiulcer activity in rats. *J. Ethnopharmacol., 29*, 25–34.

Rahman, S., (2012). *Chemical and Nutrition Evaluation of Curcuma Zedoaria (Shoti) on the Performance of Broiler Production*. M. Sc. Thesis., Bangladesh Agricultural University, Mymensingh.

Raje, K. R., Hughes, G. P., Gondhalekar, A. D., Ginzel, M. D., & Scharf, M. E., (2015). Toxicity of Turmeric extracts to the termite *Reticulitermes flavipes* (*Blattodea: Rhinotermitidae*). *J. Econ. Entomol., 108*(4), 1–7. doi: 10.1093/jee/tov109.

Ramirez, T. M. C., Mesa, M. D., Aguilera, M. C., Quiles, J. L., Baro, L., Ramirez, T. C. L., Martinez, V. E., & Gil, A., (1999). Oral administration of turmeric extract inhibits LDL oxidation and has hypocholesterolemic effects in rabbits with experimental atherosclerosis. *Atherosclerosis, 147*, 371–378.

Rani, A., (2017). Bio-efficacy of extract of turmeric and ginger as potential biopesticides on *Sitophilus oryzae*. *Adv. Sci. Res., 2*(1), 10, 11.

Ravindran, P. N., Babu, K. N., & Sivaranan, K., (2007). *Turmeric: The Genus Curcuma: Medicinal and Aromatic Plants – Industrial Profiles.* A CRC Press publication, Boca Raton, FL.

Ravindran, P. N., Pillai, G. S., & Babu, K. N., (2004). Underutilized herbs and Spices. *Handbook of Herbs and Spices, 2*, 53–103.

Sandur, S. K., Pandey, M. K., Sung, B., Ahn, K. S., Murakami, A., Sethi, G., Limtrakul, P., et al., (2007). Curcumin, demethoxycurcumin, bisdemethoxycurcumin, tetrahydrocurcumin and turmerones differentially regulate anti-inflammatory and antiproliferative responses through a ROS independent mechanism. *Carcinog., 28*, 1765–1773.

Sankari, S. A., & Narayanasamy, P., (2007). Bio-efficacy of fly ash based herbal pesticides against pests of rice and vegetables. *Curr. Sci., 92*, 811–816.

Saxena, R. C., Jilani, G., & Abdul-Kareem, A., (1988). Effects of neem on stored grain insects. In: Jacobson, M., (ed.), *Focus on Phytochemical Pesticides, the Neem Tree* (Vol. 1, pp. 97–11). CRC, Boca Raton, FL.

Soni, K. B., Rajan, A., & Kuttan, R., (1992). Reversal of aflatoxin induced liver damage by turmeric and curcumin. *Cancer Lett., 66*, 115–121.

Soudamini, K. K., & Kuttan, R. J., (1989). Inhibition of chemical carcinogenesis by curcumin. *J. Ethnopharmacol., 27*, 227.

Su, H. C. F., Horvat, R., & Jilani, G., (1982). Isolation, purification, and characterization of insect repellents from *Curcuma longa L. J. Agr. Food Chem., 30*, 290–292.

Sugiyama, Y., Kawakishi, S., & Osawa, T., (1996). Involvement of the beta diketone moiety in the antioxidative mechanism of tetrahydrocurcumin. *Biochem. Pharmacol., 52*, 519–525.

Yanai, T., Kojima, H., & Toyota, A., (1998). Essential oil constituents from Japanese and Indian *Curcuma aromatic* rhizomes. *Planta Med., 64*, 380, 381.

CHAPTER 17

Malaxis acuminata: An Important Ingredient of Ashtavarga

BANDAN THAPA, PUSHKAR SHARMA, and SUMITA PRADHAN

Regional Research Station, Hill Zone, Uttar Banga Krishi Viswavidyalaya, Kalimpong, West Bengal, India

ABSTRACT

Malaxis acuminata, an important ingredient of *Ashtavarga* is a perennial, terrestrial medicinal orchid which has been categorized as an endangered species by the IUCN as its current status in the wild. Ashtavarga is a combination of eight medicinal plants used in Ayurvedic medicine of which *Malaxis acuminata* plays an indispensable role in preparation of important Ayurvedic formulations like *Chyawanprash Malaxis acuminata* has a wide range of healing properties to cure different types of illnesses, from treating male sterility, rheumatism, dysentery, fever, emaciation, and general debility. Apart from its ethnomedicinal use, this species has also been validated for its antioxidant, anti-aging, UV-A blocking, anti-inflammatory, antiproliferative, and antimicrobial activities. This species has been of great interest for conservation and domestication because of its high demand, which suffers from indiscriminate and unsustainable harvesting, habitat degradation, construction of roads, and other human interventions and interferences in the wild, leading to decreased natural populations of this species. There has been a need for immediate remedial measures for the conservation of this medicinal orchid in its natural habitat and to promote cultivation and GACP to ensure a sustainable mode of utilization. *In situ* conservation, field gene banks, and micropropagation, a promising *in vitro* technique, show great potential in the management and conservation of plants in a sustainable way.

Bioprospecting Ethnomedicinal Plant Resources: Sustainable Utilization and Restoration.
Gopal Shukla, Jahangeer A. Bhat, A. P. Das, and Sumit Chakravarty (Eds.)

17.1 INTRODUCTION

The therapeutic value of plants in treating various diseases can be traced back over thousands of years in Indian, Chinese, Greek, and Egyptian systems of folk medicine (Mahesh & Satish, 2008). Overexploitation and illegal trade of these high-value medicinal plants have gained attention both due to their economic and conservation concerns for the past few years. Unmanaged, illicit collection and degradation of natural habitats have put these plants at severe risk of extinction. A Red Data Book of Indian plants has listed many species of Himalayan medicinal plants as endangered. Likewise, many orchid species are also used to treat various ailments as they possess special morphological, structural, and physiological characteristics (Dressler, 1990). They also occupy a wide range of habitats from tropical to alpine areas with over more than 35,000 species and 800 genera in the world and 1,141 species and 166 genera in India (Bhanwra et al., 2006). In Darjeeling Himalaya, 85 genera with 311 species of orchids have been reported (Yonzone et al., 2012). There have been reports of a high degree of threats to many of the orchid species in their natural habitat due to habitat loss, fragmentation, genetic drift, and anthropogenic pressures. The government of India has now imposed a ban on the illegal and commercial exploitation of orchids from the wild (Lohani et al., 2013).

17.2 GENERAL ACCOUNT OF THE FAMILY ORCHIDACEAE

Orchidaceae, a family of monocotyledonous, angiospermic flowering plants and masters of pollination, is arguably the largest family, with some estimates of 30,000 species distributed throughout the world. The family is further divided into five subfamilies: Orchidoideae, Epidendroideae, Cypripedioideae, Vandoideae, Apostasioideae, and Spiranthoideae (Cetzal & Savelli, 2014). The orchids basically have three forms of basic lifestyles: epiphytes, terrestrial, and semi or wholly parasites. In India, it is the second largest family of angiosperms after the family Poaceae, representing about 1,258 taxa represented by 178 genera.

17.3 THE GENUS *MALAXIS*

The *Malaxis* commonly called Adder's mouth was first described by Solander in Swartz's paper in 1788. The genus was divided into subgenera, i.e., *Microstylis* and *Malaxis* by Natall in 1818. In the later studies *Microstylis* was recognized as a synonym of the first one and the preferential genus *Malaxis* (Greuter et al., 2000). The genus *Malaxis* comprises approximately 250 species, which is considered one of the most diverse genera of Orchidaceae (Mabberley, 2008). The taxa of *Malaxis* are herbaceous and represented by subumbellate inflorescence with usually minute green to pale yellow flowers, short gynostemium, anther base below the stigma base, and rhizomatous stems with pseudobulbs usually green in color (Pérez-Escobar & Blanco, 2014). In India, there are approximately 19 species of *Malaxis* (Gogoi et al., 2012), mostly of which grows in the Himalayan region. Darjeeling Himalayas inhabits five species of *Malaxis* viz. *M. acuminata*, *M. muscifera*, *M. khasiana*, *M. ophrydis*, and *M. purpurea* (Yonzone et al., 2012).

17.4 DESCRIPTION OF THE SPECIES

Malaxis acuminata is a perennial, moisture loving, medium-sized, terrestrial orchid of about 30 cm tall consisting of pseudobulbs and fibrous roots at the base. Uma et al. (2015) reported that it arises from swollen protocorms which are often mistaken with a rhizome that is produced during embryogenesis. The plant is erect and has a small stem composed of aerial flowering axis and swollen stem at the base (pseudobulbs), bearing nodes and internodes arising from the base of mother rhizome. Several hairs like adventitious roots arise below the rhizome that support the plant to stand firmly on the soil. Sheathing leaf bases cover the rhizomatous structure and pseudobulbs which appear marginally on the nodes. Leaves are simple, alternate, three to four in number and 5–15 cm in length (Chowdhary & Wadhwa, 1984). The flowers are yellowish green in color, terminal unbranched inflorescence; about 3 mm in diameter (Chinmay et al., 2011; Uma et al., 2015). Flower bracts are 1.5 x 6 mm, narrow oval-shaped and tapering to a point. Sepals are 6 x 2 mm with a flat surface with unequal sides. Petals are 5.5 x 1.2 mm, lip 8–10 mm, erect, shaped like an arrowhead and branched at the apex, lobes acute-obtuse (Reddy et al., 2001). Fruit is a capsule, seeds very minute and powdery, and ovoid in shape. Pseudobulbs are conical in shape, 3–9 cm in length and 1–3 cm in width, fleshy, smooth, shining, greenish, covered with a membranous sheath, and slightly mucilaginous (Chowdhary & Wadhwa, 1984; Balkrishna et al., 2012). The plant flowers from July to August and fruiting takes place from September to October (Chinmay et al., 2011).

17.4.1 MORPHOLOGICAL FEATURES

Field gene bank conservation and basic features of the orchid are shown in Plate 17.1 and their qualitative morphological features, which were taken at RRS (HZ) (Regional Research Station (Hill Zone)), UBKV Kalimpong, are briefly consolidated in Table 17.1 under the NMPB sponsored project for conservation at the field gene bank and standardization of protocols for tissue culture.

a. Field view at RRS, HZ, UBKV, Kalimpong

b. Flowering at field

c. At its natural habitat

d. Rhizome with pseudobulbs

e. Fresh rhizomes

f. Flowering

PLATE 17.1 *Malaxis acuminata.* (a) Field view at RRS, HZ, UBKV, Kalimpong; (b) Flowering at field; (c) At its natural habitat; (d) Rhizome with pseudobulbs; (e) Fresh rhizomes; (f) Flowering.

TABLE 17.1 Morphological Features of Malaxis acuminata

Plant Part	Features
Habit	Erect, perennial.
Stem	Erect, modified into rhizome like structure.
Root	Adventitious arising from base of rhizome (fibrous).
Leaf	Simple, margins undulate, ovate-lanceolate, acute to acuminate, narrowed to sheathing base, petiolate.
Inflorescence	Inflorescence racemose, peduncle ribbed.
Bract	Lanceolate, sub-acute, reflexed.
Flower	Bracteates, zygomorphic, bisexual, complete, epigynous, greenish yellow tinged with purple spot.
Calyx	Sepals 3, dorsal sepal linear-oblong, sub-acute, lateral sepals' oblong, obtuse.
Corolla	Petals 3, petals linear, obtuse, margins recurved, lip, base with auricles, narrowly ovate-sagittate.
Fruit	Capsule
Seed	Microscopic, powdery.
Pseudobulb	Globular to conical.
Rhizome	Elongated and tapering.

17.4.2 DISTRIBUTION, HABITAT, AND ECOLOGY

Malaxis acuminata is found across India, Nepal, Bhutan, Myanmar, and Southeast Asia at an altitude range of 600–3,000 m. Within India, it has been recorded in temperate and subtropical Himalayan region across Himachal Pradesh, Uttarakhand to Sikkim, Darjeeling hills and North-Eastern states at an altitude of 1,200 to 2,600 m (Clarke, 1885; Balkrishna et al., 2012). The distribution of the species has also been reported in Andhra Pradesh (Reddy et al., 2001), Andaman Island, Anaimalai hills and Madhya Pradesh (Polunin & Stainton, 1977; Bose et al., 1980; Balkrishna et al., 2012). In Darjeeling Himalayas *M. acuminata* occurs in forest areas of Takdah, Maneybhangyang, Palmajua, Tonglu, Pokhriabong, Sukhiapokhri, Dhotrey, and Lamaydhura, Dello Hill, Algarah, Lopchu, Birch Hill, Durpin, and Tungsong (Yonzone, 2019). The species grows in colonies in cool, shady, and moist ground and in rock laden with mosses comprises of 5–30 individuals. In Uttarakhand, mixed oak and pine forests with a cool-moist climate were found to be the best environment for their growth (Jalal et al., 2008).

17.5 ETHNOBOTANY AND MEDICINAL IMPORTANCE

According to the WHO, medicinal plants are those plants that possess properties or compounds that can be used for therapeutic purposes or those that synthesize metabolites to produce useful drugs which have a beneficial effect on the human body. Plants produce a diverse group of bioactive molecules, making them a rich source of different medicines (Durga et al., 2009). Likewise, Paryayarathamala used the word Ashtavarga for the first time, providing the herbal benefits (Joshi, 1983). *Ashtavarga* is a combination of eight medicinal plants used in Ayurvedic medicine. The combinations are *Habenaria intermedia*, *Habenaria edgeworthii*, *Malaxis acuminata*, *Malaxis muscifera*, *Polygonatum cirrhifolium*, *Polygonatum verticillatum*, *Fritillaria roylei*, and *Lilium polyphyllum* (Dhyani, 2009). Among these eight species *H. intermedia*, *H. edgeworthii*, *M. acuminata*,and *M. muscifera* belongs to the family Orchidaceae. These eight groups of medicinal plants are useful for healing fractures, fever, diabetic conditions, and lowering body fat (Sharma & Balkrishna, 2005). *M. acuminata* is used in the preparation of various Ayurvedic formulations like *Chyawanprash*, *Astavargachurna*, *Chitraka-Ditaila*, *Vachadi-Taila*, *Mahakalyan-Taila*, *Jivaniya-Ghrita*, *Mahamayura-Ghrita*, *Vajikarma-Taila*, *Brahini-Gutika*, and *Himvana-Agada* (Balkrishna et al., 2012). This orchid species is very useful in treating tuberculosis and also acts as a great aphrodisiac (Chauhan, 1990). It is also used in treating male sterility, weakness of pitta (energy of digestion and metabolism) and vata (energy of movement), internal and external hemorrhages, dysentery fever, emaciation, rheumatism, and general debility (Charkarvarty, 1976; Jadhav, 2008).

17.6 NUTRITIONAL AND PHYTOCHEMICAL COMPOSITION

There has been an increasing demand for the use of phytochemicals because of the harmful side effects of synthetic compounds (Vikram et al., 2014). The pseudobulbs of M. acuminata have been found to contain a variety of pharmacological compounds and essential nutrients (Suyal et al., 2020). Phytochemical analysis recorded nutrients (6.48 ppm Cu, 43 ppm Zn, 35 ppm Mn, 331 ppm Fe, 21,600 ppm K, 9,000 ppm Ca, 2,800 ppm Mg, 198 ppm Al, 26.70 Ba, 55.60 ppm B, 0.30 ppm Mo, 156 ppm Cl), fatty acids, and other chemical constituents which can be used as drugs in a variety of pharmaceutical formulations (Lohani et al., 2013). Gas chromatography-mass spectrometry analysis of pseudobulbs identified the presence of long-chain saturated and unsaturated fatty acids (Rajurkar & Gaikwad, 2014). A total of 11 phenolic compounds were detected in the pseudobulbs indicating that its consumption can meet some of the basic dietary requirements (Giri et al., 2017). GC-MS analysis of leaves and stem has revealed the presence of dietary fatty acids, α-hydroxy acids, phenolic acids, sterols, amino acids, sugars, and glycosides (Bose et al., 2017).

A very little detail regarding the isolation and characterization of phytochemicals of *M. acuminata* is reported. The plant is said to possess various potent pharmaceutical compounds and is recommended as a plant of phyto-pharmaceutical importance (Arora et al., 2017a, b). A large number of alkaloids, glycosides, and flavonoids are believed to be present in *M. acuminata* (Sharma et al., 2011a, b). Sharma et al. (2009) also reported that the quality of the phytochemicals present in *M. acuminata* differed with the place of collection. GC-MS analysis of its dry root extract yielded a total of 67 compounds with different pharmaceutical activity (Raval et al., 2016). Bhatnagar et al. (1970) found β-sitosterol, cetyl alcohol, and five basic compounds like choline and sugars. Essential oil (EO) analysis of pseudobulbs by GC-MS method revealed the presence of caryophyllene (74.89%), eugenol (5.56%), humulene (5.48%), phenol (2,4-bis-(1,1-dimethyl ethyl – 5.57%), caryophyllene oxide (5.33%), 2,5-octadecadienoic acid and methyl ester 3.16% (Arora et al., 2017a, b). Thin layer chromatographic separation identified limonene, eugenol, citronellal, 1,8-cineole, piperitone, and p-cymene (Gupta et al., 1978). Pseudobulbs contain alkaloids, glycosides, flavonoids, β-sitosterol, piperitone, O-methylbatatasin, 1,8-cineole, citronellal, eugenol, glucose, rhamnose, coline, limonene, p-cymene, and ceryl alcohol (Balkrishna et al., 2012). Pseudobulbs were also reported to contain fats, alkaloids, resins, tannins, crude fiber, saponins, phenols, flavonoids, and carbohydrates (Arora et al., 2019). Cetyl alcohol, glucose, rhamnose, and diterpenes have been reported to be the constituents of *M. acuminata* in Ayurvedic Pharmacopoeia of India (2006). Pseudobulbs analyzed through HPLC-ESI-QTOF-MS/MS method yielded 11 compounds including isorhamnetin O-glycoside, bulbophythrin A, gigantol, batatasin III, lusianthrin, 2,3-dimethoxy-9,10-dihydrophenanthrene-4,7-diol, liparic acid C, and 30-O-methylbatatasin which showed high antiproliferative activity against four human cancer cell lines (Singh et al., 2017). Ethanolic and methanolic extracts of the tuber portion showed high antimicrobial activity against *Pseudomonas aeruginosa* and *Staphylococcus aureus* strains (Bharal et al., 2014). The biological activity of bioactive compounds of *Malaxis acuminata* is presented in Table 17.2.

TABLE 17.2 Biological Activity of Bioactive Compounds in *Malaxis acuminata*

Biological Activity	References
Antioxidant, anti-microbial.	Bharal et al. (2014); Raval et al. (2016); Bose et al. (2017); Giri et al. (2017)
Anti-proliferative.	Singh et al. (2017)
Anti-mutagenic.	Raval et al. (2016); Giri et al. (2017)
Photo-protective activity against UV-B and UV-A radiations, anti-aging.	Bose et al. (2017)
Anti-scorbutic, anti-leptic, anti-malarial, anti-convulsant, anti-allergic.	Raval et al. (2016)
Anti-inflammatory.	Sharma et al. (2011b); Bose et al. (2017)
Anti-fungal, neuro-protective.	Arora et al. (2017a, b)
Analgesic	Sharma et al. (2011b)

17.7 PRESENT STATUS

IUCN has categorized *M. acuminata* as endangered as it current status in the wild (Barik et al., 2018). Reports have been made by Ved et al. (2003); and Lohani et al. (2013) regarding the status of *M. acuminata* as rare and vulnerable species in the Himalayan region. Illegal harvesting, habitat degradation, construction of roads, and other human interventions and interferences in the wild have decreased natural populations of this species. It has now been categorized as endangered based on perceptions of changes in species parameters (Ved et al., 2003). The Darjeeling Himalayan region is a repository for a large number of temperate and tropical orchids. When considered from an agro-floricultural point of view, the region is known throughout the world as a botanical paradise. Horticultural interest in orchids has considerably increased in the region because of their global demand. The atmospheric humidity, rainfall, and diurnal rhythmic conditions existing in this region are also suitable for the abundance of orchid species (Yonzone et al., 2012, 2019). Over the years, the region has witnessed indiscriminate collection of orchids from wild habitat along with the destruction of habitat by multiple anthropogenic activities (Yonzone et al., 2014). Declining distribution range with habitat and huge loss of wealth of terrestrial orchids have led to sustainable conservation, universally to check and prevent it from extinction in the near future (Yonzone et al., 2012, 2019). Orchid species being very sensitive to ecological disturbances has even made it critically important to protect them in their natural habitat (Yonzone et al., 2012, 2019).

17.8 THREAT

The threat to the survival of these medicinal orchids arose due to habitat loss and its degradation and fragmentation. One major issue for their survival is these orchids having a special orchid mycorrhizal relation. There is a high dependence of orchid's seed for germination and nutrition with such association. The growth and regeneration process of orchids can be hindered and halted if there is any discrepancy in their association. Other major threats to these orchid species are:

- Growing demand for raw materials of medicinal plants by the pharmaceutical companies and their depleting resource base.
- More number of households depend on these medicinally important plants and orchids for income generation through commercial collection.
- They have a sparse occurrence and very slow growth and propagation rate.
- Lack of knowledge about sustainable harvesting and GACP.
- Habitat destruction due to construction of roads, overgrazing, and landslides.
- Forest fire is identified as the main threat to orchid diversity.
- Depletion of the natural availability of these resources has also led to intentional adulterations by *Ipomoea digitata* during drug preparation using this orchid, reducing its potency because of the high price of the drug and extreme dearth of *M. acuminata* (Chinmay et al., 2011).

Irreversible damage has already been done to the survival of *M. acuminata*, making it a rare, endangered species of the Himalayan region due to indiscriminate wild collection (Shukla & Chaubey, 2008). Wild collection has been banned in Sikkim, Uttarakhand, and other states, but with a well-defined policy for collection from the wild. There is no report of commercial cultivation of this medicinal orchid. Collection from the wild is still the main source of its availability in the market for the pharmaceutical industries. The whole plant is being collected, not even leaving small bulbs for its natural propagation for future growth, jeopardizing its survival in the wild (Puni & Singh, 2007). The species is now enlisted in the CITES Appendix-II for ensuring its conservation (Lange & Schippmann, 1999).

17.9 STRATEGIES FOR CONSERVATION

17.9.1 GOOD AGRICULTURAL COLLECTION AND CULTIVATION PRACTICES

Explicit and uncontrollable collection of numerous orchid species from the wild have led to a significant loss of habitat, resulting in the decline of many highly valued plants over the years (Kala & Sajwan, 2007). It has become very critical to conserve such plant species and provide training and conservation programs that are absolutely necessary for all stakeholders. There are no proper guidelines and sustainable harvesting practices for such medicinal orchids. We need to develop GACP (Good Agricultural and Collection Practices) and provide hands-on training to the inhabitants dwelling near the forest, who depend on their livelihoods on the collection of plants from the wild, so they can add value to the harvest over a long time and aim towards conservation. Scalable cultivation is one of the best strategies for conservation to meet the high demands of the national and international market. Developing GAP (Good Agricultural Practices) for those medicinal orchids and wild species with high commercial demand has become a must and the need of the hour. Cultivation of such medicinal plants and orchids will decrease the pressure on natural resources and wild collection and impede the use of unauthentic substitutes and adulterants, which have resulted in the decline of the standard of drugs of Indian systems of Medicine (Chauhan, 1999).

17.9.2 MICROPROPAGATION

Utilizing conventional approaches alone cannot effectively conserve important medicinal plant species, leading them to be exploited and on the verge of extinction in their natural habitat. Tissue culture, an *in vitro* technique, offers great potential for sustainable management and conservation of plants. Conventional propagation methods are too slow and inadequate in addressing the threat of extinction. Terrestrial orchids face greater difficulty in seed germination compared to epiphytic orchids in their natural environment. Therefore, *in vitro* propagation strategies are the most practical option for rapid orchid propagation. However, there is limited research on tissue culture studies in *M. acuminata.* For further information, please refer to Table 17.3, which provides a summary of *in-vitro* studies conducted on the species.

TABLE 17.3 Tissue Culture Studies on *Malaxis acuminata*

Tissue Culture Study	Explants Type	Optimum Concentration	Major Observations	References
Regeneration	*In-vitro* raised nodal segment.	NAA+BA 3 μM	Initial response time (21.6 days), number of shoots formed (18.3) and number of roots formed (3.3), 18 shoots/shoot bud.	Arenmongla & Deb (2012a)
Germination and regeneration.	Immature seeds.	MS medium + 3% (w/v) sucrose + NAA 4 μM, MS medium + 3%(w/v) sucrose + NAA+BA 3 μM + AC 0.3% w/v.	85% germination after 135 days, PLB'S developed (well rooted plantlets, expanded leaves and distinct pseudobulbs with 18 shoot buds/PLB'S).	Arenmongla & Deb (2012b)
Shoot organogenesis.	Internodal explants.	TDZ @3 mg/l + NAA@ 0.5 mg/l + 0.4 Mm @Spermidine, IBA@ 4 mg/l + AC @1.5 mg/l.	Induction of adventitious shoot bud (96%), 6.1 shoots per explants, 96% rooting.	Cheruvathur et al. (2010)
Regeneration	Pseudobulb segment.	BA 1 mg/l + NAA 1 mg/l + AC 2 g/l.	Early initiation, highest regeneration frequency, proliferation of protocorm-like body (PLB's) and plantlet development.	Kaur & Bhutani (2010)

17.10 CONCLUSIONS

Sustainable management of *M. acuminata* will play a vital role in the conservation of this species. We must look forward to the preservation of this medicinal orchid through strategic planning for utilizing it sustainably and harvesting it considering its survival and reproducing ability. The ungoverned misuse and illicit collection of *M. acuminata* is due to their soaring demand and competition amongst the collectors in the herbal market both nationally and internationally, resulting in a dwindling of their populations in their natural habitat. The scientific communities have to strive for developing proper GACP. There has also been a need for conservation efforts for resource assessment to promote the cultivation and sustainable harvesting from the wild for most of the medicinal orchids, which are on the verge of extinction. Future studies should be focused on evaluating the diversity and variability existing within the germplasm of this species at the morphological and biochemical level, studying different chemotypes of different altitudes and locations. Biotechnological resources should also be utilized for the maintenance and conservation of this orchid species.

ACKNOWLEDGMENTS

The authors thank the Ministry of AYUSH, National Medicinal Plant Board, Government of India, New Delhi (F. No. Z. 18017/187/CSS/R&D/WB-03/2017-18-NMPB-IVA) for funding the survey and study.

KEYWORDS

- **agricultural practice**
- **Ashtavarga**
- **cultivation practices**
- **endangered**
- **ethnobotany**
- **medicinal plants**
- **micropropagation**
- **orchid**

REFERENCES

Arditti, J., & Ernst, R., (1984). Physiology of germinating orchid seeds. In: Arditti, J., (ed.), *Orchid Biology: Reviews and Perspectives III* (pp. 177–222). Cornell University Press, New York.

Arenmongla, T., & Deb, C. R., (2012a). Germination of immature embryos and multiplication of *Malaxis acuminata* D. Don, an endangered therapeutically important orchid, by a symbiotic culture *in vitro*. *Ind. J. Biotechnol.*, *11*, 464–469.

Arenmongla, T., & Deb, C. R., (2012b). Studies on *in vitro* morphogenetic potential of nodal segments of *Malaxis acuminata* D. Don. *Appl. Biol. Res.*, *14*, 156–163.

Arora, M., Kaur, G., Kahlon, P. S., Mahajan, A., &Sembi, J. K., (2017a). Pharmacognostic evaluation and antimicrobial activity of endangered ethnomedicinal plant *Crepidium acuminatum* (D. Don) Szlach. *Pharmacog. J.*, *9*, S56–S63.

Arora, M., Kaur, G., Singh, S., Mahajan, A., &Sembi, J. K., (2019). Quantification of phytochemicals in the pseudobulbs of *Crepidium acuminatum* (D. Don) Szlach-a critically endangered medicinal plant. *Curr. Trend Biotech. Pharm.*, *13*, 366–375.

Arora, M., Singh, S., Mahajan, A., &Sembi, J. K., (2017b). Propagation and phytochemical analysis of *Crepidium acuminatum* (D. Don) Szlach. *IOSR J. Pharm. Biol. Sci.*, *12*, 14–20.

Balkrishna, A., Srivastava, A., Mishra, R. K., Patel, S. P., Vashistha, R. K., Singh, A., Jadon, V., & Saxena, P., (2012). Ashtavarga plants-threatened medicinal herbs of the North-West Himalaya. *Int. J. Med. Arom. Plants*, *2*, 661–676.

Barik, S. K., Rao, B. R. P., Haridasan, K., Adhikari, D., Singh, P. P., & Tiwary, R., (2018). Classifying threatened species of India using IUCN criteria. *Curr. Sci.*, *114*, 588–595.

Bhanwra, R. K., Viz, S. P., Chandel, V., Dutt, R., & Dutt, S., (2006). Development of pollinium in two epidendroid orchids. *Curr. Sci.*, *90*(10), 1384–1388.

Bharal, A., Kashyap, M., Sohpal, V. K., &Sembi, J. K., (2014). Evaluation of antimicrobial properties of terrestrial orchids (collected from Northern Himalayas) against certain human pathogens. *Int. J. Bioassays*, *3*, 3036–3039.

Bhatnagar, J., Handa, S. S., & Duggal, S. C., (1970). Chemical investigation on *Microstylis wallichii*. *Plant Medica*, *20*, 157–161.

Bose, B., Choudhury, H., Tandon, P., &Kumaria, S., (2017). Studies on secondary metabolite profiling, anti-inflammatory potential, *in vitro* photoprotective and skin-aging related enzyme inhibitory activities of *Malaxis acuminata*, a threatened orchid of nutraceutical importance *J. Phytochem. Photobiol. B.*, *173*, 686–695. doi: 10.1016/j.jphotobiol.2017.07.010.

Bose, T. K., Bhattacharjee, S. K., Das, P., & Basak, U. C., (1980). *Orchids of India* (p. 327). Naya Prakash, Calcutta.

Cetzal, W., & Savelli, N. E., (2014). *Orchidaceae*: The largest family of flowering plant. *The Encyclopedia of Earth.*

Charkarvarty, H. L., (1976). *Plant Wealth of Iraq: A Dictionary of Economic Plants* (Vol. I, p. 396). Botany Directorate, Ministry of Agriculture and Agrarian Reform, Baghdad.

Chauhan, N. S., (1990). Medicinal orchids of Himachal Pradesh. *The Journal of the Orchid Society of India*, *4*, 99–105.

Chauhan, N. S., (1999). *Medicinal and Aromatic Plants of Himachal Pradesh* (p. 632). Indus Publishing Company, Tagore Garden, New Delhi.

Cheruvathur, M. K., Abraham, J., Mani, B., & Thomas, T. D., (2010). Adventitious shoot induction from cultured Internodal explants of *Malaxis acuminata* D. Don a valuable terrestrial medicinal orchid. *Plant Cell Tissue Org. Cult.*, *101*, 163–170.

Chinmay, R., Kumari, S., Dhar, B., Mohanty, R., Dixit, R., Padhi, M., & Babu, R., (2011). Phyto-pharmacognostic studies of two endangered species of *Malaxis* (Jeevak and Rishibhak) *Pharmacognosy Journal*, *3*, 77–85.

Chowdhary, H. J., & Wadhwa, B. M., (1984). *Flora of Himachal Pradesh.* Botanical Survey of India, Calcutta.

Clarke, C. B., (1885). In: Hooker, J. D., (ed.), *Flora of British India* (Vol. V, p. 686). L Reene and Co, London.

Dhyani, A., (2009). *Lilium polyphyllum*- rarest of rare lilies. In: *Lilies and Related Plants* (pp. 85–91). Royal Horticulture Society, London.

Dressler, R. L., (1990). *The Orchid: Natural History and Classification.* Harvard University Press, U.S.A.

Durga, R. K., Karthikumar, S., & Jegatheesan, K., (2009). Isolation of potential anti-bacteria and antioxidant compounds from *Acalypha indica* and *Ocimumbasilicum*. *Afr. J. Plant Sci.*, *4*,163–166.

Giri, L., Belwal, T., Bahukhandi, A., Suyal, R., Bhatt, I. D., Rawal, R. S., & Nandi, S. K., (2017). Oxidative DNA damage protective activity and anti-oxidant potential of *Ashtavarga* species growing in the Indian Himalayan region. *Ind. Crop Prod.*, *102*, 173–179.

Gogoi, K., Borah, R. L., Sharma, G. C., & Yonzone, R., (2012). Present status of Orchid species diversity resources and distribution in Dibrugarh district of Assam of North East India. *IJMB*, *2*, 19–33.

Greuter, W., McNeilli, J., Barrie, F. R., Burdet, H. M., Demoulin, V., Filgueiras, T. S., Nicolson, D. H., et al., (2000). International code of botanical nomenclature (Saint Louis Code) adopted by the sixteenth International Botanical Congress, St. Louis, Missouri, July-August 1999. *Regnum Veg.*, 474.

Gupta, R., Agarwal, M., & Baslas, R. K., (1978). Chromatographic separation and identification of various constituents of essential oil from the bulb of *M. acuminata*. *Indian Perfumer*, *22*(4), 287–288.

Jadhav, D., (2008). *Medicinal Plants of Madhya Pradesh and Chhattisgarh*. Daya Publishing House, Delhi.

Jalal, J. S., Kumar, P., & Pangtey, Y., (2008). Ethnobotanical Orchids of Uttarakand, Western Himalayas, *Ethnobotanical Leaflets*, *12*, 1227–1230.

Joshi, V. K., (1983). Evolution of the concept of *Ashtavarga*. *Indian Journal of History of Science*, *18*(1), 9–14.

Kala, C. P., & Sajwan, B. S., (2007). Revitalizing Indian system of herbal medicines by the National Medicinal Plant Board through institutional networking and capacity building. *Curr. Sci.*, *93*, 797–806.

Kaur, S., & Bhutani, K. K., (2010). Micropropagation of *Malaxis acuminata* D. Don: A rare orchid of high therapeutic value. *J. Med. Arom. Plants*, *1*, 29.

Lange, & Schippmann, (1999). *CITES Checklist Medicinal and Aromatic Plants, Version 3.0* (p. 17). German Federal Agency for Nature Conservation, Bonn.

Lohani, N., Tewari, L. M., Joshi, G. C., Kumar, R., Kishor, K., & Upreti, B. M., (2013). Population assessment and threat categorization of endangered medicinal orchid *Malaxis acuminata* D. Don. from North-West Himalaya. *Int. J. Conserv. Sci.*, *4*, 493–502.

Mabberley, D. J., (2008). *Mabberley's Plant-Book: A Portable Dictionary of Plants, Their Classifications and Uses* (3rd edn., p. 516). Cambridge University Press, Cambridge.

Mahesh, B., & Satish, S., (2008). Antimicrobial activity of some important medicinal plant against plant and human pathogens. *WJAS*, *4*, 839–843.

Pérez-Escobar, O. A., & Blanco, M. A., (2014). Rediscovery of *Malaxis nana* (*Orchidaceae*: *Malaxideae*) in Costa Rica, with an updated description. *Lankesteriana*, *14*, 109–114.

Polunin, O., & Stainton, A., (1977). *Flowers of the Himalaya*. Oxford University Press, Calcutta, India.

Puni, L., & Singh, A., (2007). Enhancement of forest productivity through sustainable management of *Microstyliswallichii*. *ENVIS Forestry Bulletin*, *7*(2).

Rajurkar, N. S., & Gaikwad, K. N., (2014). Identification and quantification of amino acids from medicinally important plants by using high-performance thin-layer chromatography. *J. Liq. Chromat. Relat. Technol.*, *37*, 2197–2205.

Raval, S. S., Vaghela, P. G., Mandavia, M. K., & Golakiya, B. A., (2016). Phytochemical analysis of *Malaxis acuminata* D. Don (Jeevak), an ingredient of Jeevaniya group of Ashtavarga. *Ind. J. Agric. Biochem.*, *29*, 155–160.

Reddy, C., Sudhakar, Y., Reddy, K. N., & Jadhav, S. N., (2001). *Malaxis acuminata* D. Don (*Orchidaceae*): A new record for Andhra Pradesh, India. *Indian Journal of Forestry*.

Sharma, A. C., Rao, V., Tiwari, L. K., & Tyagi, M. L., (2009). Comparative study on physiochemical variation of *Microstyliswallichii*: A drug used in Ayurveda. *Academic J Plant Science*, *2*(1), 4–8.

Sharma, B. D., & Balkrishna, A. V., (2005). *Vitality Strengthening Ashtavarga Plants (Jeevaniya & Vayasthapanpaudhe)*. Divya Publishers, Divyayogmandir Uttaranchal (India).

Sharma, P. P., Dadhwal, S., Singh, G., Sharma, D., & Sharma, S., (2011a). Antimicrobial activity of butanol extract of *Malaxis acuminata*. *Journal of Pharmacy Research*, *4*(8), 2703–2704.

Sharma, P., Nipun, M., Pankaj, G., Gurukirpal, S., Sumit, D., & Sakshi, S., (2011b). *Malaxis acuminata*: A review. *Int. J. Res. Ayurveda Pharm.*, *2*, 422–425.

Shukla, P. K., & Chaubey, O. P., (2008). *Threatened Wild Medicine Plants, Assessment, Conservation and Management* (p. 105). Anmol Publications Pvt. Ltd., New Delhi.

Singh, D., Kumar, S., Pandey, R., Hasanain, M., Sarkar, J., & Kumar, B., (2017). Bio-guided chemical characterization of the antiproliferative fraction of edible pseudo bulbs of *Malaxis acuminata* D. Don by HPLC-ESI-QTOF-MS. *Med. Chem. Res.*, *26*, 3307–3314.

Suyal, R., Rawat, S., Rawal, R. S., & Bhatt, I. D., (2020). A review on phytochemistry, nutritional potential, pharmacology, and conservation of *Malaxis acuminata*: An orchid with rejuvenating and vitality strengthening properties. In: Merillon, J. M., & Kodja, H., (eds.), *Orchids Phytochemistry, Biology and Horticulture; Reference Series in Phytochemistry*. http://doi-org-443.webvpn.fjmu.edu.cn/10.1007/978-3-030-11257-8_15–12020. Springer.

Uma, E., Rajendran, R., & Muthukumar, T., (2015). Morphology, anatomy and mycotrophy of pseudobulb and subterranean organs in *Eulophiaepidendraea* and *Malaxis acuminata* (*Epidendroideae*, *Orchidaceae*). *Flora*, *217*, 14–23.

Ved, K. D., Kinhal, G. A., Ravikumar, K., Prabhakaran, V., Ghate, U., Vijayshanker, R., &Indresha, J. H., (2003). *Conservation Assessment and Management Prioritization for the Medicinal Plants of Jammu and Kashmir, Himachal Pradesh and Uttarakhand*. FRLHT, Bangaluru.

Vikram, P., Chiruvella, K. K., Ripain, I. H. A., & Arifullah, M. A., (2014). Recent review on phytochemical constituents and medicinal properties of *Kesum* (*Polygonum minus* Huds.). *Asian Pac. J. Trop. Biomed.*, *4*(6), 430–435.

Yonzone, R., (2019). Exact field status, habitat, and local distribution of orchid species of Darjeeling Himalaya of west Bengal, India. *Advances in Agricultural Technology & Plant Sciences*, *2*(1), 180021.

Yonzone, R., Bhujel, R. B., Lama, D., & Rai, S., (2012). Orchid species diversity of Darjeeling Himalaya of India. *Int. J. Pharm. Lif. Sci.*, *3*(3), 1533–1550.

Yonzone, R., Lama, D., Bhujel, R. B., Rai, S., & Gogoi, K., (2014). Orchid species of Darjeeling Himalaya of West Bengal, India: The genus Ione Lindley. *Pleione*, *8*(2), 475–477.

CHAPTER 18

Utilization of Rice Genetic Resources for Nutritional and Medicinal Benefits

LAKSHMI HIJAM, RUPSANATAN MANDAL, MOUMITA CHAKRABORTY, and BILIN MAYING

Department of Genetics and Plant Breeding, Uttar Banga Krishi Viswavidyalaya, Pundibari, West Bengal, India

ABSTRACT

Genetic diversity in *Oryza* species exceeds far more than other crops with thousands of varieties growing and spread across the globe. Approximately around 140,000 different cultivars of rice ranging from traditional rice varieties to the commercially bred elite cultivars are reported. Rice symbolizes life and prosperity for billions of people and plays a fundamental role in the world food security and socio-economic development. Till date around 80,646 accessions including over 3,000 accessions of wild rice representing 21 species of *Oryza* and 11 related genera are conserved and maintained at IRRI (International Rice Research Institute). Rice plays a vital role in Indian economy being the staple food for two-thirds of the population. India ranks first in area with 44.62 million hectares, second in production with 31% of calories to Indian diet supplied through rice. In the present context, rice is being studied beyond its stereotype of being staple food and a major source of starch or carbohydrate. Owing to the content of minerals, quality of starch, glycemic index and presence of antioxidants, rice is indeed unique among cereals. On comparing rice starch with other sources such as wheat, maize, and potato, rice starch is almost fully absorbed by the human body. Validations for the starch digestibility, the immense biological value of amino acids, high content of essential fatty acids and selenium, and anti-hypertension

Bioprospecting Ethnomedicinal Plant Resources: Sustainable Utilization and Restoration.
Gopal Shukla, Jahangeer A. Bhat, A. P. Das, and Sumit Chakravarty (Eds.)

have been reported by several scientific literatures. The use and effectiveness of Rice-based ORS (oral rehydration solution) in reducing stool output and enhancing intestinal absorption in acute diarrhea is being reported. Rice is also found to be the least allergic food substance and is highly recommended for people afflicted with diseases like the irritable bowel syndrome. An extensive study has been conducted on colored rice (red and black) and their colored pigments or anthocyanins and flavonoids are found to be associated with antioxidant properties. As red and black rice are rich in iron, zinc, other minerals, and antioxidant properties they are more nutritious. Red rice is reported have comparatively higher antioxidant and scavenging activity than black and white rice. A large number of varieties of rice with huge medicinal properties are found in India.

The need for conserving these varieties that are fast disappearing under the pressure of high yielding varieties and other cash crops is highly demanded. Very often there is hue and cry by Indian nutritionists who recommend using of green, healthy, and functional food against junk food. Various investigations and studies are being conducted on a regular basis to find out novel food-based approaches which will lower the episodes and severity of lifestyle-associated ailments with the primary focus on antioxidant properties, mineral contents, and the glycemic index of food. Rice as such holds as a promising crop with medicinal values and the need of the hour is to further extend market avenues for these varieties and promotion through greater public awareness about their importance, especially among the younger generation. To overcome and manage these kinds of lifestyle-related diseases, the conservation as well as promotion of national heritage as a health food is critical.

18.1 INTRODUCTION

Nearly half of the world's population relies on rice (*Oryza sativa* L.), the second-most significant cereal crop, as a primary food source, contributing over 20% of the total calories taken by man. In Asia, the home continent of rice, 95% is produced and consumed. Rice is by far the most economically important food crop in many developing countries, providing two-thirds of the calorie intake of more than 3 billion people in Asia. In Indian food and livelihood security system, rice occupies a pivotal role providing about 55% protein and 75% of calories in the average daily diet of the population. In India, rice is often referred to as "*Prana*" – the breath of life. Wherever rice is a staple food, such comparisons can be found for rice represents more than a source of calories in Asian culture – it is also the basis of both biological and cultural diversity. Rice also represents the pride and identity for the people of Asia from culture to history, landscape, religious, and social ideas. Every culture has its own unique set of preferences regarding the color, texture, taste, and stickiness of rice and different ways of harvesting, processing, and eating rice. Many festivals and community interactions centered on the rice production calendar and changes to rice production led to changes in social and cultural interactions. Rice indeed is one of the very few crop species endowed with rich genetic diversity accounting over one lakh landraces and improved cultivars which ultimately makes it one of the most researched crops with a wealth of scientific literature in various aspects.

Ancient Asian civilizations have always emphasized the significance of rice in supporting human health and nutrition. Our country, India, is blessed with the wealth of medicinal plants, which finds its mention in Ayurveda and Unani medicinal systems and by tribal healers for generations. Every plant on earth has uses for humans, animals, and other plants, as is also explicitly stated in ancient Indian literature (Oudhia, 1999). The importance and medicinal values of rice have been well described in Ayurveda – considered oleaginous, acrid, tonic, fattening, aphrodisiac, diuretic, and useful in biliousness (Caius, 1986). Rice as a staple crop exhibits high potential as a medicinal and healthy food. The need of the day is to enhance the marketing of these varieties and to encourage them through greater public awareness about their value, particularly among the youth. In order to stem the onslaught of lifestyle-related diseases, the promotion and conservation of this national heritage as a health food is critical. For India, the slogan 'Rice is Life' has become more appropriate as this crop serves as a source of livelihood for millions of rural people and at the same time plays a vital role in our national food security system.

18.2 RICE HISTORY, TRADITION, AND CULTURE

Rice is one of the most ancient crops cultivated for over 10,000 years and as such genetic diversity in *Oryza* species far exceeds that in other crops. At present around 117 countries cultivate rice and serve as the staple food for over one third of the world's population. There are about 140,000 varieties of cultivated rice which indicates the wide variety environmental conditions where it is grown and the associated cultural preferences and practices. In

Asia alone more than 90% of the world rice is grown and consumed where people typically eat rice twice or thrice a day. Grains vary in size and stickiness from the long and fragrant Indian Basmati and Thai Jasmine rice to the short sticky Japanese Sushi and Italian Arborio rice. The colors may be brown, black, red, green, white or the vitamin A fortified golden rice. The environment where rice is grown range from intricately irrigated terraced to rainfed lowlands. In the 'rice bowls' of Asia 2-3 crops per year is grown. According to Chang (1976), the principal cultivated species of rice, *Oryza sativa*, is believed to have been domesticated nearly 10,000 years ago in areas covering northeastern India, Bangladesh, Burma, Thailand, Laos, Southern China, and Vietnam. The only other cultivated species of rice, *O. glaberrima* Steud, is indigenous to the upper valley of the Niger River in West Africa. There are around 20 species in genus *Oryza* with a basic chromosome number of 12. The genus includes both diploid and tetraploid species with six genome groups, A, B, C, D, E, and F. The major sources of genetic diversity in rice are cultivated rice *Oryza* sativa with a gene pool consisting of primitive cultivars, land races, and improved high-yielding varieties, and wild species of *Oryza* and related species and genera comprising the primary, secondary, and tertiary gene pool. The primary gene pool contains races that usually belong to the same biological species as cultivars and can produce fertile hybrids with the other cultivars. The full spectrum of rice germplasm includes:

- Wild *Oryza* species and related genera;
- Natural hybrids between the cultivars and wild relatives and primitive cultivars;
- Germplasm generated in the breeding programs includes pure line or inbred selections of farmers' varieties, F1 hybrids, and elite varieties of hybrid origin, breeding materials, mutants, polyploids, aneuploids, etc.;
- Commercial types, obsolete varieties, and special purpose types;
- Induced mutants, obtained through physical and chemical mutagenesis;
- Somaclonal variants – obtained through tissue and cell culture procedures; and
- Transgenic plants – novel sources of genetic diversity.

All over India, rice plays an important role in festivals celebrated across the country. The harvest festivals are celebrated by preparing several delicacies from freshly harvested paddy. For instance, in Tamil Nadu, *sarkarai-pongal* is made from the combinations of raw rice, green gram, jaggery, and milk. In Assam, *ghilapitha* which is a fried rice balls are prepared and consumed. In West Bengal, traditional Bengali delicacies are made from freshly harvested rice and jaggery like *pitha* and *karpursal* or *banapuli*, and Basmati rice is also used to make a Bengali delicacy *paish*. Parboiled red rice is widely used in Kerala, including *Thondi, Kuruva, Matta, PaalThondi, Chitteni, and Chettadi. SeeragaSamba* is an aromatic variety of rice used widely in Tamil Nadu and Kerala and known as the 'Basmati of South India' and used in *biryani* preparation. Similarly, *Jatu* of Kulu valley, *Dubraj* of Madhya Pradesh, *Ambemohar* of Maharashtra, *Joha* of Assam, *Kamod* of Gujarat, *Badshahbhog* of West Bengal and Odisha, *Radhunipagal* of West Bengal, *Katrini* and *Kalanamak* of Uttar Pradesh and Bihar, *Gandhasamba* of Kerala, *Kalajira* of Odisha, and *Chakhao* varieties of Manipur are popular for their aroma (Ahuja et al., 2007, 2008a, b). In the present context the aim is to increase the production of these traditional varieties and promote consumption among the younger generations and manufacturing of nutritious and novel value-added products from colored rice.

18.3 TRADITIONAL MEDICINAL RICE VARIETIES IN INDIA

Very often one finds the mentioned rice varieties of curative properties for various diseases in the ancient records like Ayurvedic treatises (*Indian Materia Medica*) of the 15th and 16th century AD. For example, for the treatment of arthritis, rice varieties like '*Njavara*' and '*Gathuran*' were used whereas, varieties *Kalama*, *Pundarika*, *Panduka*, *Sugandhalaka*, *Kardamaka*, *Maetunaka*, and *Mahasali* have different medicinal properties. Surveys conducted in Chhattisgarh during the 1970s and 1980s had identified several traditional rice varieties with medicinal utility. To cite an instance, *Aalcha* is used for the treatment of pimples, *Baisoor* for epilepsy, *Maharaji* gives strength and stamina post-partum, and *Laicha* for healthy conception (Siddiq, 2002). In addition, the variety 'Baisoor' is used in different regions of Chhattisgarh for the treatment of headache and boils. Additionally, it is also used in dermatological diseases, ringworm infections, itching, and conjunctivitis, and as a face pack with neem leaves. Njavara, another variety, a unique landrace, is traditionally utilized for the treatment of arthritis, rheumatism, neurological disorders, and for the relaxation property as well as for rejuvenating weak muscles of old-aged persons by the Ayurvedic practitioners of Kerala (Mohanty et al., 2012). *Karkada kanjhi* is a replenishing drink and used as a supplementary diet for underweight persons. The same is also consumed during monsoon season along with some herbal medicines prepared from *Njavara* cultivar (Mohanty et al., 2012). Multiple ailments like aches, painful inflammations, spondylitis, and back pain are also treated using the oil preparations of this variety. The medicinal pouch known as *njavarakizi*, which is produced after heating it in a milk decoction of the herb *Sida cordifolia*, is an important use. By continually dipping the bag in the decoction, it is massaged over the entire body. The *Handia* rice beer, made from traditional rice varieties by fermenting moistened rice for several days, is consumed by indigenous populations in Odisha. They add an herbal mixture called "*Bakhara*" to it, which is thought to transform it into a healthy beverage that heals digestive problems. Similar behavior is seen in the indigenous populations of North Bengal, where it takes the form of the alcoholic drink *Haria*. It is an herbal remedy made from rice flour and fermented with yeast culture (Basnet, 2003).

18.4 NUTRITIONAL AND HEALTH BENEFITS OF RICE

Rice is by far the most important staple food consumed by a significant proportion of the global human population. Rice is the single most important source of protein which contributes 40–80% of calories to the Asian diet. In Asia alone, over 2 billion people derive 80% of their energy requirement from rice, which contains 80% carbohydrates, 78% protein, 3% fat, and 3% fiber (Juliano, 1985). The only things rice was regarded as were starchy foods, sources of carbs, and a small quantity of protein. Rice protein is of high nutritional value though less in amount (Chaudhary & Tran, 2001). Rice is also a good source of protein, iron, and phosphorus and also contains some amount of calcium. Since ancient times rice has enjoyed a special status in India because of its special attributes. The references about the unique and special properties of rice are found in ancient Indian texts and folklore. In the Sanskrit text *Krishi-Parashara*, the great sage Parashara wrote in praise of this food grain. "Rice is vitality, rice is vigor too, and rice indeed is the means of fulfillment of all ends in life. All Gods, demons, and human beings subsist on rice" (Majumdar & Banerji, 1960). A number of unknown properties of rice have been unraveled according to recent studies which also find its mention in ancient Indian Ayurvedic literature. In addition to being the staple food, rice is also said to have some therapeutic characteristics in Asian nations like China, Thailand, Myanmar, Malaysia, Indonesia, and India. Early oriental writings referred to whole brown rice as the ideal diet. Tables 18.1 and 18.2 enumerate the nutrient contents of brown along with comparison to other cereal and tuber crops following some earlier studies

(Eggum, 1969, 1977, 1979; Souci et al., 1987; Bradbury & Holloway, 1988). Several types of rice and their products include brown rice, parboiled rice, polished rice, basmati rice, puffed rice, organic rice, wild rice, rice bran, and rice flour. Brown rice is considered to be the most nutritious type of all rice as most of the nutrients and minerals in rice are concentrated in the outer brown layers known as husk and germs.

TABLE 18.1 Proximate Composition of Cereal and Tuber Staple Foods

FC	M	P	CF	AC	DF	WIF	L	CA	E
Brown rice	14.0	7.3	2.2	71.1	4.0	2.7	0.1	1.4	384
Wheat	14.0	10.6	1.9	61.6	10.5	7.8	0.6	1.4	375
Maize	14.0	9.8	4.9	60.9	9.0	6.8	0	1.4	396
Millet	14.0	11.5	4.7	64.6	37	2.3	0	1.5	395
Sorghum	14.0	8.3	3.9	57.4	13.8	12.4	3.0	2.6	384
Rye	14.0	8.7	1.5	60.9	13.8	8.4	14	1.8	375
Oats	14.0	9.3	5.9	63.0	5.5	39	0	2.3	392
Potato	77.8	2.0	0.1	15.4	2.5	1.9	0	1.0	70
Cassava	63.1	1.0	0.2	31.9	2.9	2.2	0	0.7	133
Yam	71.2	2.0	0.1	22.4	3.3	2.6	0	1.0	98

Abbreviations: FC: Food crop; M: moisture (in %); P: protein (gNx6.25); CF: crude fat (in grams); Ac: available carbohydrates (in grams); DF: dietary fiber (in grams); WIF: water soluble fiber (in grams); L: lignin (in grams); CA: crude ash (in grams); and E: energy (in kilocalorie).

TABLE 18.2 Vitamin and Mineral Content of Cereal and Tuber Staple Food (per 100 gm)

FC	C	T	R	N	AA	VE	Fe	Zn
Brown rice	0	0.29	0.04	4.0	0	0.8	3	2
Wheat	0.02	0.45	0.10	3.7	0	1.4	4	3
Maize	0.37	0.32	0.10	1.9	0	1.9	3	3
Millet	0	0.63	0.33	2.0	0	0.07	7	3
Sorghum	10.0	0.33	0.13	3.4	0	0.17	9	2
Rye	0	0.66	0.25	1.3	0	1.9	9	3
Oats	0	0.60	0.14	1.3	0	0.84	4	3
Potato	0.01	0.11	0.05	1.2	17	0.06	0.8	0.3
Cassava	0.03	0.06	0.03	0.6	30	0	1.2	0.5
Yam	0.01	0.09	0.03	0.6	10	0	0.9	0.7

Abbreviations: FC: Food crop; C: carotene (in mg); T: thiamine (in mg); R: riboflavin (in mg); N: niacin (in mg); AA: ascorbic acid (in mg); VE: vitamin E (in mg); Fe: iron (in mg or in %); and Zn: zinc (in mg).

18.4.1 HEALTH BENEFITS

Based on the flavor, culinary needs, availability, and potential health benefits, people select a number of rice varieties. Rice as a crop has the ability to provide rapid and instant energy. Red and brown rice are both excellent providers of fiber, vitamins, calcium, zinc, iron, manganese, selenium, and other minerals. The rich color of the red and black rice variant is due to a class of phytochemicals known as anthocyanins, which are also present in dark purple or reddish fruits and vegetables.

18.4.1.1 EXCELLENT SOURCE OF CARBOHYDRATES

Rice is an excellent source of complex carbohydrates. After being converted to glucose, carbohydrates are primarily used as fuel for the brain and as a source of energy for physical activity. Additionally, it delays starch digestion (with low glycemic index). Type II diabetics who consume rice would be better to choose slowly digesting rice variants over white rice. For instance, brown rice also has a slow rate of starch digestion, and some starch never breaks down into sugar and remains intact in the large intestine. According to Qi (2010), switching from white to brown rice or other whole grains was found to minimize the chance of developing diabetes, whereas stabilized rice bran dramatically lowered hyperglycemia and hyperlipidemia in people with Type I and Type II diabetes. In any case, Type II diabetics are recommended to refrain from eating too many carbohydrates (Frei & Backer, 2004). The pace of carbohydrate digestion and utilization varies depending on the individual's requirement for energy. Foods high in carbohydrates are crucial carriers of proteins, vitamins, and other nutritional components (Anon., 1997a, b). To maintain a varied diet, nutrition, and good health, it should be consumed alongside other foods like vegetables, fish, meat, and fruits. Its nutritional and therapeutic benefits should also be respected.

18.4.1.2 LOW FAT, LOW SALT, AND NO CHOLESTEROL

Since rice has no fat, no cholesterol, and no sodium, it is regarded as a great item to include in a balanced diet. The majority of rice lipids, which contain fatty acids, are found in the rice bran but are eliminated during milling. Rice bran oil can have up to 80% unsaturated fatty acids depending on the variety (Frei & Backer, 2004). Oleic acid and linolenic acid, which humans are unable to synthesize but are essential for sustaining the function of cell membranes and the neurological system, make up the unsaturated fatty acids found in rice.

18.4.1.3 RICE FOR PROTEIN

Up to 8% of the grain is made up of rice protein, which has eight important amino acids in precisely calibrated ratios. Internally, a complete renewal occurs when rice protein is converted into amino acids that promote health. These amino acids aid in the development of resilient muscles that can recover their original shape after being stretched and bent. Overall, it promotes the growth of healthy skin, hair, and eyesight and nourishes the glandular network, heart and lungs, tendons, and ligaments, brain, nervous system, and brain.

18.4.1.4 RICE FOR VITAMINS

The B-complex vitamins comprising of thiamine, riboflavin, and niacin, which are present in natural brown rice, help in promoting youthful energy, skin nourishment, and blood vessels. In addition, rice bran also contains anti-oxidants such as tocopherols and tocotrienols (of the Vitamin E family) and oryzanols (Lloyd et al., 2000). Tocopherols and tocotrienols were reported with anticarcinogenic properties, while oryzanols were found to play a role in reducing cholesterol absorption (Lloyd et al., 2000; Kline et al., 2004). Tocotrienols present in rice bran were reported to prevent or reverse blood clots and lesions responsible for strokes or thrombosis (Frei & Backer, 2004).

18.4.1.5 RICE FOR MINERALS

According to reports, red and black rice is filled with nutrients like iron and zinc (Ahuja et al., 2008a, b). Zinc and iron are required by the human body for enzymatic reactions and the formation of hemoglobin, respectively. Symptoms of zinc shortage include diarrhea, weight loss, and illness. If unchecked, zinc deficiency can be lethal. Similar to iron deficiency, anemia causes brittle hair and nails as well as fatigue and can be just as harmful to health. Additionally, potassium, a vital element required by the body for healthy metabolism, cellular, tissue, and organ function, muscle growth, and regular heartbeat, is proven to be a good source of potassium in rice. Furthermore, rice contains very small levels of manganese and copper. For the generation of enzymes necessary for regular physiological function, copper is required, while manganese is necessary for the normal operation of the brain and nerves. Natural brown rice contains a variety of nutrients that support healthy hormone function, wound healing, and blood pressure control. In addition, rice contains nutrients like phosphorus, potassium, and iron, which enrich the bloodstream and help to maintain internal water balance. Thus, rice aids in re-establishing physiological balance in general.

18.4.1.6 DIABETES MELLITUS

Brown rice releases sugar more gradually than white polished rice, which aids in long-term blood sugar stabilization. Because of this, brown rice is a preferable choice for those with diabetes mellitus. Studies in Asia have linked white rice consumption to an increased risk of type II diabetes. By enclosing the food and preventing the action of hydrolytic enzymes in the small intestine on food, dietary fibers are reported to inhibit the absorption of carbs and to raise the viscosity of food in the intestine (Jenkins et al., 1977). This is essential in lowering food's glycemic index and lowering the incidence of type II diabetes. Red rice contains proanthocyanidins (PAs), which protect against type II diabetes (Chen et al., 2016). Similar to this, it is claimed that black rice's anthocyanin content has a hypoglycemic effect (Tsuda et al., 2003).

18.4.1.7 CANCER

Just like brown rice, whole rice grain is rich in insoluble fiber and can possibly protect against multiple types of cancers. Brown rice is anticarcinogenic due to the abundance of manganese, selenium, and dietary fiber, which are vital against free radicals. The consumption of whole grains like brown rice is also correlated with lowered levels of colon cancer due to its high fiber content. These fibers get attached to the carcinogenic substances and toxins and help to be eliminated from the body. The presence of pro-anthocyanins in red rice modulates the inflammatory response, which also protects the human body against some cancers (Chen et al., 2016). Similarly, anthocyanins found abundantly in black rice also have anticarcinogenic properties (Pojer et al., 2013).

18.4.1.8 CHOLESTEROL

Naturally occurring bran oil is found in brown rice, which helps in reducing LDL forms of cholesterol. Due to the presence of substances including anthocyanins, flavonoids, polyphenolic compounds, phytic acid, vitamin E, and γ-oryzanol, the consumption of black rice is associated with the removal of reactive species like lipid peroxide and superoxide anion radicals as well as reducing cholesterol levels (Ichikawa et al., 2001; Nam et al., 2008). Modulation of inflammatory responses by pro-anthocyanins in red rice also provides protection from cardiovascular disease (Chen et al., 2016). They also concluded that whole grains could reduce the chances of arterial plaque buildup and thus reduce the chances of developing heart disease.

18.4.1.9 HYPERTENSION

Compared to white rice, brown and red rice have higher magnesium contents. Magnesium is a significant mineral that regulates blood pressure and the body's sodium balance, among other things (Anon., 1997b).

18.4.1.10 OBESITY

Rice varieties including brown, red, and black are high in fiber and help maintain normal metabolic and digestive processes. Red rice contains anthocyanins, which have qualities that can help with weight management (Anon., 1997b).

18.4.1.11 ALLERGY

Protein from rice is hypoallergenic. Consumption of proteins derived from additional plant sources, such as soy and peanuts, as well as from animal sources, such as eggs and milk, might occasionally result in an allergic reaction. Because it is hypoallergenic, rice protein offers a solution to this issue. Additionally, the anthocyanins included in red rice have the ability to lessen allergic symptoms (Anon., 1997b).

18.4.2 USEFUL COMPOUNDS IN RICE BRAN FOR PREVENTION OF CHRONIC DISEASE

One of the main by-products of the rice milling process is rice bran, which is made up of a mixture of protein, lipid, ash, and crude fiber. During the milling process, the outer brown coating of brown rice separates. The amount of ash in bran is increased by the discovery that it contains several minute fragments of rice hull. Rice bran has a high nutritive value. Rice bran is also an excellent source of vitamin B and E besides proteins. Small amounts of antioxidants are also found in bran which is considered to reduce cholesterol level in humans. Rice bran with its 10–23% oil also makes it an excellent binder for animal feeds. Bran oil is a high-quality vegetable oil for cooking if stabilized once and extracted. Differences in composition of rice bran occur with the differences in rice variety, geographical conditions, and processing methods. It accounts for 8–10% of the total weight of the grain. However, it contains most of the nutrients: carbohydrates (34–62%), protein (11–15%), lipids (15–20%), crude fiber (7–11%) and ash (7–10%). Lipids and bioactive components in particular are concentrated in rice bran (Oliveira et al., 2011; Gul et al., 2015). Fatty acids like palmitate (21–26%), linoleate (31–33%) and oleate (37–42%) are predominant in rice bran. Rice bran is also considered a healthy food (Lemos & Souza Soares, 2000; Oliveira et al., 2011) owing to its high content of polyunsaturated fatty acids. Rice bran contains sizable amounts of dietary fibers such as glucan, pectin, and gum as well as bioactive substances like oryzanol, tocotrienol, tocopherol, and sitosterol (Saunders, 1985; Sugano & Tsuji, 1997). The primary antioxidant in rice bran, oryzanol, has 10 times the antioxidant activity of tocopherol, but tocotrienol has 40–60 times the antioxidant activity of tocopherol. The ratios of these phytochemicals, however, differ depending on the type of rice cultivar (Iqbal et al., 2005). In addition, rice bran contains 4-hydroxy-3-methoxy cinnamic acid, which possesses antioxidant and photoprotective properties (Yan et al., 2001; Suzuki et al., 2002). Some of the bioactive compounds found in rice bran along with their properties are given in Table 18.3.

TABLE 18.3 SomeSelected Bioactive Compounds in Rice Bran and Their Properties

Rice Bran Compound	Disease Prevention Activity
Ferulic acid and γ-oryzanol.	Antioxidant, chemo-preventive, anti-inflammatory, lipid regulation.
Inositol hexaphosphate.	Blocks cancer growth and signaling.
Compesterol	Anti-angiogenic
β-Sitosterol	Blocks cholesterol
Linoleic acid	Anti-inflammatory
α-tocopherol and tocotrienol	Inhibit lipid peroxidation and intracellular signaling.
Salicylic acid	Anti-inflammatory
Caffeic acid	Gastrointestinal microbe interactions.
Coumaric acid and tricin	Anti-mutagenic, inhibit cell cycle, antioxidant, chemo-preventive.

Source: Roy & Bhattacharjya (2012); Subba Rao et al. (2015).

18.4.3 MEDICINAL USES OF RICE

Since each rice variety has distinctive qualities, using rice to treat illnesses cannot be restricted to just one variety. Due to their unique features and traits, many distinct rice cultivars are used in the treatment of diseases. Rice creates a balance to the humors (the bodily fluid of an animal) according to the practitioners of Ayurveda. Rice controls blood pressure, strengthens, revitalizes, and aids in the removal of toxic metabolites from the body. It also helps to prevent skin disorders and premature aging by enhancing the elements of the body. Rakthasali (kind of red rice) is found to be efficient in subduing disturbed humors of the body, treat fevers and ulcers, improves eyesight, and increases fertility and reported diuretic, spermatophytic, cosmetic, tonic, and anti-toxic (Krishnamurthy, 1991; Kumar, 1998; Ahuja et al., 2008a, b; Bhat & Riar, 2015). Sali, Sashtika, and Nivara rice were discovered to be utilized in Ayurveda to cure hemorrhoidal hemorrhage (piles). While Nivara rice is used to cure cervical spondylitis, paralysis, rheumatoid arthritis, neuromuscular diseases, psoriasis, skin lesions, backache, stomach ulcers, and snakebite, Sali rice is used to treat burns and fractures. Additionally, weaning food for newborns who are underweight is made using nivara rice (Sharma, 1996; Ahuja et al., 2008a, b).

Many disorders can be controlled using rice water, which is made by soaking rice in water or boiling rice in extra water. To treat conditions including vaginal and seminal discharges, diarrhea, constipation, and dysentery, Ayurvedic medicines use rice varieties such as Mahagandhak Ras, Kamdudha Ras, Sutsekhar Ras, Amritanav Ras, Swarnmalti Ras, Pradraripu Ras, Laghumai Ras, Dughdavati, Pradakn (Ahuja et al., 2008a, b). Red rice varieties are useful in the treatment of diarrhea, vomiting, hemorrhage, fever, chest pain, wounds, and burns (Hedge et al., 2013). In Himachal Pradesh, Matali and Lal Dhan rice cultivars are used for the treatment of blood pressure and fever. Leucorrhea and problems following abortion are treated with a different red rice variety from the Himachal Pradesh and Uttar Pradesh hills called Kafalya (Ahuja et al., 2007). Kari Kagga and Atikaya rice cultivars from Karnataka are consumed by breastfeeding mothers while Neelam Samba from Tamil Nadu is utilized for cooling and tonic purposes (Arumugasamy et al., 2001). Kuruvi Kar is consumed by the people for its health advantages and is drought-resistant (Anon., 2018). Raktasali is accepted in many countries for its medicinal properties. For instance, in the Philippines, to avert and cure beriberi, during rice polishing the bran is extracted and used as a very good source of Vitamin B. In Cambodia, the hulls (husk) of mature rice plants help in treating dysentery (Umadevi et al., 2012). The hulls of a three-month vintage rice vegetation are considered to be diuretic. In the Medicinal Book of Malayan Medicine, boiling of rice 'greens' is prescribed and recommended for use as an eye lotion and also for use with acute inflammation of the inner body tissues. The book also recommends the application of a mixture of dried, powdered rice on certain skin ailments. The Chinese believe rice to reinforce the spleen and stomach, increase appetite, and treat indigestion (Umadevi et al., 2012). The Pharmacopoeia of India prescribed the use of Rice water (a decoction of rice) as an ointment to counteract inflamed exterior. The nutritional composition of different rice varieties is different. Studies have depicted the importance of landraces or traditional rice varieties for their high mineral, protein, and antioxidant contents (Chaudhari et al., 2018). High biological value of amino acids, high content of essential fatty acids and selenium, and anti-hypertension effects of rice were also reported in many studies (Chaudhari et al., 2018). The majority of customers, who are conditioned to associate high-quality rice with white polished rice, are largely unaware of this fact, limiting them of the potential health benefits offered by rice diversity. Currently, several cultivars with medicinal qualities are produced and used in different regions in India (Table 18.4).

18.4.3.1 BROWN RICE

Brown rice is unpolished rice where the husk is kept and gives the kernel its usual beige color on cooking. It is high in fiber and is a very good source of thiamine, niacin, riboflavin, metal, calcium as well as proteins and small amounts of fats. It contains a lot of manganese, which is known to speed up a number of bodily functions like wound healing and blood sugar control. It is a powerful antioxidant that safeguards the body against oxidative damage. More fiber is found in brown rice than in any other refined grain, and fiber is very important for the body since it helps with digestion. It even manages diabetes and encourages a disease-free heart (https://www.nmamilife.com/food/the-versatile-rice-know-the-top-8-varieties-health-benefits/).

TABLE 18.4 Medicinal Uses of Some Rice Varieties in India

Variety	Properties and Uses
Aalcha, Nagkesar, Kari bhatta	Skin problems, pimples, small boils in infant.
Atikaya	Health tonic.
Baisoor	Epilepsy, headache, boils, ringworm, conjunctivitis.
Bhajari	Renewal in cow placenta.
Bhut moori	Cures anemia, enhance post-partum blood volume.
Chak-hao	Remedy for inflammation, anticarcinogenic.
Chinta puvvu, Pacchaganneru	Nutritional and medicinal value.
Dimarya, Kinsyalu, Sukhnandi	Good for women suffering from anemia.
Erumakkari	Cough
Gathuran	Arthritis, rheumatism.
Gathuwanor	Rheumatism
Geerakasala and Gandakasala	Rich in iron, zinc, polyphenols, antioxidant.
Kabiraj-sal	Treat convalescing patients.
Kafalya, Kaloun, BhetKalounisz	Regulates breast-feeding, strengthen bones.
Kalimooch	Skin diseases
Karhani	Paralysis
Kari Bhatta	Skin infections, increasing lactation.
Karunkuruvai	Orthopedic ailments, filariasis.
Katheri	Good for post-delivery restoration of size of reproductive organs.
Kolajoha, Kelas(rice with reddish kernel)	Served to lactating and pregnant women and also small children. Rich in iron and other minerals, used as blood booster.
Laal Dhaan(Red rice)	Beneficial for pregnant women.
Laicha	Pre-partum
Maharaji, Mehar, Saraiphool	Post-natal tonic for women.
Matali, Lal Dhan	Effective for High blood pressure, fever.
Mullarya	Coolant
Neelan Samba, Kuzhiyadichan	Regulate lactation.
Njavara	Body vitalizer, cleanser, anti-aging.
Pitchavari	Diarrhea in cattle.
Poongar	Use for gynecological issues, joint pains.
Saathi	Joint pains, arthritis.

Source: Roy & Bhattacharjya (2012); Subba Rao et al. (2015).

18.4.3.2 RED RICE

Red rice has a nutty flavor and meets iron and fiber needs which help the body to regulate weight. It contains a lot of zinc, which protects the body from free radicals that may harm any tissues or cells. Anthocyanin, the substance that gives red rice its color, also contributes to its rich nutritional content. As magnesium content is high, it also helps in preventing asthma. It is also packed with calcium, potassium, selenium, protein, vitamin B2 and B6 (https://www.nmamilife.com/food/the-versatile-rice-know-the-top-8-varieties-health-benefits/).

18.4.3.3 BLACK RICE

Also known as the forbidden rice, black rice has been consumed since ancient times and is still used in a variety of cuisines globally. Black rice has the highest protein level of all the different varieties of rice. This variety's black color is caused by anthocyanin. It also possesses anti-inflammatory and antioxidant qualities, is a good source of iron, and is high in antioxidants. Black rice also has anticarcinogenic effects, improves eye, heart, and liver health, and also regulates body weight (https://www.nmamilife.com/food/the-versatile-rice-know-the-top-8-varieties-health-benefits/).

18.4.3.4 JASMINE RICE

It is one of the rice kinds with the lowest price, rich in carbohydrates and popular in Indian cuisines like *biryani*. Brown jasmine rice is rich in dietary fiber which regulates blood sugar levels and cholesterol. Additionally, it supplies minerals and vitamin B, which improve the functioning and structure of the body.

18.4.3.5 BASMATI RICE

This long kernel Himalayan rice has a fine surface with a nutty flavor. Owing to its natural subtle perfume and distinct flavor, it finds itself to be the prevalent alternative for popular Indian and continental cuisines. It is low in fat and gluten-free. It is a wholesome and nourishing alternative for diabetics, while regulating cardiac health and body weight (weight https://www.nmamilife.com/food/the-versatile-rice-know-the-top-8-varieties-health-benefits/).

18.4.3.6 WILD RICE

It is a rice cultivar of North America and China that has gained attention recently. It is rich in protein, amino acids, and vitamin B. It strengthens the bones as it is rich in phosphorus and improves digestion and the immune system (https://www.nmamilife.com/food/the-versatile-rice-know-the-top-8-varieties-health-benefits/).

18.4.3.7 ARBORIO RICE

Arborio rice is a healthy carbohydrate source that is also high in protein, vitamins, and minerals. With a reasonable amount of vitamin C, this rice is very simple to digest and helps in developing strength and muscle. It increases the synthesis of hormones and enzymes that help the body's tissues and cells heal and grow stronger. By enhancing the immune system, it decreases the risk of numerous chronic diseases.

18.4.3.8 WHITE RICE

This Indian staple rice is obtained after milling where both hull and bran are taken. It is gluten-free and highly nutritious containing calcium and iron (carbohydrates; https://www.nmamilife.com/food/the-versatile-rice-know-the-top-8-varieties-health-benefits/).

18.5 CONSERVATION OF TRADITIONAL RICE VARIETIES

18.5.1 THREAT TO RICE DIVERSITY

Our farmers are gladly accepting the new improved high yielding rice varieties with the advent of hybrid revolution which has led to extinction of many traditional rice or farmer's varieties (Baker, 1978). Unexpected changes in this era brought about by global warming and climatic changes along with the dwindling traditional rice germplasm has necessitated the conservation of traditional rice/paddy varieties and farming to ensure sustainability, variability, and accountability in the world of food production. Unfortunately, the valuable plant genetic resources are getting rapidly eroded mainly due to their replacement by a few modern varieties and also due to habitat alteration/loss. The large-scale spread of modern, high-yielding varieties is causing concern since the genetic base of rice has narrowed down. Flood-prone systems are being replaced by irrigated systems which may eventually lead to the loss of distinct deepwater rices. Wild species of rice are threatened with extinction as a result of loss of habitat. This genetic erosion is taking place at a time when new tools of biological research enable scientists to focus as much on the diversity of genes as on the diversity of genotypes. Although there has been considerable success in protecting and preserving the biodiversity of rice during the past decades, much remains to be accomplished.

18.5.2 IMPORTANCE AND NEED TO CONSERVE

Crop genetic diversity forms the basis of our food supply and therefore the basis of our very survival. Landraces, traditionally grown primitive cultivars, and wild relatives of cultivated plants are the basic raw materials that sustain the present-day crop improvement programs. These will also be required for meeting the aspirations of future generations who may require altogether new sources of genes while facing unforeseen challenges of more virulent pathogens and pests and hostile climatic factors and abiotic stresses such as salinity, drought, and unfavorable temperatures. Traditional rice types must be preserved to prevent extinction since they are important sources of nutrients, and many have therapeutic qualities. It will be beneficial to plant and consume local seeds, as well as to conduct study and spread traditional knowledge (TK) about their importance to human nutrition and health.

In order to arrest the continued loss of traditional rice varieties, more planting and consumption of traditional rice varieties accompanied with in-depth research and promotion of TK on the value of local seeds to human health and nutrition is required. By patronizing their goods and highlighting the advantages of including different rice kinds in our diets, we may help farmers preserve and improve traditional rice varieties. One of the largest dangers to the biodiversity that provides the security, well-being, nourishment, and sustainability of the world and its inhabitants is corporate control over seeds. Because of the historic seed exchange systems and local innovation processes used by our farmers, we are able to enjoy the diversity of our food, thus we must maintain these systems and paths.

18.5.3 CONSERVATION OF RICE GERMPLASM IN INDIA

India is endowed with rich diversity of rice including the wild progenitors. Collections of approximately 200 wild rices were made from the primary and secondary centers of origin, which includes the Jeypore-Koraput region, Eastern, North-eastern hills, Chhattisgarh, and Pen India. Initiation for collection of rice germplasm started in the country by Parnell (Madras Province), Hector (Bengal), and Graham (Central Provinces) between 1910 and 1920 (Subba Rao et al., 2015). However, the systematic and organized attempt for collection of rice germplasm started in the late 50s and early 60s with the Jeypore Botanical surveys. Around 900 accessions, largely of local rice cultivars from Manipur, were added. During 1965-1980, two major collection surveys were undertaken in the North-eastern India and Chhattisgarh. The PL 480 supported scheme in Northeastern states of India led to a collection of 6,700 local rice cultivars and land races. Dr. R. H. Richaria and his colleagues in the late 70s collected about 20,000 accessions of rice from the eastern region of Madhya Pradesh. As such, recent collections from different parts of the country have led to the addition of 334 wild accessions and 700 accessions of varieties to the national collection. The total germplasm of rice on hand with the National Gene Bank is estimated to be around 94,000 (Subba Rao et al., 2015).

India has the largest network of protected areas with 553 national parks and sanctuaries providing *in situ* conservation of plant and animal biodiversity. Besides, eight biodiversity-rich areas have been designated as biosphere reserves to conserve respective ecosystems. Five of the protected areas have been designated as world heritage sites under the UNESCO's (United Nations Education, Scientific, and Cultural Organization) world heritage program. India has also taken timely action for the collection and long-term *ex situ* conservation of the rich biodiversity available in the country by *in situ* protection of habitat and ecosystem implemented by MoEF (Ministry of Environment and Forest) and *ex situ* by the establishment of NBPGR in 1976. The Bureau coordinates plant explorations for collection, facilitates the introduction and exchange, ensures medium- and long-term conservation, compiles a database on plant genetic resources, and offers training in all aspects of the management of agro-biodiversity.

18.5.4 ROLE OF LOCAL COMMUNITIES IN CONSERVATION

The indigenous communities in India are inhabitants of a region which is rich in ethnic biodiversity. In fact, these communities are the custodians of a vast genetic wealth of native genotypes. The probable reason for the richness of plant diversity is not only due to rich ecological diversity but also due to the preponderance of religious rituals and ethnic customs of the communities. The past linkages with Indo-Chinese–Indonesian, Chinese–Japanese, and the Central Asian region helped in augmenting our crop plant resources. Also, the influx of genetic material in the distant past from the Mediterranean and African regions resulted in the accumulation and diversification of enormous genetic variability. The ancient travelers, invaders, and religious missionaries have also contributed significantly towards enriching the Indian gene center. Consumer preference has played a predominant role in bringing enormous genetic diversity in rice. For a range of rice-based goods, including rice flakes, popped rice, idli, dosa, papads, appam, and murukus, only certain rice cultivars have been chosen. The tribal people of the northeastern hill states have conserved a large number of very sticky putturices for use in manufacturing rice cakes and other dishes. In addition, numerous aromatic short grain varieties with superior grain quality are unique to different states in the country. Some of the land races are still under cultivation by the farmer despite their low yield only for their quality features. Some of the *Chakuwarices* of northeastern India are "instant rices" highly suitable for ready-to-eat preparations (puffed and soft rice).

18.6 CONCLUSION

In terms of rice production, India comes in second place to China. It is also home to a number of rice types with medicinal qualities that, according to both contemporary and traditional definitions, are considered to be healthy foods. There are reportedly up to 40,000 different types of rice, although just a few are now being grown extensively. These types must be saved immediately because they are quickly going extinct as a result of high-yielding cultivars and other cash crops. Although customers favor various specialty rice cultivars, such as colored and aromatic rice variants, white rice remains the most popular type of rice consumed worldwide. Compared to variations of white rice, specialty rice has a higher nutritional profile. The current environment calls for vigorous marketing of these kinds as well as increased public awareness of their significance, particularly among the younger population. To carve out a niche in the market, the therapeutic efficacy of these cultivars needs to be empirically supported. To stop the spread of lifestyle-related disorders, it is essential to promote and preserve this national treasure as a health food. Rich in dietary fiber, rice bran is a by-product of the rice milling industry that is used to create functional foods and other goods with additional value. The bran layer of the grain contains a high concentration of phytochemicals and antioxidants due to the anthocyanin color deposition. For the benefit of mankind, rice's nutritional content still needs to be improved. Both conventional, traditional ways of crop-growing and contemporary biotechnology are being used to increase the contents of micronutrients in rice. Rice could be bio-fortified with high iron and zinc compounds, which could result in high-quality yields and be well-received by both rice growers and rice consumers.

KEYWORDS

- **chronic disease**
- **glycemic index**
- **hypertension**
- **medicinal property**
- **nutrition**
- **rice**

REFERENCES

Ahuja, U., Ahuja, S. C., Chaudhary, N., & Thakrar, R., (2007). Red rices-past, present, and future. *Asian Agri-History, 11*(4), 291–304.

Ahuja, U., Ahuja, S. C., Thakrar, R., & Shobha, R. N., (2008a). Scented rices of India. *Asian Agri-History, 12*(4), 267–283.

Ahuja, U., Ahuja, S. C., Thakrar, R., & Singh, R. K., (2008b). Rice- a nutraceutical. *Asian Agri- History, 12*(2), 93–108.

Anonymous, (1997a). *Carbohydrates in Human Nutrition.* Report of joint FAO/WHO expert consultation. Food and Nutrition Paper-66. FAO, Rome.

Anonymous, (1997b). *Standing Committee on the Scientific Evaluation of Dietary References Intakes. Dietary Reference Intakes for Calcium, Phosphorous, Magnesium, Vitamin D, and Fluoride*. Institute of Medicine, US. National Academies Press, Washington, DC.

Anonymous, (2018). *Indigenous Rice Varieties Make a Comeback.* The Hindu. The Hindu Group, India.

Arumugasamy, S., Jayashankar, N., Subramanian, K., Sridhar, S., & Vijayalakshmi, K., (2001). *Indigenous Rice Varieties*. Centre for Indian Knowledge System (CIKS), Chennai, Tamil Nadu, India.

Bakers, H. G., (1978). *Plant and Civilization* (3rdedn.). Words Worth Pub Co Inc, California, USA.

Basnet, D. B., (2003). *Ethnobotany, 15*, 64–66.

Bhat, F. M., & Riar, C. S., (2015). Health benefits of traditional rice varieties of temperate regions. *Med. Aromat. Plants, 4*, 198.

Bradbury, J. H., & Holloway, W. D., (1988). *Chemistry of Tropical Root Crops: Significance for Nutritional and Agriculture in the Pacific* (p. 201). Australian Centre for International Agriculture Research Canberra.

Caius, J. F., (1986). *The Medicinal and Poisonous Plants of India.*Scientific Publishers, Jodhpur, India.

Chang, T. T., (1976). The origin, evolution, cultivation, dissemination, and diversification of Asian and African rices. *Euphytica, 25*, 431–441.

Chaudhari, P. R., Tamrakar, N., Singh, L., Tandon, A., & Sharma, D., (2018). Rice nutritional and medicinal properties: A review article. *Journal of Pharmacognosy and Phytochemistry, 7*(2), 150–156.

Chaudhary, R. C., & Tran, D. V., (2001). Specialty rices of the world – a prologue. In: *Specialty Rices of the World: Breeding, Production, and Marketing.*

Chen, M. H., McClung, A. M., & Bergman, C. J., (2016). Concentrations of ligomers and polymers of proanthocyanidins in red and purple rice bran and their relationships to total phenolics, flavonoids, antioxidant capacity and whole grain color. *Food Chemistry, 208*, 279–287.

Eggum, B. O., (1969). Evaluation of protein quality and the development of screening technique. In: *New Approaches to Breeding for Improved Plant Protein* (pp. 125–135). Vienna IAEA.

Eggum, B. O., (1977). Nutritional aspects of cereal protein. In: Muhammad, A., Aksel, R., & Von, B. R. C., (eds.), *Genetic Diversity in Plants* (pp. 349–369). Plenum press New York.

Eggum, B. O., (1979). The nutritional value of rice in comparison with other cereals. In: *Proc. Workshop on Chemical Aspects of Rice Grain Quality* (pp. 91–111). IRRI, Los Banos, Laguna, the Philippines.

Frei, M., & Becker, K., (2004). *On Rice, Biodiversity and Nutrients.* Institute of Animal Production in the Tropics and Subtropics, University of Hohenheim, Stuttgart.

Gul, K., Yousuf, B., Singh, A. K., Singh, P., & Wani, A. A., (2015). Rice bran: Nutritional values and its emerging potential for development of functional food: A review. *Bioactive Carbohydr. Diet. Fibre, 6*(1), 24–30.

Hedge, S., Yenagi, N. B., & Kasturiba, B., (2013). Indigenous knowledge of the traditional and qualified Ayurveda practitioners on the nutritional significance and use of red rice in medications. *Indian Journal of Traditional Knowledge, 12*, 506–511.

Ichikawa, H., Ichiyanagi, T., Xu, B., Yoshii, Y., Nakajima, M., & Konishi, T., (2001). Antioxidant activity of anthocyanin extract from purple black rice. *J. Med. Food, 4*(4), 211–218.

Iqbal, S., Bhanger, M. I., & Anwar, F., (2005). Antioxidant properties and components of some commercially available varieties of rice bran in Pakistan. *Food Chem., 93*(2), 265–272.

Jenkins, D. J. A., Leeds, A. R., Gassell, M. A., Cocket, B., & Alberti, K. G. M., (1977). Decrease in post prandial insulin and glucose concentrations by gaur and pectin. *Ann. Intern. Med., 86*, 20–23.

Juliano, B. O., (1985). *Rice Chemistry and Technology.* American Association of Cereal Chemists, USA.

Kline, K., Yu, W., & Sanders, B. G., (2004). Vitamin E and breast cancer. *J. Nutr.*, *134*(12), 3458S–3462S.

Krishnamurthy, K. S., (1991). *The Wealth of Sushruta*. International Institute of Ayurveda, Coimbatore, Tamil Nadu, India.

Kumar, T. T., (1988). *History of Rice in India*. Gian Publishers Delhi, India.

Lemos, M. R. B., & Souza, S. L. A., (2000). Rice and its by-products in southern Brazil. *Vetor: Revista de CienciasExatas e Engenharias*, *10*, 21–36.

Lloyd, B. J., Siebenmorgen, T. J., & Beers, K. W., (2000). Effects of commercial processing on antioxidants in rice bran. *Cereal Chem.*, *75*(5), 551–555.

Majumdar, G. P., & Banerji, S. C., (1960). *Krsi-Parasara*. The Asiatic Society, West Bengal, Calcutta, India.

Mohanty, R. B., Panda, T., & Tripathy, B. K., (2012). Medicinal rice varieties of India need urgent attention. *Curr. Sci.*, *102*(4), 547.

Nam, Y. J., Nam, S. H., & Kang, M. Y., (2008). Cholesterol – lowering efficacy of unrefined bran oil from the pigmented black rice (*Oryza sativa* L cv. Suwon 415) in hypercholesterolemic rats. *Food Sci. Biotechnol.*, *17*, 457–463.

Oliveira, M. S., Feddern, V., Kupsk, L., Cipolatti, E. P., Badiale, F. E., & De Souza, S. L. A., (2011). Changes in lipid, fatty acids, and phospholipids composition of whole rice bran after solid state fungal fermentation. *Bioresour. Technol.*, *102*(17), 8335–8338.

Oudhia, P., (1999). Medicinal weeds in rice fields of Chhattisgarh (India). *Internat. Rice Res. Notes, 24*(1), 40.

Pojer, E., Mattivi, F., Johnson, D., & Stockley, C. S., (2013). The case for anthocyanin consumption to promote human health: A review. *Compr. Rev. Food Sci. Food Saf.*, *12*(5), 483–508.

Qi, S., (2010). White rice, brown rice, and risk of type 2 diabetes in US men and women. *Arch Intern. Med.*, *170*(11), 961–969.

Roy, S., & Bhattacharjya, S., (2012). Medicinal rice: an ignored episode. *Adv. Res. J. Crop Improv.*, *3*(1), 64–66.

Saunders, R. M., (1985). Rice bran: Composition and potential food uses. *Food Rev. Int.*, *1*(3), 465–495.

Sharma, P. V., (1996). *Classical Uses of Medicinal Plants* (p. 848). Chaukhamba Vishwabharati, Varanasi. Uttar Pradesh, India.

Siddiq, E. A., (2002). *The Hindu Survey of Indian Agriculture*, 47–52.

Souci, S. W., Fuchmann, W., & Kraut, H., (1987). *Food Composition and Nutrition Tables 1986/87* (3rd rev. edn.). Stuttgarti, Wissen Schaftloche Verlagsgeseil Schaft.

Subba, R. L. V., Chaitanya, U., & Kota, S., (2015). Rice genetic resources and their utilization for quality, nutrition, and medicinal purposes. In: *Training Manual of Short Course on Quality Improvement, Bio-fortification and Product Development in Rice for Nutritional and Financial Security of Rice Farmers* (pp. 25–29).

Sugano, M., & Tsuji, E., (1997). Rice bran oil and cholesterol metabolism. *J. Nutr.*, *127*(3), 521S–524S.

Suzuki, A., Kagawa, D., Fujii, A., Ochiai, R., Tokimitsu, I., & Saito, I., (2002). Short- and long-term effect of ferulic acid on blood pressure in spontaneously hypertensive rats. *Am. J. Hypertens.*, *15*(4), 351–357.

Tsuda, T., Horio, F., Uchids, K., Aoki, H., & Osawa, T., (2003). Dietary cyanidin 3-O-beta-Dglucoside- rice purple corn color prevents obesity and ameliorates hyperglycemia in mice. *J. Nutr.*, *133*(7), 2125–2130.

Umadevi, M., Pushpa, R., Sampathkumar, K. P., & Bhowmik, D., (2012). Rice-traditional medicinal plant in India. *Journal of Pharmacognosy and Phytochemistry*, *1*, 6–12.

Yan, J. J., Cho, J. Y., Kim, H. S., Kim, K. L., Jung, J. S., & Huh, S. O., (2001). Protection against beta amyloid peptide toxicity *in vivo* with long-term administration of ferulic acid. *Br. J. Pharmacol.*, *133*(1), 89–96.

CHAPTER 19

Pests and Diseases of Some Commercial Medicinal and Aromatic Plants and Their Management

RAKESH PANDEY[1] and SATYAJIT HEMBRAM[2]

[1]*Microbial Technology and Nematology Department, CSIR–Central Institute of Medicinal and Aromatic Plants, P.O. CIMAP, Lucknow, India*

[2]*Regional Research Station (Terai Zone), Uttar Banga Krishi Viswavidyalaya, Pundibari, Cooch Behar, West Bengal, Indial*

ABSTRACT

Global human population, especially in developing nations, still depends on medicinal and aromatic plants (MAPs) for treating various diseases through drugs, antioxidants, flavoring agents, fragrances, etc. COVID-19 outbreaks provide an opportunity to once again explore many forgotten MAPs, which also have an impact on the global market. To meet the upcoming demand for MAPs, disease-free plants and planting materials are necessary. Nematodes, fungi, bacteria, viruses, and harmful insects pose a serious threat to MAPs worldwide. Therefore, disease management is essential to increase the quality and quantity of plant-derived products or secondary metabolites. This chapter provides an overview of proven disease management practices that are of medicinal importance and commercial value. These practices will be highly useful in protecting MAPs cultivation worldwide. Furthermore, they can be utilized in future plant breeding programs to develop disease-resistant or tolerant genotypes against major pests and pathogens in MAPs.

Bioprospecting Ethnomedicinal Plant Resources: Sustainable Utilization and Restoration.
Gopal Shukla, Jahangeer A. Bhat, A. P. Das, and Sumit Chakravarty (Eds.)

19.1 INTRODUCTION

MAPs are highly useful for treating various ailments worldwide. Huge global human population, especially in developing nations, mainly depend on the traditional system of medicine for treating various human diseases. Currently, the traditional medicine derived from different medicinal plants has gained more significance and has become one of the most significant means to maintain the traditional system of medicine in most of the developing countries. Similarly, aroma compounds from botanical sources hold a promising area by being increasingly used in perfumery, nutraceuticals, food, and flavor industries due to the growing awareness in common masses about the risks involved in synthetic components in parallel products. Accordingly, nearly more than one lakh such secondary metabolites are now known to occur in about 50,000 plant species, and almost 4,000 new secondary metabolites are being discovered every year from a number of plant species (Verpoorte et al., 1999; Gomez-Galera et al., 2007; Pandey, 2017). Since the last few decades, these natural plant products have been utilized for human healthcare in the form of drugs, antioxidants, flavors, fragrances, dyes, pesticides, pheromones, etc. During the COVID-19 outbreak, there is an exhaustive search for plant-based active molecules in order to find therapy for this pandemic. In such a situation, plant-based therapy for boosting the human immune system is playing the most significant role in combating this deadly viral disease. Consequently, the current global herbal drug market has reached a level of more than US $62 billion, which is expected to grow to US $5 trillion by the year 2050 (Joshi et al., 2004). The world market for herbal medicines, including herbal products and raw materials, is actually growing at an annual rate of 10–15%. This is an indication of a possible growing demand for plant-derived drugs in the coming years. Despite this, the area under cultivation of medicinal and aromatic plants (MAPs) has been slowly increasing due to increasing demand for herbal drugs (Pandey et al., 2019). The use of MAPs plant species in the human healthcare system cannot be ignored due to the following reasons: (i) low concentration of desired molecules in definite plant tissues; (ii) medicinal crops being low-priority crops; (iii) no technical difficulty in cultivation of these crops; (iv) over-harvesting of naturally occurring herbal plants; (v) destructive collection techniques; and (vi) conversion of habitats of medicinal plants to crop-based agriculture. The business of MAP industries could be enhanced by ensuring disease-free plants and planting materials. Among the different plant pathogens, fungi, bacteria, viruses, phytoplasma, plant parasitic nematodes, etc., are the major serious menace to the future prospects of MAPs globally.

19.2 PESTS AND DISEASES OF MINTS

Mentha spp. is group of aromatic herbs belonging to family Lamiaceae, which are considered to be the most important cash crop in India. Different types of mints, which are commercially cultivated in Indo-Gangetic plains, are: Menthol mint (*Mentha arvensis*), Peppermint (*Mentha piperita*), Spearmint (*Mentha spicata*), Scotch spearmint (*Mentha cardiaca*), Bergamot mint (*Mentha citrata*), and Garden mint (*Mentha viridis*). Among different MAPs, mints come in the front line because Indian farmers grow it as a bonus crop and it also fits well in the cropping system with other crops like paddy, wheat, potato, sugar cane, etc.

19.2.1 FUNGAL DISEASES

Rust is the major disease of *Mentha arvensis* and *M. spicata* is caused by *Puccinia menthae* Pers. (Shukla et al., 1999). This fungus infests leaves and twigs of *Mentha* species and produces pustules on the upper and lower leaf surfaces, which are reddish brown and yellow orange in color. Rust fungi defoliate the crop, causing severe economic loss by reducing its oil yield in all mint growing areas of India (Ganguli & Pandotra, 1962; Shukla et al., 1999). In the Tarai region of Uttar Pradesh, the dual infection of *P. menthae* and *Alternaria alternate* has been reported (Shukla et al., 1999). The young shoots infected by rust fungi are usually twisted and most of the time broken off at the infection site (Plate 19.1(A)).

In menthol mint, typical symptoms of rust pustules develop on the lower surface of leaves between April and June. In general, there are three types of pustules found on the infected area. The first type of pustules are circular, yellowish-brown in color, whereas the second is orange in color, and the third type of pustules are of variable shape and size and are dotted tan in color. These fungi generally persist in the form of uredospores on runner and plant debris of the host plant. The rust pustules possess enough uredospores which, under favorable conditions, infect the host plant. The rust fungi also form telia at a later stage, which are thick-walled teliospores and survive in adverse conditions. The use of disease-free materials is the major preventive measure of this disease. Treatment of planting materials with hot water and

PLATE 19.1 (A) Rust of *Mentha arvensis*; (B) mosaic on leaves of *Mentha gracilis*; (C) *Alternaria* blight of patchouli; (D) yellow mosaic symptoms on patchouli; (E) root-knot nematode infested field of patchouli; (F) healthy and root-knot infested plant; and (G) roots of Egyptian henbane.

the use of resistant varieties is a common method of managing this disease. Elimination of old mint rust-infected mint plants also helps to prevent the disease from spreading and reduces the rust inoculum. The application of a few fungicides like mancozeb, propiconazole, Diclobutazol, nickel chloride, denitroamine, and Krezonit-E (DNOC) offers reliable protection against rust disease (Suab & Nagy, 1972; Margina & Zheljazkov, 1994). Himalaya,

Kalka, and Damroo menthol mint varieties/genotypes were found resistant to rust fungi, whereas genotypes MAH-3, S-10-11-45, and S-13-2-125 were moderately resistant (Shukla et al., 2008). The variety Kalka was resistant to rust, leaf spot, and leaf blight, but was found susceptible to powdery mildew, and stem blackening and rot.

19.2.2 VIRAL DISEASES

Several viruses belonging to more than 12 genera have been reported to cause diseases in various species of *Mentha spp*. World-wide, the important genera include *Alfamovirus, Begomovirus, Cheravirus, Closterovirus, Cucumovirus, Cytorhabdovirus, Hordeivirus, Nepovirus, Potexvirus, Tobamovirus, Tospovirus, Vitivirus*, etc. However, there are only few reports of viral diseases on mints from India which may be attributed that this crop has come in regular and mass cultivation lately. A light-yellow mosaic disease has been reported from Lucknow on *Mentha spicata* cv. viridis (Samad et al.,2009). The symptoms include mosaic, yellowing, leaf curling, crinkling, and retarded growth that induced drastic reduction in herb yield. An isolate of Tomato leaf curl Pakistan virus (*Begomovirus*) was identified from the infected plants. Tobacco mosaic virus (genus *Tobamo virus*)was first reported in Scotch spearmint (*M. cardiaca* Baker) from India (Samad et al., 1994). Subsequently an isolate of the virus was isolated from Scotch spearmint (*M. gracilis* Sole; Samad et al., 2000). The disease symptoms included green mosaic with vein banding, deformation of leaves, and retarded growth (Plate 19.1(B)). Another mild mosaic symptom has been recorded on peppermint (*M. piperita* L.). Initially terminal leaves of the diseased plant showed mottling, curling, and finally mild mosaic symptom which spread in whole plant giving sick appearance, associated with heavy stunting and occasional necrosis of the leaves. Flowering was delayed with small size flower.

19.2.3 NEMATODE DISEASES

19.2.3.1 MENTHOL MINT

In India, the commercial cultivation of menthol mint suffers a great yield loss due to root-knot disease caused by *Meloidogyne spp*. Both the species of root knot nematode (*Meloidogyne incognita* and *M. javanica*) occur together but the frequency of *M. incognita* is greater than *M. javanica* (Pandey et al.,

1992). In the root rhizosphere other plant parasitic nematodes viz. *Tylenchus* sp., *Tylenchorhynchus vulgaris*, *Hoplolaimus indicus*, *Helicotylenchus dihystera*, *Pratylenchus thornei*, *Rotylenchulus reniformis*, *Aphelenchoides* sp., *Longidorus pisi*, and *Xiphinema* sp. were reported (Pandey et al., 1992; Pandey, 1997a). Buhrer (1938) for the first time reported that *M. arvensis* as host of Meloidogyne species. For the first time Pandey (1989) reported the root-knot disease of menthol mint *(Mentha arvensis*) and Bergamot mint (*M. citrata*). Two major nematode species viz. *M. incognita* and *M. javanica* were found to be associated with menthol mint. Introduction of *M. arvensis* as a commercial crop into sandy loam soil in India showed that a threat was posed by the root knot nematode, *Meloidogyne incognita* (Kofoid *et* White) Chitw. (Haseeb & Pandey, 1989). As this species is sedentary endoparasitic, its spread is facilitated by infected suckers/roots that are used in the vegetative propagation of the crop (Pandey, 1998).

Whole roots and suckers are invaded by root-knot nematode. Small to medium-sized galls were noticed on *M. incognita*-infested roots. Secondary roots of suckers are also highly infested with *M. incognita,* and most of the time, whitish-colored egg masses of the pest are seen on the root system. Heavily root-knot-infested menthol mint plants are stunted, poorly tillered, and have chlorotic leaves. Infested suckers serve as a major source of infection and means of dissemination. Root-knot nematodes continue to develop after the crop has matured and been harvested. Root-knot nematodes (*M. incognita*) cause a 25–30% oil yield reduction in menthol mint. The quality of mint oil was also adversely affected due to nematode infection (Pandey et al., 1992). Eco-friendly management of plant-parasitic nematodes was reported inefficient due to the varying nature of nematodes; however, actinomycetes fungi can improve plant growth biomass and inhibit nematode infection as well (Pandey et al., 1998). Successful control of root-knot disease was achieved with carbofuran @3 kg a.i./ha, neem cake @500 kg/ha, and integration of both (Pandey, 1995, 2000, 2005). A moderate to high degree of resistance to nematodes in mint species was also reported (Pandey & Patra, 2001). Most plant-parasitic nematodes, which affect crop yield, occur in the soil or plant roots. Prevention of spreading of phytonematodes from nematode-infested fields to non-infested fields is an important part, as the nematodes that damage mints are generally endoparasitic in nature and they can easily spread through suckers or runners. Therefore, it is essential and of utmost importance to check their spread in new areas. It is recommended to use certified suckers as planting material while multiplying them in nematode-free soil. Equipment should be cleaned before moving between fields, preventing animal movement from infested to uninfected fields, and eliminating important weeds and alternate hosts of phytonematodes.

19.2.3.2 PEPPERMINT

Nematodes like *Pratylenchus thornei, P. penetrans, P. scribeneri,* and *P. minyus* are the major nematode causing root lesion in *Mentha piperita* (Pandey, 1997b, c). *Pratylenchus* species is migratory endoparasitic nematode and attack the root cortex. Roots develop dark lesions and an overall brown color. The lesions first develop on younger roots but may be found anywhere on the roots. The lesions enlarge, coalesce, and ultimately girdle the whole root, which appears constricted and then followed by secondary attack by fungi and bacteria. The affected area may be sloughed off and the overall root system is reduced. Loss of the epidermis and cortex decreases root growth, nutrient, and water uptake. Under stress, plants get yellow and stunted with reduced yields. The extent of damage depends on the phenolic content of the cells. The oxidation of these phenols causes the characteristic browning of the lesions. The aerial symptoms resemble malnutrition or water deficiency. In the initial stages, the infestation can be noticed as small patches of stunted growth of mint and yellowing of leaves. This stunting increases every year to form a large patch. Leaves may become yellowish-brown and wilt during April–May. The root system of mint species becomes very poor when severely infested with *P. thornei* and can be pulled out from the field very easily. Ethoprop @6.6 kg a.i/ha is very effective to reduce the root lesion nematode population in soil (Merrifield & Ingham, 1996). The non-chemical methods are getting more importance nowadays due to the hazardous effects of chemical pesticides (Pandey, 1993). The use of resistant crops and varieties provides the best option for farmers to control *Pratylenchus* as these crops will limit the multiplication of the nematode and reduce losses to following crops. Good soil nutrition (particularly nitrogen, phosphorus, and zinc) can reduce the effects of *Pratylenchus* on infested peppermint plants.

19.2.3.3 SPEARMINT

Species of root lesion nematode like *Pratylenchus thornei, P. penetrans*, and *P. scribeneri* are associated with spearmint (Pandey, 1997b). Pathogenicity of *P. thornei* was established on spearmint (Pandey, 1997b). The damage becomes several folds when *Pratylenchus* species interact with plant pathogenic fungi in spearmint crops. Ethoprop @6.6 kg a.i/ha is highly effective to control root lesion of spearmint (Pandey, 1993).

19.2.3.4 SCOTCH SPEARMINT, BERGAMOT MINT, AND GARDENMINT

Spearmint, Bergamot mint, and Garden mint are affected by a large number of plant parasitic nematodes other than root knot and root lesion nematodes. In India, the most important nematode pathogen of Spearmint and Garden mint is *M. incognita*, while Bergamot mint is affected by *P. thornei* (Pandey, 1997b). Among these three mint species, *M. cardiaca* and *M. viridis* are the most susceptible, while *M. citrata* is considered to be tolerant to *M. incognita* and *M. javanica* (Pandey, 1998). It is essential and of utmost importance to use nematode-free planting material to check the spread of the pathogen.

19.3 DISEASES OF PATCHOULI

Patchouli (*Pogostemon patchouli* Pellet.), (Syn. *P. cablin* Benth.) is an important aromatic plant belonging to family Lamiaceae and the major source of commercial patchouli oil. In the perfumery industries still there is no synthetic chemical which could replace the patchouli oil. It is native to the Philippines and cultivated mainly in Malaysia, Java, Brazil, West Indies, Singapore, Indonesia, China, and India. It is a perennial aromatic herb that yields fragrant leaves containing very sweet-smelling oil, attains a height of about 1.0–1.2 meters. The chemical components of patchouli oil are b-patchoulene, a-guaiene, caryophyllene, a-patchoulene, seychellene, a-bulnesene, norpatchoulenol, patchouli alcohol, and pogostol. It is propagated by stem cuttings. The commercial oil of patchouli is obtained by steam distillation of the shade-dried herbage. The oil of patchouli has a strong fixative property which helps to prevent the rapid evaporation of perfume and thereby promotes tenacity. The oil is generally blended with other essential oils (EOs). It is used in a wide range of toilet soaps, scents, body lotions, etc. At very low concentration, the oil is extensively used to add flavor to major food products including alcoholic and non-alcoholic beverages, frozen dairy desserts, candy, etc. It also possesses medicinal properties like helpful in relieving from all forms of depression, anxiety, and stress-related conditions. The other therapeutic properties of patchouli oil are antiseptic, aphrodisiac, astringent, deodorant, diuretic, febrifuge, fungicide, sedative, and also used in several other pharmaceutical preparations. It is used as a general tonic and a stimulant and helps the digestive system (Sarwar et al., 1982). In India, this plant is cultivated in the states of Assam, Madhya Pradesh, Tamil Nadu, Kerala, and Karnataka. This crop suffers from fungal, viral, and nematode diseases.

19.3.1 FUNGAL DISEASE

Wilt is the most important disease of several MAPs, but in patchouli, it is a very serious disease caused by *Fusarium solani*. The prevalence of this disease is mainly during the kharif season. The infected plants are characterized by gradual yellowing and drying of leaves followed by premature death. The plants show discoloration of roots and collar regions of fully grown plants. Severe disintegration of the secondary root surface is also observed. Wilted mature plants are characterized by blackening of roots at collar regions along with severe disintegration of the secondary root surface (Singh & Angadi, 1990). Similar yellowing of leaves is seen in Alternaria blight of patchouli (Plate 19.1(C)). EO of *Thymus vulgaris* iseffective from a minimum concentration of 50 ppm while, mint and lavender oils were only effective at higher concentrations of 400–800 ppm and in the field neem, Eucalyptus, *Ocimum basilicum*, *Prosopis cineraria,* and *Pongamia pinnata* oils are also effective (Aulerio & Zambonelli, 1997; Rai et al., 1999). Ethanol extract of *Allium cepa*, *Allium sativum*, *Lantana camara*, *Polyalthia longifolia*, *Tagetes erecta*, *Vinca rosea,* and *Withania somnifera*. sis effective against the rust (Shivapuri et al., 1997; Tripathi et al., 1999; Mamatha & Ravishankar Rai, 2004). *Bacillus subtilis* is antagonistic to *F. solani* in culture and thus seed treatment can significantly reduce seed colonization and root rot (Sarhan, 1989). Fluorescent Pseudomonads and *Trichoderma harzianum* strain THA were found to be highly effective against *F. solani* in dual culture (Zapata et al.,2001; Kavitha et al., 2004). Carbendazim, iprodione, Topsin-M, vitavax, and benomyl (500 mg a.i./ml or 10, 50, and 100 ppm) were found to be the best fungicides in disease control (Mishra & Rath, 1987; Kapoor & Kumar, 1991; Etebarian, 1992; Wahid et al., 1995).

19.3.2 VIRAL DISEASE

Patchouli is known to be infected by a number of viruses such as *Patchouli mosaic virus* (PaMV), *Tobacco necrosis virus* (TNV), *Patchouli mild mosaic virus* (PaMMV), *Patchoulimottle virus* (PaMoV), *Patchouli yellow mosaic virus* (PaYMV), and *Peanut stripe virus* (PStV) (Sastry & Vasanthakumar, 1981; Subba Rao, 1986; Zaim et al., 1999; Singh et al., 2009). The typical yellow mosaic appears when the temperature is high during April to June (Plate 19.1(D)). These symptoms generally are masked during the winter season. The Lucknow isolate of the virus, which is tentatively designated as patchouli yellow mosaic virus (PaYMV), was transmitted by sap and aphids. The host range was found to be restricted. The virus incited a typical line pattern on the leaves of *Nicotiana benthamiana*, which is a diagnostic host for this virus. The purified virus preparation showed flexuous rod-shaped virus particles about 805 nm in length. The virus was detected by RT-PCR using potyvirus-specific degenerate primers.

19.3.3 NEMATODE DISEASE

Root-knot nematode (*Meloidogyne incognita, M. javanica, M. hapla*) become most important constraint for the successful cultivation of patchouli in India. Root-knot infested plants are weak and grow slowly. Heavy galling on root system with root knot nematode on patchouli results in stunting, wilting, defoliation, and chlorosis of the plant (Plate 19.1(E)). Sometimes root galls are very small, or the surrounding galls coalesce to form large ones up to 2–5 cm or even larger. Infection of root-knot nematodes occurs when plants are in their early stage of development. Root-knot diseases of patchouli are caused by *Meloidogyne incognita, M. javanica*, and *M. hapla* (Krishna Prasad & Reddy, 1979, 1984). Good control of *M. incognita* on patchouli was obtained by nemagon, metham sodium, fenamiphos, aldicarb-sulfone, aldicarb-fensulfothion, carbofuran, phorate, cyerolan AC-92 and AC-100 @2–3.6 and 10 kg a.i./ha as pre and post inoculation treatment (Sarwar et al., 1982; Krishna Prasad & Reddy, 1984; Kumar & Nanjan, 1984). Neem oil seed cake @ 4 tons/ha and *Mucuna purita* as rotational crop or summer fallowing effectively reduce nematode population while also improving growth and yield of the crop.

19.4 DISEASES OF ASHWAGANDHA

Ashwagandha (*Withania somnifera* L. Dunal), also known as Indian ginseng, belongs to the family Solanaceae, is an important ancient medicinal plant, used in the Indian traditional systems of medicine, ayurveda, and Unani.

It is a widespread species having distributed from the southern Mediterranean area to the different parts of Africa, India, Israel, Jordan, Egypt, Sudan, Iran, Afghanistan, and Pakistan. It grows in dry parts in subtropical regions. In India, it is commercially cultivated in the states of Rajasthan, Punjab, Haryana, Uttar Pradesh, Gujarat, Maharashtra, Tamil Nadu, and Madhya Pradesh. The estimated annual production of ashwagandha roots in India is more than 2,000 tons, against an annual requirement of about 7,000 tons (Farooqi et al., 2003). Ashwagandha is regarded as the treasure of biochemical constituents serving as a remedy for many health problems. The roots contain several pyrazole alkaloids like withasomnine, withaferin A, withanolides, withaninol, and steroidal lactones, starch, and reducing sugars. Withaferin is the chief constituent (0.13–0.31%) having bacteriostatic and antitumerous properties. It is considered to be one of the best rejuvenating agents in Ayurveda (Farooqi et al., 2003). Its roots, seeds, and leaves are used in ayurvedic and Unani medicines. Roots are used for treatment of rheumatic pain, inflammation of joints, nervous disorders, and epilepsy, and also used as tonic for hiccup, cold, cough, and female disorders. The leaves are also useful for carbuncles, inflammation, swellings, and conjunctivitis.

19.4.1 FUNGAL DISEASES

Root-rot and wilt are one of the most important diseases of this crop in the nurseries and commercial field ashwagandha growing areas of Uttar Pradesh. The diseases occur during April-May and cause 30–50% mortality of plants (Gupta et al., 2004). The disease is caused by *Fusarium solani*. The major symptom of the disease is withering and drooping of plants whereas, at an advanced stage, it causes severe wilting leading to decomposing and death of underground parts. The cottony growth appears around the infected roots under moist conditions. The infected roots become soft and brown in color. Mature leaves turn yellow with necrotic margins, wither, and shed off. In addition to killing plants and reducing the quantity of saleable crop, root rots can also slow or stop plant growth and thus suppress plant quality. Root-rotted plants are usually smaller, less vigorous, produce fewer and/or smaller leaves, flowers, and fruit than healthy plants of equal age. Flowering may be delayed when the roots are rotted. As a result, the crop quality is very uneven. Root rots must be managed early in the disease if the losses are to be avoided. The disease root rot and wilt complex of ashwagandha is caused by soilborne pathogens. Before any action is taken, a diagnosis must be made of the actual cause of the symptoms. If the damage is due to the activity of fungi, chemicals are sometimes available which can check the fungus and allow the plant to grow.

19.4.2 PHYTOPLASMA DISEASE

In CIMAP, Lucknow, the Witches broom disease was first observed in 1988 and more than 30% of plants were affected by this disease (Zaim & Samad, 1995). The succession of symptom appearance includes little leaf, phyllody, dense clusters of highly proliferating branches with shortened internodes resulting in witches' broom and bushy appearance of infected plants (Plate 19.2).

PLATE 19.2 Phytoplasma infected *W. somnifera* plant along with healthy plant (A); electron microscopy of phytoplasma infected plant (B); and field view of naturally infected plants (C).

19.4.3 NEMATODES

Root-knot disease of ashwagandha caused by the root-knot nematode *Meloidogyne incognita* (Kofoid and White) Chitwood, is very widespread affecting more than 80% of plants in India (Pandey & Kalra, 2003). A large number of other phytoparasitic nematodes were found to be associated in the root rhizosphere of ashwagandha. The nematode-infected plants typically show chlorotic, stunted, less branched with fewer and smaller leaves, and poor response to fertilizer and irrigation. Such symptoms usually are not noticeable until severe damage to the root system has been done by the nematodes. Roots of such plants were severely galled. If the stem touches the soil, it also gets infested with root-knot nematode. It was also noticed that root-knot nematode infected plants are more likely to be killed early with adverse effects on the environment than healthy non-infested plants. Neem, *Artemisia annua marc*, *Mentha*, and *Murraya koenigii* distillates were found to be highly useful in suppressing root-knot development and *M. incognita* population (Pandey & Kalra, 2003). Bio-agents, however, were found to be less effective compared to organic materials due to host reactions towards the bio-agents. Integration of bio-organics like FYM (farmyard manure), cow urine, neem cake, and vermicompost with *Trichoderma harzianum* and *Mentha* distillates with *Glomus aggregatum* were found to be nematode suppressive and enhanced the growth of the crop (Pandey & Kalra, 2010).

19.5 DISEASES OF HENBANES

Henbanes (*Hyoscyamus muticus*, *H. niger*, and *H. albus*) are important tropane alkaloid bearing plants belonging to family Solanaceae and one of the chief sources of tropane alkaloids like hyoscine or scopolamine, hyoscyamine, and atropine which are obtained from the dried leaves and other plant parts. The hyoscine and its derivatives are used in pharmaceutical preparations, since they have anticholinergic, antispasmodic, and mydriatic properties. Leaves, seed, and flowering tops of the plant are used in respiratory diseases and old cough. These plants originated in Egypt and spread to the rest of the tropics. The two species of henbane, i.e., black henbane (*H. niger*) and Egyptian henbane (*H. muticus*) are widely cultivated for pharmaceutical purposes.

19.5.1 FUNGAL DISEASE

Leaf spot caused by *Ascochyta kashmiriana* is an important disease and first time reported in the state of Jammu and Kashmir (Ganguli & Pandotra, 1962) Leaf spot on Egyptian henbane is caused by *Alternaria alternata*. This disease appeared during cultivation in the months of January-March. On infected leaves, several irregular brownish spots occur and later on these spots enlarge in size and become big spots. The infected leaves generally defoliate prematurely and consist of spores. These spores are the major source of perpetuation of the disease. Spraying of Dithane M-45 (0.3%) or Kavach (0.2%) at an interval of 15 days minimizes the disease severity and incidence.

19.5.2 VIRAL DISEASE

Among the first viruses reported to cause diseases on hebane are Hyoscamus virus II, Hyoscamus virus III, and Hyoscamus virus I (Hamilton, 1932). Later, a number of diseases were reported caused by viruses belonging to the genus *Tobamovirus*, *Cucumovirus*, *Tospovirus*, *Potexvirus*, and *Tobravirus* (Samad et al., 1999, 2000; Zaim, 1999).

19.5.3 NEMATODES

Large number of plant parasitic nematodes viz. *Tylenchorhynchus vulgaris, Pratylenchus thornei, Hoplolaimus indicus, Longidorus pisi* and *Meloidogyne* sp. Were reported associated with different species of henbane but the nematode which has been reported to cause serious damage to the crop is root-knot nematodes, i.e., *Meloidogyne incognita* and *M. javanica* (Pandey, 1990). Root-knot nematode causes poor growth of the seedling establishment and leads to a significant decrease in the herbage yield of the plant (Plate 19.1(F) and (G)). The infested plants show chlorosis, stunted plant growth, and a patchy appearance with fewer smaller leaves and flowers, while the infested roots have severe galling of various degrees (Haseeb & Pandey, 1989). In the control program of root-knot nematode, the major attention has been paid to protecting the young seedlings. Reduction of nematode population in the soil prior to sowing is very important to avoid poor plant growth. Pre-sowing measures to suppress root-knot nematode population include rotation with non-host crops (*Cymbopogon flexuosus*, *C. winterianus*, and *C. martinii*), fallowing, green manuring, organic amendments, and application of some nematode antagonistic organisms. Marigold (*Tagetes spp.*) has the ability to suppress nematode population with antagonistic phytochemical in root exudates, i.e., polythienyls. Therefore, prior to sowing henbanes, growing *Tagetes minuta* at the distance of 30 cm x 30 cm can drastically reduce the nematode population. Removal of weed hosts can avoid the infestation of root-knot nematodes as these weeds serve as a reservoir for the nematodes. Avoiding the general practice of irrigating the noninfested field with water from an infested field will also restrict *Meloidogyne sp.* population spreading from one field to another. Chemicals such as carbofuran (@2 kg a.i./ha soil) and monocrotophos @ 0.1% SS) have been used to prevent root-knot nematode infestation of henbane. Monocrotophos can be used for seed soaking whereas carbofuran is preferred prior to sowing the seeds of the crop. The harvesting period of this winter crop is only four months, so sowing daicha (*Sesbania bispinosa*) as a green manuring crop can be used, which decreases nematode population below the economic damage levels. Inoculum of different mycorrhizal fungi showed better crop growth due to suppression of nematode population in comparison to untreated ones (Pandey et al., 1999). The best results were obtained with *Glomus aggregatum* for increasing growth/biomass of plants and reduction in root-knot nematode population. Experimental findings indicated that application of bio-inoculants have not only enhanced the total biomass yield of *H. niger* but it has also significantly decreased the multiplication of nematode, however, a significantly higher reduction was recorded in the treatment where all bio-inoculants are combined. Thus, mixed inoculation of rhizobacterium with VAM fungi could be considered as biological management instead of nematicides for reducing the deleterious effect of root knot disease in black henbane (Pandey, 1997c; Pandey et al., 2000a). Oils of *Cymbopogon martinii*, *C. winterianus*, *Ocimum basilicum* and *Mentha arvensis* were effective in reducing *M. incognita* population and improving the growth of plant; the oil of *C. martinii* @ 2 ml/pot being most effective (Pandey et al., 2000b).

19.6 CONCLUSION

The cultivation of MAPs has considerably increased in tropical and subtropical countries of the world to meet the requirements of pharmaceutical and perfumery industries. The diseases caused by a variety of pests and pathogens are major limiting factors for its production. Although chemical control of pests and pathogens has proved effective, it cannot be recommended to farmers because of its cost, harmful effects on human and animal health, and its percolation into groundwater resources. Other alternatives are available in medicinal plants, such as resistant germplasms, useful organic materials, effective biocontrol agents, and other cultural/physical methods of disease management. Such resistant and tolerant germplasms could be exploited in future plant breeding programs to develop resistant or tolerant genotypes against major pests and pathogens. Various organic materials are available that have been proven useful in decreasing disease incidence and enhancing MAP yield, which could be used in ongoing programs for healthier plants. A large number of bio-agents are available that could be used on a large scale to protect MAPs against various pests and pathogens. Based on the contributions made to the science of mycology, virology, and nematology in relation to MAPs, CIMAP has been placed in national and international records. However, much attention is also needed to study and develop new strategies to manage major diseases in an eco-friendly way that is cost-effective and environmentally friendly.

ACKNOWLEDGMENTS

Authors are grateful to Prabodh Trivedi, Director CSIR-CIMAP, Lucknow for his kind help and encouragement. Authors are also grateful to all colleagues of Crop Protection Division of CSIR-CIMAP for their help in all the ways during preparation of this manuscript.

KEYWORDS

- **Commercial MAP**
- **Disease**
- **Management**
- **Mentha**

REFERENCES

Anonymous, (2013). *PQR Database.* European and Mediterranean plant protection organization, Paris, France. Available at: http://www.eppo.int/DATABASES/pqr/pqr.htm (accessed on 18 December 2023).

Attique, M. R., Rafiq, M., Gaffar, A., Ahmad, Z., & Mahyuddin, A. I., (2003). Hosts of *Bemisiatabaci* (Genn.) (*Homoptera: Aleyrodidae*) in cotton areas of Punjab, Pakistan. *Crop Protection, 22*, 715–720.

Aulerio, A. Z., & Zambonelli, A., (1997). Fungal diseases of officinal plants: Spread, effects and defense. *Informatore-Agrario., 53*, 91–93.

Ayyar, T. V. R., (1984). *Handbook of Economic Entomology for South India* (p. 528). International Books & Periodicals Supply Service, New Delhi.

Ben-Dov, Y., (2020). *Coccidohystrix Insolita.* http://scalenet.info/catalogue/Coccidohystrix%20insolita/ (accessed on 18 December 2023).

Blackman, R. L., & Eastop, V. F., (2007). Taxonomic issues. In: Van, E. H. F., & Harrington, R., (eds.), *Aphids as Crop Pests* (pp. 1–29). CABI, UK.

Buhrer, E. M., (1938). Addition to the list of plants attacked by root-knot nematode (*Heterodera marioni*). *Plant Dis. Reptr.*, *22*, 216–234.

Chandranath, H. T., & Katti, P., (2010). Management of epilachna beetle on ashwagandha. *Karnataka Journal of Agricultural Sciences*, *23*(1), 171.

Chaudhary, V., (2013). Arthropods associated with ashwagandha, (*Withania somnifera* Dunal) in semi-arid region of Gujarat. *Karnataka J. Agric. Sci.*, *26*(3),433–435.

Etebarian, H. R., (1992). Studies on *Fusarium* wilt of tomato and its chemical control in Varamin area. *Iranian J. Agric. Sci.*, *23*, 1–14.

Farooqi, A. A., Iqbal, A., Aquil, F., & Ahmad, L., (2003). Broad-spectrum antibacterial and antifungal properties of certain traditionally used Indian medicinal plants. *World J. Microbiol. Biotech.*, *19*, 653–657.

Ganguli, D., & Pandotra, V. R., (1962). Some of the important commonly occurring diseases of important medicinal and aromatic plants in Jammu and Kashmir. *Indian Phytopath.*, *15*, 50–54.

Ghosh, S. K., & Senapati, S. K., (2001). Biology and seasonal fluctuation of *Henosepilachna vigintioctopunctata* Fab. on brinjal under Terai region of West Bengal. *Indian Journal ofAgricultural Research*, *3*, 149–154.

Gomez-Galera, S., Pelacho, A. M., Gene, A., Capell, T., & Christou, P., (2007). The genetic manipulation of medicinal and aromatic plants. *Plant Cell Rep.*, *26*, 1689–1715.

Gotyal, B. S., Babu, V. R., Satpathy, S., & Selvaraj, K., (2019). Biological control of key pest of jute, *Spilosoma obliqua* walker (*Lepidoptera: Arctiidae*): a case study *Journal of Biological Control*, *33*(3), 222–224.

Gupta, M. L., Mishra, H. O., Kalra, A., & Khanuja, S. P. S., (2004). Root-rot and wilt: A new disease of ashwagandha (*Withania somnifera*) caused by *Fusarium solani*. *Journal of Medicinal and Aromatic Plants Sciences*, *26*, 285–287.

Gupta, S. K., & Karmakar, K., (2011). Diversity of mites (Acari) on medicinal and aromatic plants in India. *Zoosymposia*, *6*, 56–61.

Hamilton, M. A., (1932). On three new virus diseases of *Hyoscyamus Niger. Ann. Appl. Biol.*, *19*, 550–557.

Haseeb, A., & Pandey, R., (1989). Observation on *Meloidogyne* spp. infesting Japanese mint – New disease records. *Nematropica*, *19*, 93-97.

Hilje, L., Costa, H. S., & Stansly, P. A., (2001). Cultural practices for managing *Bemisia tabaci* and associated viral diseases. *Crop Prot.*, *20*, 801–812.

Hirekurubar, R. B., Patil, S., Jayappa, J., Tatagar, M. H., & Masuthi, D. A., (2018). Evaluation of new insecticides and Acaricides against red spider mite (*Tetranychus urticae* Koch) in Ashwagandha (*Withania somnifera* Dunal). *Journal of Pharmacognosy and Phytochemistry*, *SP3*, 319–322.

Joshi, K., Chavan, P., Warude, D., & Patwardhan, B., (2004). Molecular markers in herbal drug technology. *Curr. Sci.*, *87*, 159–165.

Kapoor, I. J., & Kumar, B., (1991). Relative efficacy of systemic and non-systemic fungicides against *Fusarium oxysporum* and *Fusarium solani* affecting tomato. *Indian Phytopath.*, *44*(1), 87–93.

Kavitha, M., Gopal, K., Anandam, R. J., & Prasad, B. G., (2004). Evaluation of native isolates of *Trichoderma* in the control of dry root-rot in acid lime. *J. Mycol. Pl. Pathol.*, *34*, 384–386.

Kedar, S. C., Saini, R. K., & Kumaranag, K. M., (2017). Seasonal abundance and predatory potential of *Serangium parcesetosum* Sicard (*Coleoptera: Coccinellidae*) against whitefly, *Bemisiatabaci* on cotton. *Journal of Entomological Research*, *41*(1), 63–68.

Kedar, S. C., Saini, R. K., & Ram, P., (2011a). Biology of *Phenacoccus solenopsis* Tinsley (Hemiptera: *Pseudococcidae*) on potato sprouts. *Journal of Insect Science, 24*, 30–34.

Kedar, S. C., Saini, R. K., & Ram, P., (2011b). Record of coccinellid predators of solenopsis mealybug, *Phenacoccus solenopsis* Tinsley (*Hemiptera: Pseudococ cidae*) in Haryana, India. *Journal of Entomological Research, 35*(3), 245, 246.

Kedar, S. C., Saini, R. K., & Ram, P., (2013a). Bionomics of solenopsis mealybug, *Phenacoccus solenopsis* Tinsley (*Hemiptera: Pseudococ cidae*), on cotton in Haryana. *Journal of Cotton Research and Development, 27*(1), 99–103.

Kedar, S. C., Saini, R. K., & Ram, P., (2013b). Relative population density of *Phenacoccus solenopsis* Tinsley on some host plants. *Bioinfolet., 10*(2), 586.

Khan, S. M., (1976). *Biology and Control of Agrotisspp.* M.Sc. Thesis. University of Peshawar.

Khanuja, S. P. S., Shasany, A. K., Dhawan, S., Darokar, M. P., Satapathy, S., Santha, K. T. R., Saikia, D., et al., (2004). *Mint Plant Named Sambhav*, USPP14538P2.

Krishna, P. P. R., & Reddy, D. D. R., (1979). Study on the root-knot nematode infecting patchouli (*Pogostemon cablin* Benth.). *Maysore J. Agric. Sci., 13*(3), 367, 368.

Krishna, P. P. R., & Reddy, D. D. R., (1984). Pathogenicity and analysis of crop losses in patchouli due to *Meloidogyne incognita. Ind. J. Nematol., 14*(4), 36–38.

Kumar, A., Sharma, P. C., & Mehta, P. K., (2009b). Insect pests on Ashwagandha, *Withania somnifera* Dunal in Himachal Pradesh. *Journal of Insect Science, 22*(2), 210–211.

Kumar, A., Singh, C. P., & Pandey, R., (2009a). Insect pests of ashwagandha, *Withania somnifera* Linn. in Tarai region of Uttarakhand. *Entomon., 34*(2), 115–118.

Kumar, R., Nitharwal, M., Chauhan, R., Pal, V., & Kranthi, K. R., (2012). Evaluation of ecofriendly control methods for management of mealybug, *Phenacoccus solenopsis* Tinsley in cotton. *Journal of Entomology, 9*,32–40.

Kumar, S., & Nanjan, K., (1984). Control of spiral nematode in patchouli. *Fungicide & NematicideTests, 39*, 93.

Mamatha, T., & Ravishankar, R. V., (2004). Evaluation of fungicides and plant extracts against *Fusarium solani* leaf blight in *Terminalia catappa. J. Mycol. Pl. Path.,* 34.

Margaritopoulos, J. T., Tsitsipis, J. A., Goudoudaki, S., & Blackman, R. L., (2002). Life cycle variation of *Myzuspersicae* (*Hemiptera: Aphididae*) in Greece. *Bull. Entomol. Res., 92*, 309–319.

Margina, A., & Zhelijazkov, V., (1994). Control of mint rust (*Puccinia menthae* Pers.) on mint with fungicides and their effect on essential oil content. *Journal of Essential Oil Research, 6*, 607–615.

Mathur, A. C., & Srivastava, J. B., (1964). *Epilachna vigintioctopunctata* as a defoliator of some solanaceous medicinal plants. *Indian Journal of Entomology, 26*, 246.

Meena, N. K., Singh, B., Lal, G., Kant, K., & Meena, R. D., (2016). Sustainable management of aphid in coriander (*Coriandrum sativum* L.) through botanicals and bio-pesticides. *International J. Seed Spices, 6*(1), 25–31.

Merrifield, K. J., & Ingham, R. E., (1996). Population dynamics of *Pratylenchus penetrans, Paratylenchus* sp. and *Criconemellaxenoplax* on Western Oregon peppermint. *Journal of Nematology, 28*, 557–564.

Mishra, D., & Rath, G. C., (1987). Some pathological aspects of *Fusarium solani*(Ment.) Sacc. causing post-harvest decay of *Colocasia* and yam. *Indian J. Mycol. Pl. Path.*, 16.

Nagrare, V. S., Kranthi, S., Kumar, R., DharaJothi, B., Amutha, M., Deshmukh, A. J., Bisane, K. D., & Kranthi, K. R., (2011). *Compendium of Cotton Mealybugs* (p. 42). Central Institute for Cotton Research, Nagpur, India.

Newton, H. C. F., (1934). On the biology of *Psylliodeshyoxcyami* Linn. (*Chrysomelidae, Coleoptera*), The henbane flea beetle with descriptions of the larval stages. *Annals of Applied Biology, 21*(1), 153–161.

Oliveira, M. R. V., Henneberry, T. J., & Anderson, P. K., (2001). History, current status and collaborative research projects for *Bemisiatabaci. Crop Protection, 20*, 709–723.

Pandey, R., & Kalra, A., (2003). Root-knot disease of ashwagandha, *Withania somnifera* and its ecofriendly and cost-effective management. *Journal of Mycology and Plant Pathology, 33*(2), 240–245.

Pandey, R., & Kalra, A., (2010). Inhibitory effects of vermicompost produced from agro-waste of medicinal and aromatic plants on egg hatching in *Meloidogyne incognita* (Kofoid and White) Chitwood. *Curr. Sci., 98*, 833–835.

Pandey, R., & Patra, N. K., (2001). Screening mint (*Mentha* spp.) accessions against root-knot nematode infection. *Journal of Spices and Aromatic Crops, 10*(1), 55, 56.

Pandey, R., (1989). *Studies of Root-Knot Nematode, Meloidogyne Incognita Infesting Japanese Mint (Mentha Arvensis) Linn Subsp. Haplocalyx Briquet var. piperascens Holmes) and its Control* (p. 163). Ph. D. Dissertation, Kanpur University Kanpur, India.

Pandey, R., (1990). Pathogenicity and reproduction of *Meloidogyne incognita* on *Hyoscyamus albus* and rhizospheric association of other phytoparasitic nematodes. *International Nematology Network Newsletter, 7*(4), 34-37.

Pandey, R., (1993). Conceptual models for the management of root-knot and lesion nematodes in important medicinal and aromatic plants. *Afro-Asian Journal of Nematology, 3*(1), 103-106.

Pandey, R., (1995). Application of oil seed cake, pesticides, and dry leaf matter on the reproduction potential of *Meloidogyneincognita* and yield of Japanese mint. *Indian J. Nematol., 25*(1), 12, 13.

Pandey, R., (1997a). Seasonal variation in population of phytonematodes associated with peppermint (*Mentha piperita* L.) and spearmint (*Mentha spicata* L.). *Journal of Aromatic & Spices Crops, 6*(2), 149-151.

Pandey, R., (1997b). Comparative potentiality of *Pratylenchus thornei* on different mint species. Indian *Journal of Plant Pathology, 15*(1, 2), 24–30.

Pandey, R., (1997c). Ecofriendly approaches for the management of root-knot nematode on black henbane. *Indian J. Nematol., 27*(2), 175-178.

Pandey, R., (1998). Phytopathological impact of root-knot nematode on some medicinal and aromatic plants. *Journal of Medicinal and Aromatic Crop Sciences, 20*, 67–84.

Pandey, R., (2000). Additive effect of three organic materials and pesticides on the reproduction of *Meloidogyne incognita* and yield of *Mentha arvensis. Nematropica, 30*, 155–160.

Pandey, R., (2005). Field application of bio-organics in the management of *Meloidogyne incognita* in *Mentha arvensis*. *Nematologia Medit., 33*(1), 51–54.

Pandey, R., (2017). Diseases of medicinal and aromatic plants: Insights in nematode biomanagement. *Indian Phytopathology, 70*(1), 12–21.

Pandey, R., Gupta, M. L., Singh, H. B., & Kumar, S., (1999). Interaction potentialities of Vesicular-arbuscular mycorrhizal fungi with root knot nematode, *Meloidogyne incognita*on black henbane. *Bioresource Technol., 69*, 275–278.

Pandey, R., Gupta, M. L., Singh, H. B., Singh, H. N., & Kumar, S., (2000a). Disease management potentiality of VAM fungi to *Meloidogyne incognita* on *Hyoscyamus muticus*. In: *Proc. International Conference on Integrated Plant Disease Management for Sustainable Agriculture* (pp. 1239, 1240). IARI, New Delhi.

Pandey, R., Haseeb, A., & Husain, A., (1992). Distribution, pathogenicity, and management of *Meloidogyne incognita* on *Mentha arvensis* cv. MAS-1. *Afro-Asian Journal of Nematology, 2*(1), 27-34.

Pandey, R., Kalra, A., & Kumar, S., (2000b). Efficacy of various essential oils in the management of root-knot disease in black henbane. In: *Proc. Centennial Conference on Spices and Aromatic Plants* (pp. 263–266). Calicut.

Pandey, R., Mishra, A. K., Singh, H. B., Kalra, A., & Singh, D., (2019). *Diseases of Medicinal and Aromatic Plants and its Management* (p. 366). Today & Tomorrow Press, New Delhi.

Pandey, R., Singh, H. B., Gupta, M. L., Singh, H. N., Sharma, S., & Kumar, S., (1998). Efficacy of VAM fungi, green manure, organic materials, and pesticides in the management of root-knot nematode in menthol mint. In: Dhawan, S. C., & Kaushal, K. K., (eds.), *Proc. National Symposium on Rational Approaches in Nematode Management for Sustainable Agriculture* (pp. 47–50). GAU Anand.

Rai, M. K., Qureshi, S., & Pandey, A. K., (1999). *In vitro* susceptibility of opportunistic *Fusarium* spp. to essential oils. *Mycoses, 42*, 97–101.

Ramanna, D., Prasad, K., & Basavanagoud, K., (2010). Pest complex of Medicinal Plants. *Karnataka Journal ofAgricultural Sciences, 23,* 197–199.

Rathour, K. S., Ganguly, S., Das, T., Singh, P., Kumar, A., & Somvamshi, V. S., (2014). Biological management of subterranean termites (*Odontotermes obesus*) infesting wheat and pearl millet crops by entomopathogenic nematodes. *Ind. J. Nematol., 44*, 97–100.

Ravikumar, A., Rajendran, R., Chinniah, C., Irulandi, S., & Pandi, R. J., (2008). Induction of resistance through organic amendments for the management of spotted leaf beetle, *Epilachna vigintioctopunctata* Fab. on ashwagandha. *Journal of Biopesticides, 1*, 23–27.

Rehaman, S. K., & Pradeep, S., (2016). Effect of biopesticides on management of *Helicoverpa armigera*in Ashwagandha (*Withania somnifera* Linn.). *Pest Management in Horticultural Ecosystems, 22*(1), 99–102.

Rehaman, S. K., Pradeep, S., & Dhanapal, R., (2018a). Incidence of fruit borer *Helicoverpa armigera* (Hubner) on ashwagandha in Shivamogga. *International Journal of Current Microbiology and Applied Sciences,7*(3), 1060–1066.

Rehaman, S. K., Pradeep, S., Dhanapal, R., & Chandrashekara, G. V., (2018b). Survey studies on insect-pests associated with important medicinal plants in Shivamogga, Karnataka. *Journal of Entomology and Zoology Studies, 6*(1), 848–857.

Richards, A. M., (1983). The *Epilachna vigintioctopunctata* complex (*Coleoptera: Coccinellidae*). *International Journal ofEntomology, 25*, 11–41.

Rita, P., & Animesh, K., (2011). An updated overview on peppermint (*Mentha piperita* L.). *International Research Journal of Pharmacy, 2*(8), 1–10.

Sagar, P., & Ramji, (1990). Natural enemies of cabbage semilooper on Japanese mint crop at Ludhiana. *Indian Perfumer, 34*(1), 44–46.

Sagar, P., & Ramji, (1991). A study on the host range of cabbage semilooper, *Thysanoplusia orichalcea* (Fab.) in Punjab. *Indian Perfumer, 35*(1), 61, 62.

Sagar, P., & Reddy, R. O., (1985). Host range and preference of Mentha leaf Webber, *Syngamia abruptalis* (*Lepidoptera: Pyralidae*) in Punjab (India) *Journal of Research Punjab Agricultural University, 22*(3), 488–492.

Sagar, P., & Reddy, V. O., (1987). Nature and extent of damage of Mentha leaf Webber, *Syngamia abruptalis* Walker, a pest of Japanese mint in the Punjab *Indian Perfumer, 31*(4), 379–382.

Sagar, P., & Singh, D. P., (1981). Chemical control of the aphid, *Aphis affinis* (*Homoptera: Aphididae*) a pest of Japanese mint *Mentha arvensis* in the Punjab India. *Entomon.*, *6*(1), 73–80.

Sagar, P., (1984). Comparative efficacy of some insecticides against leaf Webber, *Syngamia abruptalis* Walker infesting Japanese mint in the Punjab; *Inter. Pest Control.*, *26*(4), 97–105.

Sagar, P., (1985). Note on outbreak of the cabbage semilooper, *Thysanoplusia orichalcea* (Fab.) = (*Plusia orichalcea*) on Japanese mint, *Mentha arvensis* Linn. in Punjab. *Indian Perfumer, 29*, 167, 168.

Sagar, P., (1988). Efficacy of some synthetic pyrethroids and other insecticides against infestation of Japanese mint in Punjab *Indian Perfumer*, *32*(3) 186–189.

Sahu, C., Ganguli, J. L., & Joshi, S., (2017). Seasonal incidence of the mealybug, *Phenacoccus solenopsis* Tinsley (*Hemiptera: Pseudococcidae*) on ashwagandha in Raipur (Chhattisgarh). *International Journal of Plant Protection*, *10*(2), 270–274.

Samad, A., Mahesh, K., Gupta, A. K., Shasany, P. V., Ajayakumar, & Alam, M., (2009). Begomovirus related to *Tomato leaf curl Pakistan virus* newly reported in *Mentha* crops in India. *Plant Pathology, 58*,404.

Samad, A., Zaim, M., & Ajayakumar, P. V., (1994). An outbreak of mosaic disease on mint (*Mentha cardiaca* Baker) in India. *Ind. J. Plant Pathol.*, *12*, 1–4.

Samad, A., Zaim, M., & Ajayakumar, P. V., (1999). White mosaic: A new virus disease on Egyptian Henbane in India. *Ind. J. Virol.*, *15*, 39–42.

Samad, A., Zaim, M., Ajayakumar, P. V., & Garg, I. D., (2000). Isolation and characterization of a TMV isolate infecting scotch spearmint (*Mentha gracilis* sole) in India. *J. Plant Dis. Prot.*, *107*, 649–657.

Sanjta, S., & Chauhan, U., (2018). Incidence and diversity of thrips and its associated natural enemies in medicinal plants. *Journal of Medicinal Plant Studies*, *6*(1), 1, 2.

Sarhan, A. R. T., (1989). Biological control of *Fusarium* root rot of broad bean. *Acta Phytopathologica*, *24*, 271–275.

Sarwar, M., Parameswaren, T. N., Shanamugham, C., & Narayana, M. R., (1982). Appraisal of patchouli in India, root-knot problem and control. *Indian Perfumer*, *26*, 117–121.

Sastry, K. S. M., & Vasanthakumar, T., (1981). Yellow mosaic virus of Patcouli (*P. patchouli*) in India. *Curr. Sci.*, *50*, 767, 768.

Senthil, K. C. M., Jeyaram, K., & Singh, H. B., (2011). First record of the entomopathogenic fungus *Entomophagaaulicae* on the Bihar hairy caterpillar *Spilarctia obliqua* in Manipur, India. *Phytoparasitica*, *39*, 67–71.

Sharma, A., & Pati, P. K., (2012). First record of the carmine spider mite, *Tetrany chusurticae*, infesting *Withania somniferain* India. *Journal of Insect Science*, *12*, 50.

Sharma, A., & Pati, P. K., (2013). First record of Ashwagandha as a new host to the invasive mealybug (*Phenacoccus solenopsis* Tinsley) in India. *Entomological News*, *123*(1), 59–62.

Shivapuri, A., Sharma, O. P., & Jhamaria, S. L., (1997). Fungitoxic properties of plant extracts against pathogenic fungi. *J. Mycol. Pl. Pathl.*, *27*, 29–31.

Shukla, R. S., Alam, M., Singh, V. P., & Kumar, B., (2008). Multiple disease resistance in menthol mint genotypes. *Curr. Sci.*, *95*(7), 836–838.

Shukla, R. S., Singh, H. B., Kalra, A., Singh, V. P., & Kumar, S., (1999). Occurrence of dual infection by *Puccinia menthae* and *Alternaria alternate* on *Mentha arvensis* in Tarai region of Uttar Pradesh. *Journal of Medicinal and Aromatic Plant Sciences*, *21*, 311–315.

Singh, D. P., & Sagar, P., (1981). Biology of the aphid (*Aphis affinis* Del Guercio; *Homoptera, Aphididae*) on the Japanese mint in the Punjab. *Indian Perfumer*, *25*(1), 105–109.

Singh, D., & Bali, R., (1993). New record of coccinellid predators on aphids (*Aphis affinis* and *Myzus persicae*) in Japanese mint (*Mentha arvensis* subsp. *haplocalyx* var *piperascens*) and Egyptian henbane (*Hyoscyamus muticus*). *Indian Journal of Agricultural Science, 63*(5), 313, 314.

Singh, D., Singh, D. K., Bahl, J. K., & Tiwari, R. K., (2004). Evaluation of different genotypes of *Mentha* species for resistance potential to sweet potato whitefly *(Bemisiatabaci) Indian J. Genet., 64*(3), 253, 254.

Singh, J. P., & Gupta, P., (1970). The red pumpkin beetle *Raphidopalpa foveicollis* (Lucas), as a pest of the Japanese mint. *Journal of the Bombay Natural History Society*, *67*(1), 123–144.

Singh, M. K., Chandel, V., Hallan, V., Ram, R., & Zaidi, A. A., (2009). Occurrence of *Peanut stripe virus* on patchouli and rising of virus-free patchouli plants by meristem tip culture. *Journal of Plant Diseases and Protection, 116*(1), 2–6.

Singh, O. P., & Rawat, R. R., (1980). Alternative food-plants of the soybean looper *Plusiaorichalcea* F. (*Lepidoptera: Noctuidae*) in Jabalpur, India. *Anzeiger fur Schadingskundepflnzenschutz Umweltschutz, 53*(2), 25.

Singh, R., & Angadi, S. P., (1990). Wilt of patchouli (*Pogostemon patchouli* Pellet) and reaction of some patchouli strains to the disease. *Indian Perfumer*, *32*(2), 81, 82.

Suab, J., & Nagy, F., (1972). Results of mint rust control experiments. *Mycopathol. Mycol. Appl.*, *31*, 66, 67.

Subba, R. B. L., (1986). Patchouli (*Pogostemon patchouli*) mosaic caused by a potyvirus in India. *Phytopathology*, *76*, 1109.

Szezeponek, A., & Mazur, S., (2006). Occurrence of fungal diseases on lemon balm (*Mentha officinalis* L.) and peppermint (*Mentha piperita* L.) in the region of Malopolska. *Communications in Agricultural and Applied Biological Sciences*, *71*(3), 1109–1118.

Tripathi, A. K., (2018). Pests of medicinal and aromatic plants. In: Omkar, (ed.), *Pests and Their Management* (pp. 719–822).Springer Nature Singapore Pte Ltd.

Tripathi, M. N., Shukla, D. N., & Dwivedi, B. K., (1999). Effect of aqueous extracts of leaves of some medicinal plants on spore germination of *Fusarium* spp. *Bioved., 10*, 43, 44.

Triveni, B., (2017). *Bio-ecology and Eco-friendly Management of Insect Pests of Patchouli, Tulsi, and Mint (Family: Lamiaceae).* M. Sc. (Ag) Thesis, University of Agricultural Sciences, Bangalore.

Varma, S., & Anandhi, P., (2008). Seasonal incidence of brinjal hadda beetle *Henosepilachna vigintioctopunctata* (Fab.) and its natural enemies. *Indian Journal of Entomology, 70*, 31–39.

Venkatesh, M. G., (2006). Seasonal occurrence of *Henosepilachna vigintioctopunctata* (F) (*Coleoptera: Coccinellidae*) and its parasitoids on ashwagandha in India. *Journal of Asia andPacific Entomology, 9*, 265–268.

Vennila, S., Prasad, Y. G., Prabhakar, M., Kumar, R., Nagrare, V., Amutha, M., Dharajyothi, et al., (2011). *Spatio-temporal Distribution of Host Plants of Cotton Mealybug, Phenacoccus Solenopsis Tinsley in India* (p. 50). Technical Bulletin No. 26. National Centre for Integrated Pest Management, New Delhi, India.

Verpoorte, R., Van, D. H. R., Ten, H. H. J. G., & Memelink, J., (1999). Metabolic engineering of plant secondary metabolic pathways for the production of fine chemicals. *Biotechnol. Lett.*, *21*, 467–479.

Wahid, A., Jawed, M. S., & Idress, M., (1995). Chemical control of *Fusarium* root rot, wilt, and collar rot of soybean. *Pakistan J. Phytopathol.*, *7*, 21–24.

Williams, D. J., (2004). *Mealybugs of Southern Asia*. Southdene SDN, BHD, Kuala Lumpur.
Yuksel, E., Taskesen, Y. E., Erarslan, D., & Canhilal, R., (2018). Effectiveness of different entomopathogenic nematode species against the variegated cutworm, *Peridroma saucia* (Hubner) (*Lepidoptera: Noctuidae*). *Egyptian Journal of Biological Pest Control, 28*, 8.
Zaim, M., & Samad, A., (1995). Association of phytoplasmas with a witches-broom disease of *Withania somnifera* (L.) Dunal in India. *Plant Science, 109*, 225–229.
Zaim, M., (1999). *Studies on Transmission, Purification and Some Properties of Virus Strain Causing Mosaic Disease in Hyoscyamus Muticus L. and its Effect on Alkaloid Production*. Phytopathological Society, Zonal Meeting, Gorakhpur University, Gorakhpur.
Zaim, M., Joseph, J., Samad, A., Khaliq, A., & Savithri, H. S., (1999). Mixed viral infection in patchouli (*Pogostemon cablin* (Blanco) Benth.). In: *Proc. National Symposium on Changing and Prospects of Plants Pathology in Coming Millennium*. NBRI Lucknow.
Zapata, R. L., Palmucci, H. E., & Blanco, M. V., (2001). *Fusarium solani*, causal agent of wilt and root rot of eggplant (*Solanum melongena*) in Argentina. *Fitopat., 36*, 15–18.

CHAPTER 20

Changing Marketing Scenario of Medicinal and Aromatic Plants in India

MD. SHAMSHER AHMAD,[1] MD. WASIM SIDDIQUI,[1] M. P. DEVI,[2] S. EDA,[2] and P. K. PAUL[2]

[1]*Department of Food Science and Postharvest Technology, BAC, Sabour (Bihar), India*

[2]*Department of Pomology and Postharvest Technology, Faculty of Horticulture, UBKV, Pundibari, Cooch Behar (WB), India*

ABSTRACT

MAPs are one of the important segments of horticultural produce in India. It is supposed to be the backbone of Indian herbal-based pharmaceutical industries. The indigenous systems of medical treatments found in India are mainly based on MAPs. The majority of Indian producers have small land holdings and are grouped as small and marginal farmers. The horticultural marketing system in the country is dominated by middlemen in the name of service providers. These persons play a role in the supply chain between producers and consumers and take a bigger margin out of the consumer's rupee. For the last few years, there has been an entry of organized retailers of established private firms under wholesale and retail business sectors dealing with horticultural produce. Another motivating force for the marketing of MAPs is an increase in per capita income and healthy herbal treatment consciousness among Indian populations, increasing the demand for herbal products. With this increase in demand, opportunities for earning more for local farmers have also increased. The marketing of horticultural produce in India is one of the most unorganized sectors of Indian agricultural industries. The growers mostly sell their produce to the middlemen, who make more

Bioprospecting Ethnomedicinal Plant Resources: Sustainable Utilization and Restoration.
Gopal Shukla, Jahangeer A. Bhat, A. P. Das, and Sumit Chakravarty (Eds.)

profit than the growers. But slowly the situation is changing and now the growers are selling their produce directly to the consumers, organized retail chains, and modern formats. This shift may be termed as the retail marketing revolution of horticultural produce in India.

20.1 INTRODUCTION

Herbal plants, which mainly include medicinal and aromatic plants (MAPs), are used as raw materials for manufacturing herbal medicines like Ayurvedic, Unani, and to some extent, homeopathic. These herbal medicines are in demand in both the national and international markets. Medicines based on MAPs are traditionally used all over the world. Most of the Ayurveda, Unani, and Siddha medicines are manufactured from plant species grown in India (Shankar & Majumdar, 1997; Prajapati et al., 2003). MAPs, a major segment of flora, are used as raw materials ranging from pharmaceuticals, cosmetics, F&F (flavor and fragrance) to drug industries. According to an estimate by Schippmann et al. (2005), about 2,500 species of MAPs are marketed internationally. India is assumed to be the birthplace of many old systems of medical treatments based on medicinal plants categorized as ethnomedicines. About 7,500 species of plants are used in the preparation of such medicines (Shankar & Majumdar, 1997). The indigenous systems of medicine developed in India for centuries are based on MAPs and their derivatives. People in developing and underdeveloped countries still use medicines of plant origin for their disease treatment (De Silva, 1997). About 30% of the drugs and cosmetic products marketed globally contain compounds of plant origin. The demand for herbal medicines in India is increasing every year. According to Ved & Goraya (2008), the annual demand for Indian herbal products was 250–1,000 tons. This demand provides an opportunity for rural populations in India to grow and gather MAPs. The MAPs cultivated or gathered provide a sure source of cash income for many small and marginal farmers. MAPs-based industries create employment opportunities for local populations as well as act as a source of their permanent income (Lacuna-Richman, 2002).

20.2 MARKETING OF MAPS

Marketing of MAPs in India is more or less similar to fruits and vegetables where the supply chain is dominated by middlemen, popularly called traders or commission agents. These people have nothing to do with cultivation but are prominent in the market and influence market price. Their main job is to sell the produce brought by farmers on a commission basis or purchase directly from farmers at a low price at sourcing points and sell at a high price in the markets or to manufacturers. They are very much aware of the sale prices of different markets located in different parts of the country. After deducting commission, labor charges of loading and unloading, and other charges if any, the net payment is given to the producers. The result of the whole transaction reflects the fact that a very small amount of the consumer's price is paid to the farmers. The actual share of farmers in the price paid by the consumers depends on the length of marketing channels of that particular commodity (Anantia, 2008). The longer the channel, the smaller will be the farmer's share in the consumer's price and vice versa. Many surveys have been conducted from time to time to estimate the farmer's share in the consumer's rupee. One such survey conducted by the Government of India in 2001 is on horticultural produce which concluded that the farmer's share in the consumer's rupee is only one rupee when INR Rs 3.50 is paid by the consumers (Anon., 2001). The remaining Rs 2.50 is distributed among intermediate people (middlemen) such as Rs. 0.50 to the wholesaler, 0.75 to the retailer, and Rs 1.25 in between the wholesaler and the retailer. Based on many surveys, it is concluded that farmers receive only 33-75% share in the consumer's price. In another survey by Olsen-Smith (1996), the share of collectors or producers of MAPs is 40-50% of the sale price in the Delhi wholesale market. The producer's net margin may go high if they are in direct contact with the traders. Prasad & Bhatnagar (1991) noticed a remarkable price fluctuation of a few selected medicinal plants at three different markets (Table 20.1). It is also reported that in a few markets, farmers are not allowed to sell their produce directly to the buyers. Jairath (2008) also revealed a similar opinion and reported that many farmers face many problems while doing sales without the help of any trader or commission agent.

There are two groups of people associated with MAPs. One group of people is called collectors, as they used to collect MAPs from forests and mountains and the other groups are producers or cultivators. There are only two supply sources of MAPs first is source is via collection from the wild (uncultivated source) and second is via cultivated source (Kuipers, 1997; Anon., 2005). The wild source is prominent over the cultivated one. The volume of trade is directly connected with the volume of medicinal plants collected from wild sources (Gusainet al., 2009). The majority of both groups are poor and marginal people. Direct marketing to manufacturers or consumers is difficult for them. They used to sale their produce either to village level or district level wholesalers or private agents at a very cheap rate.

TABLE 20.1 Prices of Few Important Medicinal Plants in Jabalpur (Madhya Pradesh) and Delhi Markets

Sl. No.	Common Name	Scientific Name	Price (Rs/Kg)		
			Village Level Market	Jabalpur Market	Delhi Market
1.	Aonla	*Ernbilica officinalis*	60	140	600
2.	Bahera	*Terminalia bellirica*	20	200	250
3.	Bach	*Acorus calamus*	70	NA	1,200
4.	Baibidang	*Embeliaribes*	125	NA	1,800
5.	Baelguda	*Aegle rnarmelos*	25	NA	500
6.	Bhilwa	*Semecarpus anacardium*	10	900	NA
7.	Chirayata	*Swertia chirayita*	NA	500	NA
8.	Nagamotha	*Cyperus rotundus*	20	900	NA
9.	Safed musli	*Chlorophyturnborivilianum*	500	25,000	2,000
10.	Sarpgandha	*Rauwolfia serpentine*	NA	2,500	6,000

Source: Prasad & Bhatnagar (1991); Ghate & Dube (2014).

Indian industries purchase MAPs from the suppliers and wholesalers. There are no or very few direct sourcing from collectors and producers. The reason is obvious that the bulk quantities and a wide range of raw materials needed by the industries can't be supplied by individual farmers. A survey conducted jointly by Pragya and Whitley Fund for Nature in 2006-2007 reveals the dominance of intermediate people (middlemen) in marketing channels of MAPs (Anon., 2007a). During the survey, about 47% of the manufacturers interviewed responded that they source over 75% of their raw materials from middlemen and only a small quantity from direct producers and collectors. But when retailers interviewed, it was found that about 77% retailers procure from smaller regional markets and collectors (Mills & Lee, 2011). Only 23% of all traders source their produce from cultivators directly. This makes product traceability nearly impossible. At present, contract farming and buy-back arrangements are the only practical solution for exporters and manufacturers who want to maintain traceability. In the present scenario, MAPs are marketed in both raw as well as in processed form. Demand for both forms is increasing every year. This increase is mainly due to its use in traditional systems of medicine by developing countries of the world. Roughly 80% of the population in most of the developing countries uses traditional medicines of plant origin (De Silva, 1997). Nowadays, many products of MAPs based origin are available in the market (Table 20.2). Many reputed firms are already manufacturing drugs, cosmetics, food supplements, and beauty products. One of the main ingredients of their products is MAPs based. A few reputed firms are also listed in Table 20.3.

TABLE 20.2 Marketing Based on MAPs

Sl. No.	Types of MAPs Products	Product Form	Example
1.	Phyto-pharmaceuticals.	Isolated pure active compounds are used to treat diseases.	Torex Herbal, Honitus, Patanjali products.
2.	Dietary supplements.	Extracts of plants or plant-parts are used for health benefits.	Dabur Chyawanprash, Immuno-care Craft Fit.
3.	Nutraceuticals	Plant or plant-part extracts with a specific health benefit in prevention and treatment of disease.	Animi-3, Divista, fish oil, lipoicin, thioctan.
4.	Cosmetics/beauty care.	Beauty products which contain biologically active molecules and compounds.	Beauty balm, mask Moha, Patanjali beauty cream.
5.	Flavor and fragrance.	Used for imparting flavor/masking unpleasant flavor in foods and fragrance in cosmetics.	Vanilla, rose essence.

TABLE 20.3 Reputed Firms Engaged in Manufacturing MAPs-based Herbal Products

Sl. No.	Name of Firms	Establishment	Mode of Marketing	Major MAPs Products
1.	Patanjali Ayurveda	Across India	Franchise shops and organized retail shops.	Ayurvedic products, herbals, nutrient supplement, and cosmetics.
2.	Dabur India	Across India	Pharmacies and general retail shops.	Ayurvedic products, herbals, and cosmetics.
3.	Baidynath Ayurveda Bhavan	Across India	Pharmacies and general retail shops.	Ayurvedic and herbal products.
4.	Himalaya Wellness	Across India	Pharmacies	Ayurvedic and herbal products.
5.	Zandu Ayurveda	Across India	Pharmacies and general retail shops.	Ayurvedic and herbal products.
6.	Vicco Laboratories	Across India	Pharmacies and general retail shops.	Ayurvedic and herbal products.
7.	Kottakal Arya Vaidyasala	South India	Authorized Dealers and clinical hospitals.	Healthcare
8.	Vaidyaratnam Oushadhasala	South India	Authorized dealers and nursing homes.	Ayurvedic and healthcare products.
9.	Nagarjuna Herbal Concentrates	South India	Organized retail shops and franchise clinics.	Ayurvedic medicines and healthcare products.

It is evident from the last two decades, there are few changes happening on both national and international level in the marketing patterns of horticultural produce including MAPs. The changing patterns are aimed for proper post-harvest management and quality control of MAPs. The ultimate aim is focused to satisfy consumers' or manufacturers' demand of quality and convenience. In other words, all the changes made are business oriented. Now marketing of MAPs has been transformed into a multi-million-dollar business. This is also reflected in the convention on biological diversity (CBD). As per the secretariat of the CBD, the marketing of only herbal products all over the world was estimated about US $60,000 million in 2002 (Anon., 2003). According to an estimation by National Medicinal Plant Board (NMPB), India, the annual growth in several herbs is between 15% and 30% and the annual exports of MAP plants and its derivatives are in the range of INR Rs. 600–700 million (Arun & Madhav, 2004).

Use of information technology and cashless transactions is one of the factors contributing to the increased sale of all commodities, including MAPs. Today, there are many electronic platforms that are actively dedicated to horticultural produce and MAPs. All these platforms operate online, such as the electronic national agriculture marketing (e-NAM) and electronic negotiable warehouse receipts (e-NWRs). The National Medicinal Plants Board (NMPB) of India has also launched an electronic market (www.e-charak.in) and a mobile application called e-CHARAK for the trade of MAPs. It also serves as an information exchange center among various stakeholders involved in the MAPs sector. On this online platform, farmers and collectors can display their produce, and buyers such as traders, manufacturers, and exporters can purchase their requirements. There are always a few quality issues with MAPs, both online and offline, but these can be addressed through improved grading, packing, storage, and transportation. With the success of the online business model, many big wholesale markets in India have now joined each other and are connected with other small wholesale markets as well. Similarly, all organized retail chains are also connected with the nearest wholesale market in their region and major wholesale markets in India.

20.2.1 MAP MARKETING CHANNEL

Marketing of horticultural produce and MAPs are similar in many ways. These similarities are due to several factors starting with the nature of commodities, socio-economic conditions of producers, supply, and demand fluctuations, and marketing channels, to name a few. Among these factors, the marketing channel plays the most important role in deciding the retail or final price of the raw material. In the majority of cases, the produce passes through four to six hands or stages to reach manufacturers. The stages may include primary aggregators or producers, local traders, village markets, district or regional wholesale markets, and large wholesale markets. The long marketing chain means a low price for primary collectors and farmers of their products. Most of the farmers of MAPs are small and marginal with poor economic conditions. Generally, MAPs are sold in raw form. In fact, most of the growers are not aware of the actual sale price of their raw products, and they receive only a small part of the final sale. Another reason for getting a lower price might be that farmers are unable to market it in the form wanted by the buyers or are unable to sell to these buyers directly (Gularia & Gupta, 2020).

Studies reported five stages and actors involved in the marketing channels for medicinal plants starting from the forest or farm to their final export or consumption or utilization in drugs, cosmetics, or herbal preparations (Olsen-Smith, 1996; Gularia & Gupta, 2020). These actors are collectors or farmers, local traders, private agents, wholesalers, and consumers and the channels are:

Collectors/farmers → Local traders → Consumers
Collectors/farmers → Local traders → Private Agents → Industry
Collectors/farmers → Local traders → Private Agents → Wholesaler→ Industry
Collectors/farmers → Local traders → Private Agents → Wholesaler→ Export
Collectors/farmers → Local traders → Private Agents → Industry

However, in many cases marketing of medicinal plants will not move in the same way as given above. Even in a single market, there might be multiple sales of the same commodity or from one market to another or from one private agent to another until it reaches the industry. All established wholesalers always manage a few people of their confidence to get price, arrival, and quality information in the sourcing areas. In many cases, advance orders are placed and advance payment is also paid by wholesale dealers to private agents. These private agents then hire local traders who visit the farmer's fields on a daily basis. They also visit these markets every day and buy directly from the collectors/farmers. The sale and purchase of MAPs are generally based on fresh weight basis, although some traders also prefer to use dry weight basis. After purchasing at the local level, these traders then send it to the wholesaler of district-level markets, for sale to private agents or commission agents popularly known as '*arhatiyas*.' These '*arhatiyas*' charge a fixed commission and possess a valid registration for auctioning and transaction on a commission basis. They are well aware of the expected supply and demand in other markets throughout the state as well as with the prevailing rates in the larger markets of other states. This capability attracts a greater number of buyers and sellers towards the commission agents. At present, the major known markets for MAPs are found in Amritsar, Delhi, Kanpur, Kolkata, Chennai, Cochin, Mumbai, and Tuticorin (Olsen-Smith, 1996). High-volume buyers purchase the products from wholesalers or service providers and supply to manufacturing units or the international market.

20.2.1.1 DIRECT MARKETING

In this system, no middleman or intermediary exists between producers or collectors and consumers. So, the whole profit goes to the producers. Few examples are given below (Plate 20.1(a) and (b)):

- Producer/Collector → Consumers;
- Farmer's market → Consumers (*Ryathu* market);
- Farm stall/garden sale → Consumers;
- Contract farming → Industry (Consumers);
- Reverse marketing → Industry (Consumers);
- Cooperative marketing → Bulk buyer (Consumers).

PLATE 20.1(A) Roadside stall farmers markets.

PLATE 20.1(B) Own farm/garden sales or own shop.

20.2.1.2 INDIRECT MARKETING

In this system, there is always a middleman or an intermediary between producers or collectors and consumers. So, the profit money of producers always gets reduced. The reduction in money is directly related to the length of intermediary. As the length of this chain will increase, the profit money of the producer will decrease and vice versa. Few examples are.:

- Producer/Collector→ Local traders) → Consumers;
- Producer/Collector → Local trader → Private agent → Big wholesaler/Industry → Consumers;
- Producer/Collector → Local trader → Wholesalers → Industry → Consumers;
- Producer/Collector → Local medical man/*Vaidh* → Consumers;
- Producer/Collector → Street/Roadside shop owner → Consumers;
- Producer/Collector → Markets owned by municipality/multi-outlets shop → Consumers.

20.2.2 MODE OF SALE AND PURCHASE IN MAPS

Many growers in India have few modes or alternatives for marketing their produce. Each alternative has its own advantages and disadvantages. A producer may use one or more than one alternative based on their knowledge and opportunities. Some important factors like quantity or volume, location, and quality of produce are considered while choosing a marketing mode. Producers are not very aware of marketing alternatives. However, intermediate people are more capable of using or developing more alternatives.

20.2.2.1 RURAL HAAT OR VILLAGE MARKET

Village markets are weekly markets of Indian villages and very popular among villagers. In general, its occurrence is weekly but may be twice in a week. Rural *haats* (markets) are the only markets easily accessible to village people. There are no high-volume sale and purchase as only local people are involved.

20.2.2.2 MUNICIPAL HAAT

As the name indicates, this market comes under municipality areas of small and big cities. The shops are permanent but may or may not be under permanent structure and always regulated by municipal authorities. Such shops are dominated by fruits, vegetables, and flowers with few exceptions like MAPs.

20.2.2.3 WHOLESALE

It is the biggest market of horticultural produce located generally in an urban area. Here, producers, traders, and all stakeholders gather every day, particularly in the morning hours. The sale commodities are auctioned and sold as a whole. In this way, a very big transaction occurs every day. The commission agents or traders do auction in the morning so that buyers can transport it in the morning hours for their destinations. All other shops directly or indirectly connected with agriculture or horticulture are present in or around all wholesale markets of India. All inputs like insecticides, pesticides, boxes, and other packing materials are available to cater to farmers' needs. Like any other commodity, the price of MAPs also fluctuates based on demand and supply situations. More supply and less demand will result in a low price, and vice versa leads to a high price at the whole market.

20.2.2.4 ROADSIDE SHOPS

In all cities, it is a common sight. It is always temporary in nature and mostly unauthorized but opens early morning and continues till evening. Change of place is also a feature of roadside shops. Only a temporary structure is erected with the help of wooden planks and bamboo just to protect from rain and sunshine.

20.2.2.5 STANDING CROP SALE AND PURCHASE

In few MAPs, crops are sold to buyers well before harvesting. It simply means the rate of produce is fixed during the growing period. The buyers see the crop condition and offer a price. If the farmer agrees to the offered price, they take some amount as advance and the rest after harvesting. A time period is given by the farmer and within that period, the buyer harvests the crop.

20.2.2.6 PRE-CONTRACT SALE

In this system, producers are forced to sell their produce to a fixed buyer. In most cases, the producers borrow money from that particular buyer well ahead of the crop harvesting. So, the farmers have the obligation to sell to that particular buyer from whom money has been taken in advance. This model of sale is mainly dependent on the socio-economic condition of producers.

20.2.2.7 LOCAL TRADERS

Local traders are active in all growing areas and well connected with both producers and buyers. They are functioning like a connecting link between producers and buyers. A good amount of MAPs is sold through local traders only.

20.2.2.8 SERVICE PROVIDER

Those wholesalers or traders have sorting, grading, packing, or storage facilities that act as service providers for processing and manufacturing companies. Service providers have the capability to fulfill the demand of companies.

20.3 CHARACTERISTICS OF MAP MARKET

There are some special characteristics of Indian MAP market which are described in subsections (Mills & Lee, 2011).

20.3.1 UNSTRUCTURED MARKET

There is no single supply chain in MAPs. The produce passes through many stages or hands before it falls into the hands of manufacturers or processors. In fact, the hands are multiple, including producers, collectors, locals, small traders, big traders, and so on. But in most cases, it is the local and small traders that act as suppliers to the big traders and wholesalers of cities, and from there it is finally passed to manufacturers.

20.3.2 DOMINANCE OF THE INTERMEDIARIES

There is a dominance of middlemen in the marketing channel of MAPs and other horticultural produce. About 47% of the processors purchase over 75%

of their requirement of raw materials from intermediate people and only 23% from producers or collectors directly.

20.3.3 REVENUE SHARING

The revenue sharing in terms of producer's share in consumer's rupee is reported very low in all horticultural produce including MAPs. Only a small part of the final sale price is shared by the producers ranging from 18-30% of the final sale.

20.3.4 FLUCTUATING SUPPLY AND DEMAND

Like any other horticultural produce, MAPs are also harvested in bulk during peak harvesting season. Therefore, maximum supply is expected during the harvesting season and very little or no supply during the off-season when the price of produce generally remains high. This supply pattern results in very high fluctuation in demand and price. Round the year supply needs storage facility which is a limiting factor for the majority of the producers. They have no storage facilities, hence unable to store their harvest for off-season sale at a higher price.

20.3.5 PRICE TRENDS

It is difficult to assume any particular price trends in MAPs due to seasonal harvesting, perishability, and lack of storage facility. However, a survey finds the price of 82% of the major Himalayan plant species are on increasing trends, i.e., a continued price increase was observed for the last three years while, remaining 18% registered a constant price but none of them registered a decreasing pattern.

20.3.6 QUALITY FACTOR

Quality is the most important factor which determines the price of any commodity. Similarly, traders and retailers dealing with MAPs put the highest priority on quality and accordingly determine the price of raw material. Installation of primary processing and packaging facilities is beyond the scope of growers. Some of the traders have limited post-harvest processing, storage, and packaging facilities who act as service providers. Quality assessment of raw materials by buyers is unique. Visual appearance plays (color, odor, and appearance) the most important role in judging raw material quality. Mainly the exporters and the manufacturers use laboratory-based, scientific chemical analysis of the raw materials.

20.3.7 ADULTERATION

Adulteration is very common in MAPs. It may start at the producer level or at any stage of the supply chain. Experienced traders are capable of identifying adulterants and eliminating them during marketing. In general, three kinds of adulterants are found, i.e., plants as a whole or parts of a plant belonging to different species but of the same genus, coloring of low-quality roots or stems in order to hide blemishes and create a feel of high-grade material, and the addition of foreign materials such as dried straw, grass, dried branches, and other similar-looking or smelling materials to increase the overall quantity or volume of the raw material.

20.3.8 FACILITIES

Most of the MAP markets lack basic infrastructure such as sorting, grading, and packing. Since the supply chain is dominated by intermediaries, Indian farmers solely depend on these middlemen for the sale of their produce. Always there is a risk in the supply chain like heavy rain, transit losses, more supply, and fewer buyers. All these risks before sale are also borne by the growers/collectors.

20.3.9 TRANSACTIONS

Most of the transactions take place in cash and in informal ways due to lack of any clear policies of marketing of MAPs. The regulation in licensing and registration is established in few commodities but enforcement is not strict and generally ignored by the traders.

20.4 MARKETING INTERVENTIONS AND DEVELOPMENT

The most common intervention in the marketing channel of MAPs is the establishment of cooperatives to secure a fair price for the welfare of collectors and producers. Cooperatives are established in many states of the country with the aim of improving the life and livelihood of producers or collectors by protecting their interests on one hand and promoting conservation and marketing on the other hand. This is achieved by covering a wide range of forest products, which also include MAPs. The GSCS, TDCC, FDC, and MFPD are some of the noted cooperative societies working independently. Some NGOs like Pragya India (www.pragya.org), BAIF Institute for Rural Development, BIRD-K (www.birdk.org.in), and FRLHT (Foundation for Revitalization of Local Health Tradition; envis.frlht.org; frlhtenvis.nic.in) are reputed ones working in the field of MAPs. The state-level federations or cooperatives are also formed in many states. The central government also passed a Cooperative Act in 1984 with an attempt to link these cooperatives together for the benefit of the indigenous communities or rural people involved in the collection of MAPs. Under this act, a TRIFED (tribal federation) is formed. The main function of this federation is facilitating communication and cooperation between the cooperatives and also developing an MIS (marketing information system) at the national level. TRIFED works in collaboration with other government organizations by forming SHGs. In 2018, TRIFED signed an MoU with NMPB and established more than 1200 SHGs across 22 states of India. These SHGs, in the name of VDVK (Van Dhan Vikash Kendra), work to identify potential medicinal plant species in forests and rural areas while supporting skill development, standardization of marketing norms for medicinal plants, and facilitating and providing financial assistance for plantation and value addition of forest produce. As per TRIFED, the number of these SHGs is increasing and currently 18,000 SHGs are working across 22 states of India which provides employment opportunities to 3.6 lakh tribal gatherers (https://timesofindia.indiatimes.com/india/trifed-to-link-tribal-shgs-to-medicinal-plant-cultivation-to-bolster-tribal-livelihood/articleshow/76450679.cms?from=mdr).

During the past few years, some changes have been registered in the marketing pattern of MAPs. The changes are mainly confined to the diversity of products manufactured by the F&F industries and FMCG (fast moving consumer goods) companies. The domestic and global demand for fragrances is required in detergents, soaps, perfume products, cosmetics, and packaged food industries. Indian manufacturers like M/s., S. H. Kelkar & Company Pvt. Ltd., Oriental Aromatics Ltd., Ultra International Ltd., Khattri Fragrances and Flavors Limited, Gupta, and Company (P) Ltd., Sachee Aromatics Limited and Aarav Fragrances are major players that use MAPs as raw material. M/s. S. H. Kelkar and Company Pvt. Ltd. leads the sales among Indian F&F companies among Indian players. There is also a growing demand under pharmaceuticals for natural medicinal products, and a growing market for aromatherapy. Thus, there is a huge demand for processing of raw materials into finished products such as essential oils (EOs), cosmetic products or herbal beauty products and a need for quality assurance in order to ensure the marketability of these products. The Indian F&F market is highly fragmented ranging from multinational companies and large Indian industrial houses to small-scale industrial units and local manufacturers. The flavor industries are more interested in savory, bakery, and confectionery segments. On the other hand, the fragrance market is more organized with larger players like Unilever, Godrej, Dabur, Reckitt Benckiser, Wipro, Henkel, and ITC generating most of the sales.

The idea of organized retail came in India in the early to mid-80s with the launch of Mother Dairy for distribution of milk in Delhi NCR (National Capital Region). But a quantum jump occurred only in the last two decades when companies like Reliance Industries, Bharti Walmart, Future Group, Aditya Birla Group, and other corporate houses entered the retail business in FMCG sectors covering a vast range of products of daily use. Many studies on the growth of the retail business predicted faster growth in India and other developing countries. Studies predicted faster growth of organized retail business in developing countries in the next few decades due to increasing income and population in cities and towns (Anon., 2007b). Reardon & Gulati (2008) found that the Indian organized retail sector is growing at an annual rate between 16 and 50% over the past few years. Similar trends were also observed in less developed countries also (Timmer, 2005). The increased growth in retail sectors is a boon to the majority of the farmers though with varying benefits. Gularia & Gupta (2020) reported comparative prices of a few medicinal plants as mentioned in Table 20.4. The authors mentioned that this increase in price is due to the direct transportation of MAPs from production sites to improved marketing models such as Herbal mandi located in Amritsar (Punjab), Delhi, and other retail outlets.

The purchase models adopted by retail firms are indirect sourcing through either a wholesaler or service provider, but not direct with the producers except for a few larger ones. The success of this model lies in the fact that wholesalers and service providers have the capacity to provide bulk to retail chains (Alam & Verma, 2007; Mangala & Chengappa, 2008). The present retail chains like Reliance Fresh, Big Bazaar, More, Heritage, Food World, Big Apple, Easy Day, and Cash and Carry are owned by established corporate firms of India and abroad and are flooded with herbal-based cosmetics, health booster foods, and beauty products that are directly or indirectly associated with MAPs (Ahmad et al., 2014; Table 20.5).

TABLE 20.4 Comparative Prices of Few MAPs between 1997 and 2019

Sl. No.	Scientific Name	1997 Price (INR Rs/Kg)	2019 Price (INR Rs/Kg)	Price Difference (INR Rs/Kg)
1.	*Jurineamacrocephala*	40–100	240	140–200
2.	*Valeriana jatamansi*	50	1,200	1,150
3.	*Valeriana hardwickii*	30–50	250	200–220
4.	*Dactylorhizahatagirea*	400–750	4,500	3,750–4,100
5.	*Parmelia*spp.	20–40	140	100–120
6.	*Viola odorata, V. serpens*	80	1,400	1,320
7.	*Aconitum heterophyllum*	800–1,100	6,000	4,900–5,200
8.	*Picrorhizakurrooa*	60–110	1,350	1,240–1,290
9.	*Morchella esculenta*	1,800–2,500	2,500	0–700
10.	*Hypercum perforatum*	75–100	1,500–2,000	1662.50

TABLE 20.5 Major Organized Fruit and Vegetable Retails in India

Sl. No.	Name of Group	Company/Brand Name	Area of Operation
1.	NDDB	Safal	Delhi NCR, Bangalore
2.	RPG	Food World	South India
3.	RPG	Spencer	South, Central, and East India
4.	Reliance	Reliance Fresh	All over India
5.	ITC	Choupal Fresh	North, South, and Central India
6.	Aditya Birla	More	All over India
7.	Heritage	Fresh@	South India
8.	Future Group	Food Bazaar	North, South, and Central India
9.	Bharti Wall mart	Easy Day	North, South, and Central India
10.	Mahindra & Mahindra	Mahindra Shubhlabh	North
11.	TATA	Khet Se	North and South

The overall horticultural sectors, including MAPs, are benefited by the entry of organized retail sectors in India. Farmers dealing with medicinal plants are supposed to equally benefit, but not equally as claimed by many researchers. According to research, the farmers who supply their produce, mainly fruits and vegetables, directly to retail chains receive higher prices and higher net profit (Alam & Verma, 2007; Mangala & Chengappa, 2008), and lower transaction costs (Singh & Singla, 2010). The quality of the produce was found to be the main criteria upon which prices were fixed. Farmers are also aware of this fact, and they are trying to increase the level by adopting some scientific approaches (Jairath, 2008). Almost all the corporate firms dealing with fresh horticultural produce and MAPs appointed experts to train and guide the farmers about good agricultural practices (GAPs) and quality production management. The training programs conducted by these firms were mainly focused on awareness among farmers on quality, cost cutting, and primary processing (Singh & Singla, 2010).

20.5 CONCLUSION

India is rich in MAPs. Here cultivation is scattered and dominated by small and marginal farmers and gatherers. Marketing of MAPs in India is dominated by intermediaries and mainly unorganized. This system results in poor returns to growers/collectors in terms of producers' share in the consumer's rupee. Recently, there have been some changes or improvements in the marketing of MAPs in developing countries like India. These changes are due to a change in marketing systems from traditional middlemen-dominated unorganized marketing to organized herbal markets (*mandis*) and modern retail chains of multinational companies. Most of these herbal markets and retail chains have direct access to farmers' organizations. They work with many groups of farmers. Most of the multinationals also have specialists who advise farmers in pre- and post-harvest management of MAPs. This guidance not only raises the quality bar of their produce but also increases their net profit over the past 10–15 years. All these changes indicate a change in the marketing patterns of MAPs in the future.

KEYWORDS

- **adulteration**
- **consumer**
- **MAP market**
- **marketing interventions**
- **middleman**
- **producer**
- **retail chain**
- **trader**

REFERENCES

Ahmad, M. S., Nayyer, A., & Siddiqui, M. W., (2014). Changing scenario of fresh fruit and vegetable marketing in India. *Processed Food Industry, 17*(5), 21–26.

Alam, G., & Verma, D., (2007). *Connecting Small-Scale Farmers with Dynamic Markets: A Case Study of a Successful Supply Chain in Uttarakhand.* Centre for Sustainable Development, Dehra Dun, India.

Anantia, (2008). *What is India's Share in Global Vegetable & Fruits Market? Economic Survey 2007–08.* Ministry of Finance, Government of India, New Delhi. Available at https://www.managementpedia.com/threads/what-is-indias-share-in-global-veg-fruits-market.25024/ (accessed on 18 December 2023).

Anonymous, (2001). *Report of the Working Group on Horticulture Development for Tenth Five Year Plan (Main Report).* Planning commission, Government of India, New Delhi.

Anonymous, (2003). *WHO Guidelines on Good Agricultural and Collection Practices (GACP) for Medicinal Plants.* World Health Organization, Rome.

Anonymous, (2005). *Changes in Food Retailing in Asia: Implication of Supermarket Procurement Practices for Farmers and Traditional Marketing Systems.* Agricultural Management, Marketing and Finance Occasional Paper 8. Food and Agricultural Organization, Rome.

Anonymous, (2007a). *International Market Prospects for Sustainably Sourced Medicinal and aromatic Plants.* Whitley Fund for Nature-Pragya.

Anonymous, (2007b). *Global Powers of Retailing.* Deloitte Touche Tohmatsu and Stores Magazine, Deloitte-Stores.

Arun, N., & Madhav, K., (2004). *A Study on Marketing Opportunities for Medicinal, Aromatic, and Dye Plants in South Asia.* Medicinal and Aromatic Plants Program in Asia (MAPPA). IDRC/FORD Foundation/IFAD-Supported Research Network.

De Silva, T., (1997). Industrial utilization of medicinal plants in developing countries. In: Bodeker, G., Bhat, K. K. S., Burley, J., & Vantomme, P., (eds.), *Medicinal Plants for Forest Conservation and Healthcare, Non-Wood Forest Products 11* (pp. 38–48). FAO, Rome.

Gularia, M., & Gupta, P., (2020). Marketing channel of medicinal and aromatic plants (MAPs) in the great Himalayan National Park (GHNP), Kullu, Himachal Pradesh India. *J. Medicinal Plants Studies, 8*(5), 107–120.

Gusain, M. S., Sultan, Z., Singh, C. J., & Bisht, N. S., (2009). Potential of marketing medicinal plants in Uttarakhand. *J. Non-Tim. For. Prod., 16*, 239–244.

Jairath, M. S., (2008). Enhancing farmers' linkage to markets. *Indian Journal of Agricultural Economics, 63*(3), 355–356.

Kuipers, S. E., (1997). Trade in medicinal plants. In: *Medicinal Plants for Forest Conservation and healthcare, Non-Wood Forest Products 11.* Food and Agriculture Organization of the United Nations, Rome.

Lacuna-Richman, C., (2002). The socio-economic significance of subsistence non-wood forest products in Leyte, Philippines. *Environmental Conservation, 29*, 253–262.

Mangala, K. P., &Chengappa, P. G., (2008). A novel agribusiness model for backward linkages with farmers: A case of food retail chain. *Agricultural Economics Research Review, 21*, 363–370.

Mills, S., & Lee, J., (2011). *International Market Prospects for Sustainably Sourced Medicinal and Aromatic Plants from India* (pp. 35–41). A project by Pragya and Whitley Fund for Nature in partnership.

Olsen-Smith, C., (1996). *Medicinal Plants, Markets, and Margins: Implications for Development in Nepal*. Paper presented at the INBAR/IDRC/IPGRI/ICIMOD Workshop on the Role of Bamboo, Rattan, and Medicinal Plants in Mountain Development. IDRC, New Delhi.

Prajapati, N. D., Purohit, S. S., Sharma, A. K., & Kumar, T., (2003). *A Handbook of Medicinal Plants* (p. 553). Agribios, India.

Prasad, R., & Bhatnagar, P., (1991). *Socio-Economic Potential of Minor Forest Produce in Madhya Pradesh*. Madhya Pradesh State Forest Research Institute Bulletin, No. 26. Jabalpur.

Reardon, T., & Gulati, A., (2008). *The Rise of Supermarkets and their Development Implications-International Experience Relevant for India*. IFPRI Discussion Paper 00752. International Food Policy Research Institute, Washington, D.C.

Schippmann, U., Leaman, D. J., Cunningham, A. B., & Walter, S., (2005). Impact of cultivation and collection on the conservation of medicinal plants. *Acta Horticulturae*, *676*, 31–44.

Shankar, D., & Majumdar, B., (1997). Beyond the biodiversity convention: The challenge facing the biocultural heritage of India's medicinal plants. In: Bodeker, G., Bhat, K. K. S., Burley, J., & Vantomme, P., (eds.), *Medicinal Plants for Forest Conservation and Health Care; Non-Wood Forest Products 11* (pp. 87–99). FAO, Rome.

Singh, S., & Singla, N., (2010). *Fresh Food Retail Chains in India: Impacts on Small Primary Vegetable Producers and Traditional F&V Retailers*. Study Report. Indian Institute of Management, Ahmedabad.

Timmer, C. P., (2005). *Agriculture and Pro-Poor Growth: An Asian Perspective*. CGD Working Paper No. 63. Center for Global Development, Washington DC.

Ved, D. K., & Goraya, G. S., (2008). *Demand and Supply of Medicinal Plants in India* (p. 216). National Medicinal Plants Board, New Delhi.

Index

B

O

P

S

W

X

Y

Z